CAMBRIDGE LIBRARY COLLECTION

Books of enduring scholarly value

Technology

The focus of this series is engineering, broadly construed. It covers technological innovation from a range of periods and cultures, but centres on the technological achievements of the industrial era in the West, particularly in the nineteenth century, as understood by their contemporaries. Infrastructure is one major focus, covering the building of railways and canals, bridges and tunnels, land drainage, the laying of submarine cables, and the construction of docks and lighthouses. Other key topics include developments in industrial and manufacturing fields such as mining technology, the production of iron and steel, the use of steam power, and chemical processes such as photography and textile dyes.

Life of Richard Trevithick

Cornishman Richard Trevithick (1771–1833) was one of the pioneering engineers of the Industrial Revolution. Best remembered today for his early railway locomotive, Trevithick worked on a wide range of projects, including mines, mills, dredging machinery, a tunnel under the Thames, military engineering, and prospecting in South America. However, his difficult personality and financial failures caused him to be overshadowed by contemporaries such as Robert Stephenson and James Watt. This two-volume study by his son Francis, chief engineer with the London and North-Western Railway, was published in 1872, and helped to revive his neglected reputation. It places its subject in his historical and technical context, building on the work of his father, Richard Trevithick Senior, and the Cornish mining industry. It contains much technical detail, but is still of interest to the general reader. Volume 1 covers his predecessors, and early life, before examining his work thematically.

Cambridge University Press has long been a pioneer in the reissuing of out-of-print titles from its own backlist, producing digital reprints of books that are still sought after by scholars and students but could not be reprinted economically using traditional technology. The Cambridge Library Collection extends this activity to a wider range of books which are still of importance to researchers and professionals, either for the source material they contain, or as landmarks in the history of their academic discipline.

Drawing from the world-renowned collections in the Cambridge University Library, and guided by the advice of experts in each subject area, Cambridge University Press is using state-of-the-art scanning machines in its own Printing House to capture the content of each book selected for inclusion. The files are processed to give a consistently clear, crisp image, and the books finished to the high quality standard for which the Press is recognised around the world. The latest print-on-demand technology ensures that the books will remain available indefinitely, and that orders for single or multiple copies can quickly be supplied.

The Cambridge Library Collection will bring back to life books of enduring scholarly value (including out-of-copyright works originally issued by other publishers) across a wide range of disciplines in the humanities and social sciences and in science and technology.

Life of Richard Trevithick

With an Account of his Inventions

VOLUME 1

FRANCIS TREVITHICK

CAMBRIDGE UNIVERSITY PRESS

Cambridge, New York, Melbourne, Madrid, Cape Town, Singapore,
São Paolo, Delhi, Dubai, Tokyo, Mexico City

Published in the United States of America by Cambridge University Press, New York

www.cambridge.org
Information on this title: www.cambridge.org/9781108026673

This edition first published 1872
This digitally printed version 2011

ISBN 978-1-108-02667-3 Paperback

Richard Trevithick

Engraved by W. Holl, from a Painting.

Published by E and F. N. Spon. 48. Charing Crofs. 1872

LIFE

OF

RICHARD TREVITHICK,

WITH AN ACCOUNT OF HIS INVENTIONS.

BY FRANCIS TREVITHICK, C.E.

ILLUSTRATED WITH ENGRAVINGS ON WOOD BY W. J. WELCH.

VOLUME I.

LONDON:
E. & F. N. SPON, 48, CHARING CROSS.
NEW YORK:
446, BROOME STREET.
1872.

PREFACE.

THE events in a man's life and the progress of the steam-engine are so dissimilar, that the reader is solicited to pass with the writer over the numerous breaks in the thread of the story. The good Trevithick did in his generation is found in the extended use of steam-engines: in tracing their various forms in their applicability to numerous purposes, the labour of years has to be reviewed, and each idea may be followed, from its first becoming a useful form to its perfect growth, and acceptance by the public as a good thing.

The overflow of Trevithick's practical designs has caused a difficulty in fairly defining and estimating the facts bearing on each of the particular kinds selected, as belonging to understood classes of the steam-engine; while the great importance of his inventions caused many engineers to labour on their improvement, both during and since his time.

Engines originating or made commercially serviceable by Trevithick are given as his, but other engineers may have a fair claim as improvers of the same, or of nearly similar inventions; therefore, at the risk of wearying the reader, repetitions of evidence are given, that, if possible, a true knowledge may be formed.

The history of the day-by-day life of the man has been made subservient to that of the steam-engine: to partially restore this break, the reader is asked to bear with the frequent introduction of dates.

Trevithick's correspondence during a great portion

of his life with his scientific friend, Mr. Davies Gilbert, written with the freedom of impulsive genius, longing for some one person to whom his thoughts might be made known, together with a rough draft letter-book, used only in his Cornish home during four or five years, constitute the ground-work of my history. His practical engineering works, nearer perfection in their first movements than those of other engineers, and the after success and eminence of many who in early life were his working assistants, are evidences of his strength of character.

Very little has been written of Trevithick, though Mr. Richard Edmonds, the late Lord Brougham, Sir Edward Watkin, Mr. Enys, Mr. Bennet Woodcroft, Mr. Hyde Clarke, Professor Pole, and others, have published outlines, or have collected information; and though they may not have succeeded in making him fully known, their labours have lessened mine.

To the Hon. Mrs. Gilbert I am indebted for the numerous letters written to Mr. Davies Gilbert; and among many who assisted Trevithick by friendly acts, I observe mention in his papers of Mills, Gittens, Wyatt, Banfield, Potter, and the Dartford mechanics.

FRANCIS TREVITHICK.

THE CLIFF, PENZANCE,
April, 1872.

CONTENTS OF VOLUME I.

CHAPTER IV.

EARLY LIFE OF TREVITHICK.

CHAPTER V.

PLUNGER-POLE PUMP AND WATER-PRESSURE ENGINE.

CHAPTER VI.

HIGH-PRESSURE WHIM-ENGINE.

CHAPTER VII.

Camborne Common Road Locomotive.

CHAPTER VIII.

Patent of 1802, and London Locomotive.

CHAPTER IX.

Tram and Railway Locomotives.

CHAPTER X.

Partnership, and Early High-pressure Engines.

CHAPTER XI.

High-pressure Steam-dredger.

CHAPTER XII.

Thames Driftway.

CHAPTER XIII.

Iron Tanks.

CHAPTER XIV.

Ships of Wood and Iron.

CHAPTER XV.

Propelling Vessels by Steam.

CHAPTER XVI.

Recoil Engine and Tubular Boiler.

ILLUSTRATIONS TO VOLUME I.

CHAPTER I.

EARLY CORNISH ENGINES.

DURING many centuries the wants of the mining interests of Cornwall led to improvements in pumping machinery.

The ancient tin streamer in the Cornish valleys used the fall of the running stream to separate the grains of tin from a mass of sand and gravel.

The tin sand of the valleys having all been washed, the miner followed the mineral vein in its downward course, and successive miners, working a lower strata, were obliged to use pumps or other means for removing water from their excavations. Adits were driven from the lowest convenient surface level, to assist in draining the workings. Every day added to the depth, and called for improved machinery.

The open launder of the tin streamer had been converted into a close launder or pipe, having a square leather bag, attached to a long wooden handle, serving as a bucket in the primitive hand-pump, but increased depths overtaxed the strength of the square-sided pump, and cylindrical wooden pumps were made, bound by iron bands. The rag and chain pump then came into use, being a revolving or endless chain, moving upwards through the pump, having at every two or three feet of its length a piston made of a ball of rags bound together, or of wood edged with cloth or leather, for up to this time valves had not been used.

The observed fact that water would rise to a certain height in a pump, following a well-constructed piston, suggested the use of a bottom or stop valve, preventing the descent of the water, and also the use of a valve in the piston or bucket. In 1663 the Marquis of Worcester applied steam to force water up a pipe; other inventive men varied the methods and even constructed model steam-engines, but failed to help the working miner.

The ever-increasing depth and size of the pumps required the power of water-wheels and horse-whims until about the year 1702 Savery is said to have erected the first steam pumping-engine in Cornwall, of which he wrote thus in the 'Miner's Friend':—

"I have known in Cornwall a work with three lifts of about 18 feet each, lift and carry a 3½-inch bore;

that cost forty-two shillings a day. I dare undertake that my engine shall raise you as much water for eight-pence as will cost you a shilling to raise the like with your old engines in coal pits."

Savery's boiler was of cast iron. A vacuum was formed in a receiver by admitting steam to drive out air, and then condensing this steam by pouring cold water on the outer surface of the receiver, caused the water in the shaft to rise by the weight of the atmosphere into the void in the receiver; a valve prevented its return, while another valve opened a passage to the upcast pipe from the receiver towards the surface of the mine; a supply of steam again passed from the boiler to the receiver, forcing the contained water upwards through the pipes. The only moving parts of the machine were the valves.

This engine illustrated three leading principles of the modern steam-engine,—the use of steam to expel the atmosphere, its condensation by cold forming a vacuum, and the more direct use of steam as a strongly expansive manageable agent.

Savery's engine, requiring to be fixed near the bottom of the shaft, or within 30 feet of the level of the water to be raised, never came into general use, and absence of moving parts made it unsuitable for any other purpose than the raising of water, and even for that there are not many traces of its practical application.

In 1705 Newcomen, from Devonshire, combining the ideas of others with his own great mechanical genius, constructed, or it may fairly be said, invented the means of giving motion to a beam by using a cylinder and piston. The steam pressure in the boiler was 1 or 2 lbs. on the square inch above the pressure of the atmosphere, sufficient to drive the air out of the cylinder and prepare it for the vacuum. This was known as the "Atmospheric Engine," its power being measured by the weight of the atmosphere on a piston, whose under side was in vacuum.

In 1712 such an engine was erected at Griff in Warwickshire which raised a load equal to 10 or 11 lbs. on each square inch of the piston. Others followed, and many improvements were made. The first steam-cylinder was about 23 inches in diameter.

In 1720 Newcomen erected at Ludgvan-lez, in Cornwall, a pumping engine with a cylinder of 47 inches in diameter, working at the rate of fifteen strokes a minute.

The best working engine in Cornwall in 1746 was at the Pool Mine, interestingly described by Borlase:[1]—

"The most powerful as well as constant engine hitherto invented is the fire-engine.

"This engine is now well known to the learned; but as their books do not reach everywhere, and this machine is especially serviceable for the working of deep mines, and of great advantage to the public revenue, a general explanation of its principal parts, its powers, and profit to the Government, may not be improper. The principal members of this engine are exhibited in the engraving annexed; the cistern or boiler T, the cylinder P, and the bob O I, turning on an axis which rests in the middle of the wall Y. The following is the process of its several operations:—The cistern T, full of boiling water, supplies steam (by means of an upright tube and valve which shuts and opens) to fill the hollow cylinder P, and expel the air through a horizontal tube S, placed at its bottom. As the steam rises, the piston, which plays up and down in the cylinder, rises, and when it is got near the top opens a clack, by which cold water is injected, and condenses the vapour into nearly the twelve thousandth space which it

[1] See Borlase's 'Natural History of Cornwall,' pub. 1758.

before occupied, and the cylinder being then nearly empty the piston of iron edged with tow and covered with water (to prevent any air from above getting into the cylinder), is driven down by the pressure of the atmosphere (with the force of about 17 lbs. on every square superficial inch) nearly to the bottom

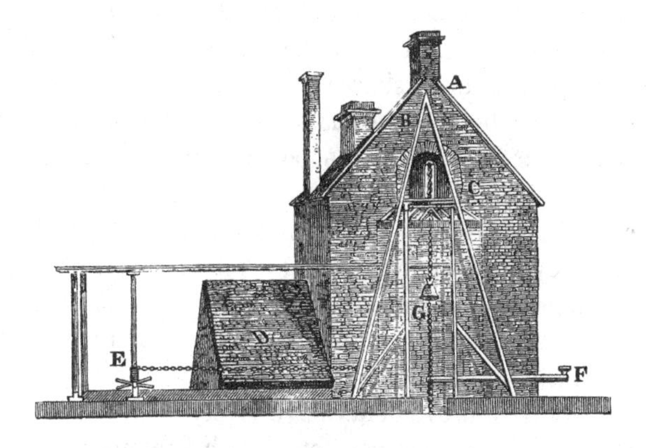

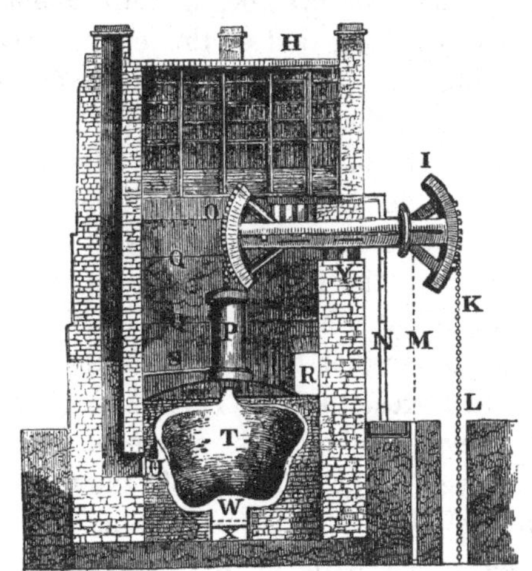

NEWCOMEN ATMOSPHERIC ENGINE, POOL MINE, 1746.

A, south front of fire-engine house; B, triangle for tending the engine-pumps; C, arch for main bob to play in; D, coal-house and fire-place; E, capstan and cable for the triangle; F, balance-bob to assist the draught; G, the bell.

H, section from the west; I, south end of the main bob; K, main chain, to draw up the water from the bottom; L, end of balance-bob, marked F in Fig. 1; M, a small chain, drawing from the adit to a cistern; N, force-pump to supply the cistern for the boiler T, &c.; O, north end of the main bob; P, the cylinder; R, the eastern door; S, pipes to let out the air and steam from the cylinder; T, the boiler which supplies the steam; U, the damper to moderate the fire; W, the fire-place; X the ash-pit; Y, the axis of the main bob.

of the cylinder; at this instant it opens the valve which lets in the steam from the boiler T, and then the piston ascends till it opens the condensing clack above, which brings it down again to open the under clack and admit the steam, and thus continues ascending and descending as long as the managers think proper; this process is quick or otherwise, as the steam is by increase or subtraction of fire made more or less violent, to drive the engine faster or slower. To this piston the end of the bob O is fastened by an iron chain, and as the piston descends in the cylinder P this end of the bob is drawn downwards, and *vice versâ.* As the end O is drawn down, the other end of the bob I ascends, and by a chain I K draws up with it, from an iron or brass cylindrical tube, called a pit-barrel, through a tyre of wooden pumps, a column of water out of the mine equal in diameter to the bore of that tube, and in height to each stroke or motion of the piston in the cylinder P, and the sweep of the bob I K. Many improvements have lately been added to this excellent piece of mechanism, among which I cannot but mention one in particular, which is, that as this engine stood formerly, if the firemen chanced to nod, the violence of the motion increasing with the fire, the weighty bob O I beat, shocked, and endangered the whole machine, and the fabric it is enclosed in; but now, when the fire is at the extreme height, and the bob begins to beat and strike the springs, it lets fall a trigger into a notch and stops the injection-cock, and the whole movement is stopped till the injection of the cold water into the cylinder is restored; so that this engine is now brought to such perfection that in a great measure it tends, regulates, frees, and checks itself. Several subordinate members, wires, clacks, and valves are all moved, opened, and shut by the force of the steam and the motion of the piston; inasmuch as that by enlarging the cylinder and other parts in proportion, few Cornish mines are subject to more water than this engine will master. Its power is in proportion to the diameter of the cylinder principally, the strength of the steam, and the depth it draws. This is the fire-engine which in the year 1746 belonged to the Pool Mine and the cylinder's diameter from the outer edge was but 3 feet; but they make them much larger now, and it is imagined that if they were still to increase the diameter of the cylinder and make it also shorter than they do now, the force would be augmented, and though the column of water exhausted would be shorter, yet might this be well remedied by increasing the number of tubes, which the greater pressure on the piston would easily manage. A cylinder of 47 inches bore at Ludgvan-lez-work, in the parish of Ludgvan, making about fifteen strokes in a minute, usually drew through pit-barrels of 15 inches diameter, from a pump 30 fathoms deep. The cylinder at Herland (or Drednack) Mine, in the parish of Gwinear, is 70 inches in diameter, and will draw a greater stream of water at any equal depth, in proportion to the square of its diameter.

"The only objections to this engine are the great expenses in erecting and vast consumption of coals in working it. To obviate these expenses several methods have been suggested of increasing the elasticity of the steam and reducing the size of the boiler, which can be decided only by experience, and to that we must refer them."

The cylinder, 36 inches in diameter, rested on a large slab of stone, which seems to have served as the top of the steam-boiler. A hole cut in the stone, having a valve in it, was the steam-pipe. The piston was packed with tow; a layer of water on its top prevented leakage of air or steam. A piston-rod and chain connected the piston with the arched head of the wood beam. The steam pressure was 1 or 2 lbs. on the square inch above the pressure of the atmosphere, and the valves were self-acting, enabling the engine to make fifteen strokes a minute. If the engineman fell asleep and the steam becoming stronger made the engine work too rapidly, the self-acting gear, by shutting off the injection, caused the engine to cease working.

The boiler is not particularly described, but the drawing represents a large caldron or pot of metal with a hollowed bottom, under which the fire was placed and circulated through mason-work flues around the sides. The top of the boiler, being nearly flat, was its weakest part, and was strengthened by a stone cover, or possibly this stone cover, in the earlier engines, formed the boiler-top.

In 1756, several were at work; about a dozen are specified; one of them at the Herland Mine having a cylinder of 70 inches in diameter.

The only objection to the engine is the cost of the coal. To lessen this, several methods had been suggested for increasing the elasticity of the steam, and reducing the size of the boiler.

Such were the Newcomen atmospheric open-top cylinder steam-engines, among which Trevithick, sen., lived, depending mainly on the vacuum and weight of the atmosphere for their power, but yet having caused the discovery that an increased pressure of steam in the boiler added to the speed and power of the engine, they only needed an inventor who should design a smaller and stronger boiler, to give steam of greater elastic force.

This Pool Mine, now called North Wheal Crofty, adjoined Dolcoath Mine, and was within half a mile of the house in which Richard Trevithick, sen., lived, then a boy eleven years of age. It is more than likely that in early manhood he exercised an authority over it, for shortly after that period, in 1765, he was the manager of Dolcoath, and resided just midway between the two mines, both of which were worked under the Dedunstanville interest for which Trevithick was the agent. In 1746 the Cornish pumping engine worked fifteen strokes a minute, and was so under control from its well-contrived gear and valve work, that the engine regulated its own movements, and its power was only limited by the diameter of the cylinder and the strength of the steam. In 1758 Borlase wrote:—"Several methods have been suggested for increasing the elasticity of the steam, and reducing the size of the boiler, which can be decided only by experience, and to that we must refer them." And we shall find that shortly after that period Richard Trevithick, sen., took the first step in overcoming the difficulty.

In 1830 the writer saw at the Weith in Camborne a floor about 12 feet square of blocks of granite, known as the old Moor-stone boiler, and conversed with several old men, who, when boys, played in it, when it had sides or walls three or four feet high of blocks of granite.

Captain Joseph Vivian recollected hearing, when a boy, his uncles Simon and John Vivian talk of having taken a contract to break up this boiler, and cut out the copper pipes in the inside.

Mr. C. E. Edwards, a smith at Perran, near Marazion, has the screw-plate and taps used in constructing a granite boiler for Gwallon Mine, near the present Wheal Prosper.

His great-grandfather, Edwards, was a smith in Ludgvan-lez when Newcomen put his great engine there, about 1720. His grandfather was married in 1764, and set up the smith's shop C. E. Edwards still works in, bringing as a marriage present those old taps and screw-plate, said to be the first used in Cornwall. When a boy he often heard his grandfather talk of the old smiths who failed to clamp together by lambs' tails[1] the blocks of granite forming the boiler at the Gwallon Mine; and one of these smiths said, go over to Edwards at Ludgvan-lez, he can make screws that will draw them up beautiful.

[1] Bands of iron tightened by cutters in lieu of bolts.

Gwallon Mine was less than a mile from Newcomen's early engine at Ludgvan-lez.

James Banfield, for many years the principal smith in the engineering works of Harvey and Co., at Hayle, says:—"In 1813 I was rivet-boy at the making of Captain Trevithick's high-pressure boilers at Mellinear Mine. The largest boiler-plates then to be had in Cornwall were 3 feet by 1 foot. My father served his apprenticeship in 1784 with uncle Jan Hosking, a famous smith, on Long-stone Downs in Lelant parish. Father has often told me how the work used to be made when he was a young man. The only wrought iron they could get was Spanish bar, 2 inches square, hammered thin in the middle that it might be bent for the convenience of carriage. It was red-short iron difficult to work; and Swedish or Danish bar, said to come from Siberia, 3¼ inches wide, and ⅝ths of an inch thick. Whatever was wanted in the mine had to be made with such bars."

In 1746, the only wrought iron of commerce in Cornwall being small bars, to make a wrought-iron boiler was a difficult and almost hopeless undertaking.

Small boiler-plates and boilers were made in Staffordshire or Shropshire; but the Cornish roads did not admit of the easy conveyance of heavy weights. For fifty years after the erection of these early engines, the coal and mineral from the ports and mines were conveyed on mules, the roads being unsuitable for wheel carriages.

The traditions relating to granite boilers may not absolutely lead to the conclusion of their use; but the many difficulties in procuring a good boiler are evident. Borlase's drawing shows a close intimacy of boiler and stonework; the steam-pipe leading from the top of the boiler into the bottom of the cylinder being merely a hole in the large stone resting on the top of the boiler, on which the cylinder was supported.

The drawing, coupled with tradition, implies, almost to a certainty, the use of granite in the construction of early steam-engine boilers; and the cutting out of copper pipes from a boiler during the childhood of a person still living, seems to prove that even in those early days we had tubular boilers; for in the granite boiler it is impossible to imagine any other way of getting up steam than by some internal fire-place, as suggested by those tubes. Such is the history of the steam pumping engine in Cornwall up to 1758, enabling the miner to follow to a greater depth the glittering tin, sparkling in its bed of hard rock, from which its minute pin-point particles were separated by stamping the lode-stone into sand, and washing it in a gentle stream, on the higher portion of which the heavier tin settled, while the lighter refuse went with the stream, except in the chance eddies and hollows where again the greater gravity of the particles of tin settled them at the bottom, while the waste earth still floated away, on and on to the sea.

"If the ore be very full of clammy slime, it is turned from the area C into a pit near by, called a buddle, L I, to make it stamp the freer without choking the grates, and brought back to C. If the ore is not slimy, it is shovelled forward from C into a sloping channel of timber E, called the pass, from whence it slides by its own weight and the assistance of a small rill of water, D, into the box at Y; then by the lifters *a b c* falling on it after being raised by the axletree *d*, which is turned round by the water-wheel B, it is pounded or stamped small; to make the lifters more lasting, and fall upon the ore with the greater force, they are armed at the bottom with large masses of iron of 140 lbs. weight each, called stamp-heads; and to assist the

STAMPING AND WASHING TIN ORE, POOL MINE, 1746.

attrition, the rill of water D keeps the ore perpetually wet, and the stamp-heads cool, till the ore in the box Y is pulverized, and small enough to pass through the holes of an iron grate at Y. The grate is a thin plate of iron, no more than $\frac{1}{10}$th of an inch thick, 1 foot square, full of small holes punched in it about the bigness of a moderate pin, not always of the same diameter, but as the different size of the tin granules requires; for the larger the crystals enclosing the metals are, the larger must be the holes, and *vice versâ*, so that in suiting the grate to the nature of the tin, the skill of the dresser appears. From this grate the tin is carried by a smaller gutter *e*, into the fore pit F, where it makes its first and purest settlement, the lighter parts running forwards with the water through holes made in the partition *f*, into the middle pit A (much of the same shape and size as the fore pit), and thence into the third pit H; what settles in A and H is called the slimes, and what runs off from them is good for nothing. The fore pit F, as soon as full is emptied, and the contents carried to the buddle I, a pit 7 feet long, 3 wide, and 2 deep; the dresser standing in the buddle at I, spreads the pulverized ore at K, called the head of the buddle, in small ridges parallel to the run of the water which enters the buddle at L, and falling equally over the cross-bar M, washes the slime from the ridges (which are moved to and fro with a shovel) till the water, permeating every part, washes down the whole into the buddle I: whilst the dresser's hands are employed in stirring the ridges at K, he keeps his foot going always, and moves the ore to and fro so as the water may have full power to wash and cleanse it from its impurities; the buddle fills, and the tin is sorted into three divisions; that next the head, at *g*, is the purest; the middle, at *h*, is next in degree; that at *i* most impure of the three; and each of these divisions goes through a different process. The fore part at A is taken out first, and carried to a large tub N, called the *keeve;* there immersed in water, it is moved round with a shovel for a quarter of an hour, by which means the impurities rise from the ore, and become suspended in the water; the tin ore is then sifted in a sieve purposely constructed, and if it needs, must be sent to be buddled again, then returned to the keeve and worked as before with a shovel, which they call *tozing* the tin; the keeve is then *packed*, that is, beat with a hammer or mallet on the sides, that the ore within may shift and shake off the waste, and settle the purer to the bottom. The foul water then on the top of the keeve is poured off, and the *sordes* which settles above the tin is skimmed off, and what remains is pure enough to be sent to the melting house, and is then called black tin.

"The waste skimmed off is carefully laid by to undergo another washing: whilst the fore part of the buddle I is thus manufacturing at the keeve, another hand is moving forth that part of the buddle *h* in the same manner as *g* was before; and in its turn that, and the settlement at *i*, is promoted to the keeve, and thus what is deposited in the fore pit F, is *brought about*, as the tinners term it, that is, undergoes all the necessary lotions.

"What runs off from F into G and H must be dealt with in another manner. The contents of these pits consist of the small and lighter parts of the ore, and are intimately mixed with a greater quantity of earth and stone bruised to dust by the mill. These are called the *slimes*, and are carried to the trunk O to be again reworked."[1]

The same principles remain in operation even to the present day, but the work of the man's foot is performed by the more effective movements from the steam-engine, and the plane surface of the old buddle has generally given place to a circular buddle, the tin stuff taking the form of a flat conical mound, the heavy particles of tin remaining near the centre, while the lighter particles are carried by a film of running water toward the circumference of the mound; revolving arms having bits of rag or brushes attached slightly disturb the surface of the tin stuff, inducing by the running water the constant change of position of the particles, the lighter or refuse portions being washed farthest from the centre of the mound.

Not only does Borlase give explicit detail on the mine

[1] Borlase's 'Natural History of Cornwall,' pub. 1758.

mechanism of his day, but also indirectly shows when Cornish streams were first poisoned by mineral water. "About fifty years since there were plenty of fine trout in the river Conar in Gwythian"[1]—this would be about 1708; it now sweeps to the sea yearly many thousands of tons of sand from the tin stamps, and from its colour is called the red river; no fish swim in it, though about 1820 the writer caught trout in a tributary stream, from

CROWLISS STREAM, 1871. [W. J. Welch.]

Roseworthy. The destruction of fish one hundred and sixty years ago in the Conar, or valley from the Great Dolcoath mining district, was from tin streaming or shallow mining; but when in after years the steam-engine drained the mines and copper ore was worked, the rivers became still more poisonous. "At Crowliss, a village of Ludgvan, in the year 1739, a flock of geese belonging

[1] Borlase's 'Natural History of Cornwall.'

to James George, tailor, went into the river as usual, and drinking heartily of the water (a very strong mundic), upon their return to the bank nine of them lay down immediately and died."[1] The water fatal to the thirsty geese came from Ludgvan-lez copper mine; it is now a red stream from tin mines where geese may safely seek for grains of tin, readily taking a taste when passing a heap of craze, or tin ore and sand, to allay the cravings of the gizzard. The writer believes but three geese recovered from the Crowliss flock. Cattle are apt to prefer this mine water, but it gives a rough coat and sometimes causes death. Large quantities of arsenic are taken from the flues of tin-calcining furnaces: in 1868 a field of apparently well-grown oats ripe for the sickle was valueless, from the absence of grain in the pods, caused by arsenical smoke from Stray Park Mine, while the straw was unfit even for bedding for cattle, because they might eat it; cabbages in an adjoining field were uninjured, the numerous little flakes of white arsenic resting on them could be washed off; a grass field near by was utterly untrustworthy, but potatoes thrived well and were said to be comparatively free from disease. The smoking chimney in Cook's Kitchen[2] is attached to one of these calciners: on going into the flues when the workman was shovelling the arsenic from their sides and bottom into a barrow, the writer breathed a palpable atmosphere of arsenic; on coming out the man took a pasty from a bench, having a suspiciously thick layer of arsenic on it; brushing off with his hand the superfluous poison, he dined heartily. A heap of arsenic of several tons lay by the furnace at the road-side, to be prepared for commercial wants.

---

[1] Borlase's 'Natural History of Cornwall.

[2] Cook's Kitchen Water-wheel, chap. iii.

CHAPTER II.

RICHARD TREVITHICK, SEN.

GILBERT, in his 'Survey of Cornwall,' says :—

"The name of Trevithick is certainly of great antiquity in the county of Cornwall, and the family is supposed to have been resident at Trevimider for many descents before the seventeenth century.

"A monument in St. Eval Church, to the memory of the Rev. William Trevithick, in 1692, and one with the impalement of the arms of Leach and Trevithick, in 1672; and several places or properties in that neighbourhood called Trevithick, point to their Cornish origin. Arms—Argent, a unicorn rampant.

"Anne Trevithick, the sole heiress of William Trevithick, of Trevimider, married Francis Leach Llewellan, Sheriff of Cornwall, in 1740. The arms of Llewellan seem to denote their descent from the ancient Princes of Wales.

"Trevithick, formerly a seat of the Arundells; and Trevithick, formerly a seat of the Polomounters.

"Trelissick, formerly a seat of the Tremaynes. The house appears to have been built by the Hookers, and in the glass of the windows are preserved the family arms, with the letters J. V. H. Trevithick."

The 'Parochial History of Cornwall' says of the St. Eval monument to Trevithick :—

"In remembrance of William Trevithick, of Trevimider, in St. Eval, Gent., who died the 3rd day of Nov., 1731, aged 52.

"'Farewell kind friends,
Farewell dear wife and brother;
Peace be your ends,
United to each other.'"

Polwheles' 'History of Cornwall' has the following:—

"I have first to remark that Alfred devised Cornwall to his eldest son Edward, and that he devised it under the name of Triconshire.

"The natives that occur as men of property, or who probably held lands here before the Conquest, have been distinguished by Carew under the appellations of Tre, Pol, and Pen, and it seems worthy of remark that as representatives of Tre and Pol, if not of Pen, there exist several families, who have possessed lands from all antiquity.

"By Tre, Pol, and Pen, you shall know the Cornishmen.

In olden time property and birth gave power and influence to the family of Trevithick. History traces the last male Trevithick, of Trevimider, down to 1731; and shortly after his death, his daughter and sole heiress married Francis Leach Llewellan, who believed himself a descendant of the Princes of Wales. Some poor branch of the male line survived unnoticed, for Richard Trevithick, sen., was born in 1735. He is said to have walked a day's journey to make the acquaintance of one known by his almost obsolete name; but little trace remains of where he was born or how he was brought up, though the writer has talked with those who knew him well in his old age, and who spoke of him with respect, as of one above his fellows.

The writer's first reliable evidence of his acts is from his old account-books, some of which, after his death, were fortunately retained by his son, that their unfilled leaves might serve for his rough draft letters. Three of these books still remain, showing that when thirty years of age, Richard Trevithick, sen., was the manager of the leading mines in Cornwall; Dolcoath Mine, the oldest, richest, and most famous in Cornish history, being at that time his head-quarters, while his place of residence was in Illogan parish, midway between the mine and Carn Brea Hill.

How long he had filled the position of leading man in Cornish mines, prior to the dates in the writer's possession, is an unanswered question—certainly for several years, for to be manager of the great Dolcoath at the age of thirty implies unusual ability; but to be at the same time manager of several other of the leading mines, proves that at that early age he was a man of eminent practical experience. Borlase thus speaks of Cornish mining when Trevithick, sen., was a child in 1738:—"All these are easily performed when the workings are near the surface; but the difficulty increases with the depth, and skill and care become still more and more necessary; and, indeed, all the mechanical powers, the most forcible engines, and the utmost sagacity of the chief miner, are often too little and vain when the workings are deep and many."

In the year 1760 Richard Trevithick, sen., when twenty-five years old, married Miss Anne Teague, whose family were mine managers in the Redruth district; their forefathers were said to have been driven from Ireland during a rebellion. Richard Trevithick, sen., and his wife, their four daughters, and one son, averaged 5 feet 11 inches in height.

In 1765 he was the manager of Dolcoath Mine, and constructed "the deep adit," a work of difficulty and importance in those days, still serving as the lowest practical drain for numerous mines, and reducing the adit level for the exit mine-water by 60 or 70 feet.

Arthur and John Woolf were in his pay; the father and uncle of Arthur Woolf the well-known engineer, and of his brother who mutinied on board the fleet at the Nore, under an assumed name.

The Dolcoath Mine at that time used two atmospheric steam pumping engines, the water from which, on its way to the lower level of the new adit, by its passage over two water-wheels with cranks on their axles attached to beams, worked pumps.

RESIDENCE OF TREVITHICK, SEN., IN 1760, AS IT APPEARED IN 1871. [W. J. Welch.]

The mineral from the mine was raised by horse-whims, and water was elevated by the same means in buckets or kibbals from the shallow levels, showing that the steam pumping apparatus had not driven the horse quite out of the field in 1765, nor had the pump-barrel entirely superseded the original tub.

The merciful care in supplying wholesome air to the underground miner, exercised by Richard Trevithick a hundred years ago, is now enforced by the strong arm of the law. Those who carried out the detail in sending air into the mine by the power of the steam-engine, were the fathers of the men who, in after years, prominently helped in the detail manufacture of the first locomotive.

In Trevithick's accounts of 1765 is the following entry:—

"To William Jeffry and partner for driving north from Bullan Garden (a part of Dolcoath) fire-engine shaft.

"To clearing and timbering from the western end of Tonkin's Ground to the western water-engine shaft.

"To bringing air from the little engine to the ends.

"To winding kibbals (of water or stuff), at 5*d.* per hundred kibbals."

The notable re-erection by Richard Trevithick, sen., of the old Carloose engine, just before Watt's first engine in Cornwall, forms an era in the history of the steam-engine. The following extract from his account-book, closing in 1775, gives the items of chief interest:—

"DOLCOATH NEW ENGINE COST ACCOUNT.

	£	s.	d.
Carloose adventurers, for materials	414	12	3
John Commins, for boiler-top, &c...	93	8	9
John Jones and Co. (Bristol), for iron pumps	118	6	10
Dale company, for iron pumps	131	9	4
Mr. Budge, for erecting the engine	63	0	0
Carriage of the boiler, cylinder, &c., from Carloose, including attendance, &c., &c.	50	0	0
Arthur Woolf, per month	1	14	4
John Harvey and partners, for putting in the boiler and building the shed-walls, &c.	33	1	9
To new ironwork, as per account	187	10	4
To timber boards, &c., as per account	255	10	10"

This worn-out Carloose engine, bought for 414*l.* 12*s.* 3*d.*, had been one of Newcomen's early erections, of perhaps forty or fifty years before, but in 1775 the gear and valves were improved and made self-acting, and it became a new engine, with the exception of a few large parts, such as cylinder, large bob, &c., and was named Dolcoath New Engine.

The greatest improvement, however, was Trevithick's new semicircular boiler-top, which, at a cost of 93*l.* 8*s.* 9*d.*, took the place of the original flat top, weighted down by slabs of granite. The cylindrical sides, and indented or curved bottom, of the Newcomen boilers gave strength to those parts; but its flat weak top prevented the use of steam of a higher pressure. Newcomen's steam-boilers were simply the ordinary household boilers used in cooking, on an enlarged scale, with the lid or top weighted down, to enable them to retain steam of one, or from that to two pounds on the square inch. Richard Trevithick, sen., removed the objectionable flat top; every part was made more or less circular, giving uniform and greater strength. The increased pressure of steam in the stronger boiler, by only a pound or two on the inch, materially increased the effective force of the engine, and practically pointed out the true source of the further increase in the power and usefulness of steam.

A story of Trevithick, sen., when at Bristol ordering cast-iron work for this engine, is still told by Captain Joseph Vivian.

"When dining at the inn the waiter remained in the room after the dinner had been placed on the table. Trevithick not wanting him, said, 'What are you doing here?' 'Oh, sir! it is my business to wait upon you, sir!' 'Well, but I do not want you here; peeping upon every bit I put in my mouth. Will you be off now?' 'Oh no, sir, I am ordered to remain!' 'You won't be off, won't you? We'll soon see then!' and striding towards the waiter with an evident inclination to shake him, he drove him out."

Trevithick, sen., also went to Coalbrookdale, which supplied much of the cast-iron work for the manufacture of the early steam-engines, for the art of casting pumps was not then known in Cornwall. The John Harvey who worked to fix the improved boiler in its house was then a country smith, at Carnhell Green, a small village a few miles from Dolcoath. He having discovered how to cast iron pipes, established the now famous Harvey and Co.'s engineering foundry at Hayle.

The improvement and re-erection of this engine by Trevithick, sen., was an important event. Fifty-five years had passed since Newcomen had worked his first Cornish pumping engines. The Carloose engine was one of them; and having done its share of work, old age and growing improvements had caused it to be set aside by the Carloose adventurers; the manager of Dolcoath Mine, however, determined to give a new life to this discarded engine. When re-erected in its improved form it cost 2040*l.* A man then earned from 1*s.* to 1*s.* 6*d.* a day. With the present rate of wages the engine would have cost 6000*l.* or 7000*l.* It was, therefore, of great importance in those rude times, and in its new form was esteemed as the most perfect steam-engine in Cornwall.

Mr. Budge was the working engineer erecting it. The cylinder was removed from its objectionable seat on the granite-boiler top, and placed on cross beams from wall to wall of the engine-house; thus avoiding the jar of the steam-cylinder, resting on the weak boiler and masonry, which had been a source of trouble in the early Newcomen.

The wrought-iron and the timber work were prepared in the mine, Arthur Woolf, sen., working on it as mine carpenter at 1*s*. 5*d*. a day.

This 45-inch open-top cylinder engine, erected by Trevithick, sen., is shown and described by Pryce.

"Mr. Newcomen and Mr. J. Cawley contrived another way to raise water by fire, where the steam to raise the water from the greatest depths of mines is not required to be greater than the atmosphere; and this is the structure of the present fire-engine, which is now of about seventy years' standing.

"B is a large boiler, whose water, by the fire under it, is converted into an elastic steam. The great cylinder C, C, is fixed upon it, and communicates with it by the pipe *d*; on the lower orifice of which, within the boiler, moves a broad plate, by means of the steam-cock, or regulator E 10, stopping or opening the passage to prevent or permit the steam to pass into the cylinder, as occasion requires. The diameter of the pipe *d* is about 4 inches.

"The steam in the boiler ought always to be a little stronger than the air, that, when let into the cylinder, it may be a little more than a balance to the external air, which keeps down the piston to the bottom *d*, *n*. The piston being by this means at liberty, the pump-rod will, by its great weight, descend at the opposite end to make a stroke, which is more than double the weight of the piston, &c., at the other end. The end of the lever at the pump, therefore, will always preponderate and descend when the piston is at liberty. The handle of the steam-cock E 10, being turned towards *n*, opens a pipe *d* to let in the steam; and being turned towards O, it shuts it out, that no more can enter. The piston is now raised towards the top of the cylinder at C, and the cylinder is full of steam. The lever O, I, must then be lifted up to turn, by its teeth, the injecting cock at N, which permits the water, brought from the cistern *g* by the pipe *g* M N, to enter the bottom of the cylinder at *n*, when it flies up in the form of a fountain, and striking against the bottom of the piston, the drops, being driven all over the cylinder, will, by their cold-

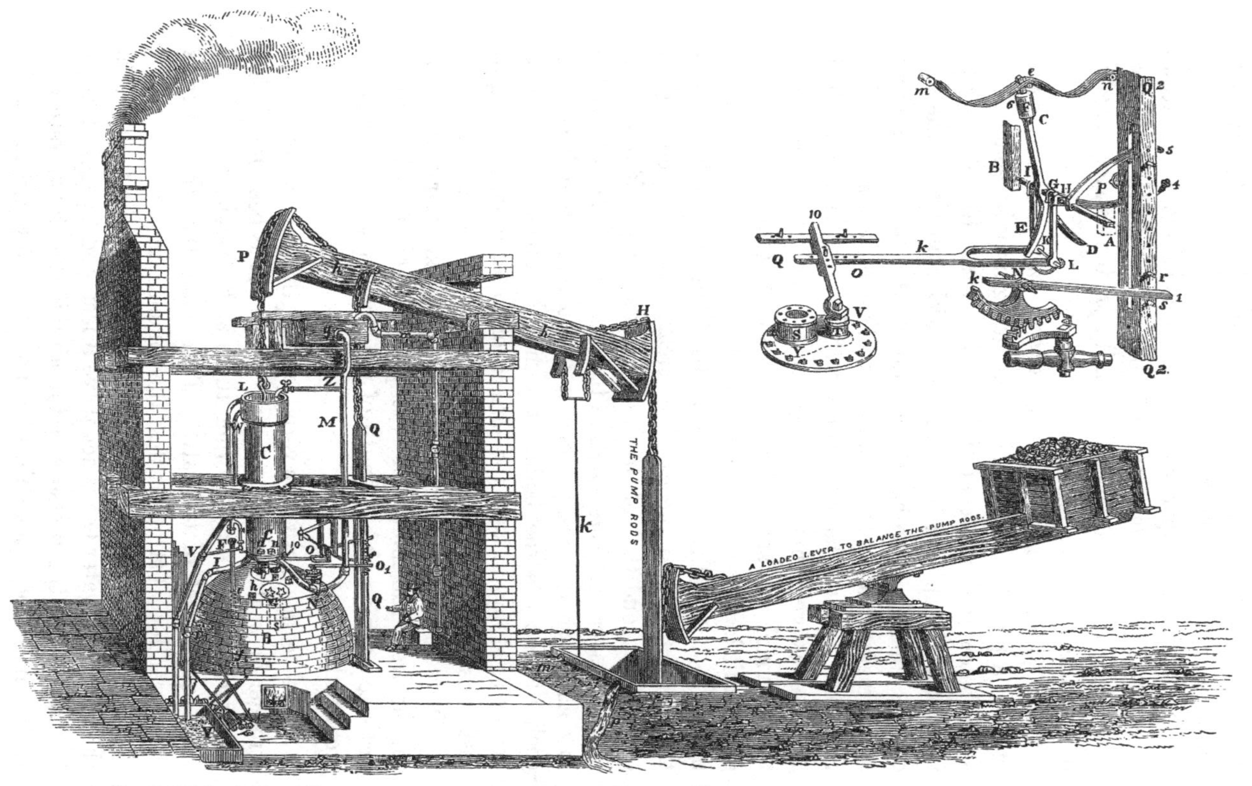

NEWCOMEN ATMOSPHERIC ENGINE, IMPROVED AND RE-ERECTED BY RICHARD TREVITHICK, SEN., AT BULLAN GARDEN, DOLCOATH, IN 1775.

ness, condense the steam into water again, and precipitate it to the bottom of the cylinder.

"Now this whole operation of opening and shutting the steam-regulator and injection-cock will take up but little more than three seconds, and will therefore easily produce sixteen strokes in a minute.

"The water in the boiler which wastes away in steam, is supplied by a pipe I *i*, about 3 feet long, going into the boiler about a foot below the surface of the water. On the top of this pipe is a funnel I, supplied by the pipe W with water from the cup of the cylinder, which has the advantage of being always warm, and therefore not apt to check the boiling of the water. That the boiler may not have the surface of the water too low, which would endanger bursting, or too high, which would not have room enough for steam, there are two gauge-pipes at G, one going a little below the surface of the water when at a proper height, and the other standing a little above it. When everything is right, the stop-cock of the steam-pipe, being open, gives only steam, and that of the long one water; but if otherwise, both cocks will give steam when the surface of the water is too low, and both give water when it is too high; and hence the cock which feeds the boiler at I may be opened to such a degree as always to keep the surface of water to its due height, lest the steam should grow too strong for the boiler and burst it. There is a valve fixed at *h*, with a perpendicular wire standing up from the middle of it to put weights of lead upon, in order to examine the strength of the steam pushing against it from within.

"The steam has always a variable strength, yet never one-tenth stronger or weaker than common air; for it has been found that the engine will work well when there is the weight of 1 lb. on each square inch of the valve. This shows that the steam is one-fifteenth part stronger than the common air.

"Now as the height of the feeding pipe from the funnel F to the surface of water G *s* is not above 3 feet, and 3½ feet of water is one-tenth of the pressure of the air; if the steam were one-tenth part stronger than air, it would push the water out at F.

"Among the great improvements of this engine, we may reckon that contrivance by which the engine itself is made to open and shut the regulator and injection-cock, and that more nicely than any person attending could possibly do it. For this purpose, there is fixed to an arch 12, at a proper distance from the arch P, a chain, from which hangs a perpendicular piece or working beam Q Q, which comes down quite to the floor and goes through it in a hole which it exactly fits. This piece has a long slit in it and several pin-holes and pins for the movement of small levers destined to the same office of opening and shutting the cocks, after the following manner: between two perpendicular pieces of wood, on each side, there is a square iron axis A, B (upper Fig., p. 25), which has upon it several iron pieces of the lever kind. The first is the piece G, E, D, called the Y, from its representing that letter, inverted by its two shanks E and D; on the upper part is a weight F to be raised higher or lower, and fixed as occasion requires. This Y is fixed very fast upon the said iron axis A, B. From the axis hangs a sort of iron stirrup I, K, L, G, by its two hooks I, G, having on the lower part two holes K, L, through which passes a long iron pin L, K, and keyed in the same. When this pin is put in, it is also passed through the two holes in the ends E, N, of the horizontal fork or spanner E, Q, N, joined at its end Q to the handle of the regulator V 10. From Q to O are several holes, by which the said handle may be fixed to that part of the end which is most convenient. Upon this axis A, B, is fixed at right angles to the Y, a handle or lever G 4, which goes on the outside of the piece Q 2, Q 2, and lies between the pins. Another handle is also fastened upon the same axis, *viz.* H 5, and placed at half a right angle to the former G 4; this passes through the slit of the piece Q 2, Q 2, lying on one of the pins. Hence we see that when the working beam goes up, its pin in the slit lifts up the spanner H 5, which turns about the axis so fast as to throw the Y, with its weight F, from C to 6, in which direction it would continue to move after it had passed the perpendicular, were it not prevented by a strap of leather fixed to it at Q *e*, and made fast at the ends *m* and *n*, in such a manner as to allow the Y to vibrate backwards and forwards about a quarter of a circle, at equal distances, on this side and that of the perpendicular."[1]

---

[1] Pryce's 'Treatise on Minerals, Mines, and Mining, in Cornwall,' 1778.

The Newcomen Pool engine of 1746, followed by the Trevithick Bullan Garden engine with improved working gear, allowing the engineman to sit at ease, while the engine moved at the increased speed of sixteen strokes a minute, and the improved boiler, give a good idea of the state and progress of the steam-engine from its first really useful working days, up to Trevithick, sen., in 1775, or even to Watt, who erected his first engine in Cornwall shortly after that date.

Trevithick's, sen., account-books, commencing in 1765, prove the use, ten years before the erection of his engine, of two steam-engines in the Dolcoath Mine, the Bullan Garden fire-engine, and the little engine; and many others were then at work, for Borlase said in 1746 "there are several other very considerable mines now worked by the fire-engine in Cornwall,—Huelrith in Godolphin Hill, Herland, Bullan Garden, Dolcoath, The Pool, Bosprowal, Huel-ros, and some others."

Feed-water was supplied from a cistern three or four feet above the boiler, serving also as a safety-valve should the steam become of greater pressure than the weight of the head of water, thus allowing the steam to escape on the water becoming lower than the bottom of the feed-pipe; two gauge-cocks and a weighted safety-valve showed the water level and the steam pressure.

A plug-rod, so called from its shifting plugs or pins, worked the gear-handles for moving the steam and injection valves or cocks; and Y shafts, named from the shape of the levers fixed on them, moved the valves. A cataract regulated the time of rest of the piston at the top of the cylinder; water from a regulating cock ran into a balanced tub, which when full descended, and capsizing its contents, returned to its former level, having during its movement opened the injection-valve, and caused the piston to descend.

These contrivances are used in the present day, in an improved form, but still retaining their old names.

Borlase wrote hopefully of a coming boiler to give more highly elastic steam; Pryce, with the accumulative improvements of twenty years, said, "The steam has always a variable strength, yet never one-tenth stronger or weaker than common air; for it has been found that the engine will work well when there is the weight of 1 lb. on each square inch of the valve."

The last words in Pryce's book relate to Watt and his expected improvements; and his view of the best steam-pressure was just that which Watt adopted and acted on.

It is unimportant whether Trevithick, sen., used the increased pressure of the steam; he certainly constructed a boiler on scientific principles that allowed of its use. Stuart, fifty years after Pryce, gave a copy of Pryce's drawing of Trevithick's engine erected in Bullan Garden, a part of Dolcoath Mine; but Stuart in error calls it "a view of the atmospheric engine as improved by Beighton." The detail accounts of its construction by Trevithick, sen., make no mention of the name of Beighton. The reader who cares to examine the drawings by Pryce and Stuart will observe that this copy is in perspective, while the other drawings by Stuart are not. Pryce states that his drawing of the engine was made at his own expense and was dedicated to his kinsman John Price, late High Sheriff for the county of Cornwall. Trevithick, sen., was the mineral agent for the great mining property of Lord Dedunstanville and Basset, in which were situated the leading mines of that day, retaining the post for twenty-two years, to the time of his death.

In 1776 he received a grant from Sir Francis Basset, of the mine sett of Roskear (or Wheal Chance), which became a large mine, and is still at work.

Another of his account-books, commencing January, 1777, contains a list of sixty-four different Cornish copper mines then at work; the greater number of which must have used steam-engines. It also gives particulars of the sale of copper at the Cornish ticketings for the two years 1777 and 1778, showing a yearly produce of about 24,000 tons of copper ore, worth 156,000*l.*

These account-books show that he was the manager of Dolcoath, Wheal Chance, now Roskear, Wheal Treasury, and Eastern Stray Park, at 2*l.* a month from each, the greatest pay in the cost-books of the time: being the lord's agent he had the leading authority in the principal Cornish mines situated on the Basset property.

In the year following Trevithick's engine, Watt erected his first working engine.

"In 1776 Watt, after much difficulty, erects his first working engine; and in 1777 erects his first engine in Cornwall at Wheal Busy Mine. Here he met the Hornblowers, who had been erecting engines for fifty years, and Bonze, who had five engines at work, with cylinders of 60 to 70 inches in diameter.

"In 1778 Watt is again in Cornwall, and says the Chacewater engine goes satisfactorily, making fourteen strokes per minute, and others are ordered. Even the infidels of Dolcoath are now obliquely inquiring after our terms.

"He would almost as soon have wrestled with the Cornish miners as higgled with them. They were shrewd, practical men, rough in manner and speech, yet honest withal.

"In 1779 Murdoch joined Watt in Cornwall. Watt, in writing to Boulton, says:—'At Wheal Union account our savings were ordered to be charged to the interest of Messrs. Edwards and Phillips; but when to be paid, God knows! Bevan said in a month. After all this was settled, in came Captain Trevithick, I believe on purpose, as he came late, and might have heard that I was gone there. He immediately fell foul of our account, in a manner peculiar to himself; laboured to demonstrate that Dolcoath engines not only surpassed the table, but even did more work with the coals than Wheal Union did, and concluded with saying that we had taken or got the advantage of the adventurers. I think he first said the former, and then edged off by the latter statement. Mr. Phillips defended, and Mr. Edwards, I thought, seemed staggered, though candid. Mr. Phillips desired the data, that he might calculate it over in his way. Mr. Edwards slipped away; but I found afterwards that he was in another room with Captain Gundry (who, and Hodge also, behaved exceedingly well; I believe Gundry to be a very sincere, honest man). I went out to speak to Joseph, and on my return found only Trevithick, Bevan, Hodge, and some others. Soon after, Mr. Edwards called out Trevithick to him, and Gundry. I heard them very loud, and waited their return for an hour; but they not seeming ready to return, night coming on, and feeling myself very uncomfortable, I came away, so know not what passed further. During all this time I was so confounded with the impudence, ignorance, and overbearing manner of the man, that I could make no adequate defence, and indeed could scarcely keep my temper, which, however, I did, perhaps to a fault; for nothing can be more grievous to an ingenuous mind than the being suspected or accused of deceit. To mend the matter, it had been an exceedingly rainy morning, and I had got a little wet going thither, which had rather hurt my spirits. Yesterday I had a violent headache and could do nothing. Some means must be taken to satisfy the country, otherwise this malicious man will hurt us exceedingly. The point on which Mr. Edwards seemed to lay the most stress was the comparing with a $77\frac{1}{10}$ cylinder, as he alleged they would not have put in so large an engine; and in this there is some reason, as I do not think they believe that the engine would be so powerful as it is. Add to this that the mine barely pays its way. Trevithick made a great noise about short strokes at setting on, &c. The Captains seemed to laugh at that; and I can demonstrate that, were it allowed for, it would not come to 2*s.* 6*d.* per month. I believe they can be brought to allow that they would have put in a 70-inch. Now query if we ought to allow this to be calculated from a 70 (at which it will come to near 400*l.* a year), and on making this concession insist on our having a good paymaster to pay regularly once a month, and not be obliged to go like beggars to their accounts to seek our due, and be insulted by such scoundrels into the bargain. As to Hallamanin, they have not met yet, and when they do meet I shall not go to them. I cannot bear such treatment; but it is not prudent to resent it too warmly just now. I believe you *must* come here. I think fourteen days would settle matters. Besides my inability to battle such people, I really have not time to bestow on them.'

"In the autumn of 1780 Boulton went into Cornwall for a time, to look after the business there; several new engines had been ordered, and were either erected or in progress, at Wheal Treasury, Tresavean, Penrydel, Dolcoath, Wheal Chance, Wheal Crenver, and the United Mines.

"One of the principal objects of his visit was to settle the agreements with the mining companies for the use of these engines. It had been found difficult to estimate the actual savings of fuel, and the settlement of the accounts was a constant source of cavil."[1]

In 1779 Watt, with his low-pressure steam vacuum engine, had declared war against Trevithick, sen., and his improved higher pressure steam-engine. Watt's first impression of "the infidels of Dolcoath" was that they were "obliquely inquiring"; but after a year's unfruitful negotiation, he "was so confounded with the *impudence, ignorance, and overbearing manner of the man*" (Trevithick, sen.), that he could make no adequate defence.

In the autumn of 1780 several new engines had been

[1] 'Lives of Boulton and Watt,' by Smiles, p. 270

ordered from Watt for Cornwall, for Wheal Treasury, Tresavean, Penrydel, Dolcoath, Wheal Chance, Wheal Crenver, and the United Mines.

The spirit of rivalry and even of mistrust arose of necessity between Trevithick, sen., and Watt; but it is much to the credit of the former that, within three years of the erection of Watt's first engine in Cornwall, he, as manager, ordered his rival's engine for Wheal Treasury, Dolcoath, and Wheal Chance, being three out of the seven mines that at that time had agreed to try Watt's engine.

The Watt Dolcoath engine had a cylinder of 63 inches in diameter, with a cylinder cover, and an air-pump and condenser, and was called the great 63-inch double engine. It was erected near to a 63-inch Newcomen engine, mentioned by Watt as Bonze's, with open-top cylinder, and known as the fire-engine, and was also close alongside of the 45-inch open-top cylinder Carloose, erected by Trevithick, sen., known afterwards as the Shammal engine, because those two engines pumped from the same shaft, which was called Shamalling.

The oldest account-book now in Dolcoath Mine going back to 1783, has the entry of 52*l*. paid to Boulton and Watt for a month's saving by their patent engine.

As these three engines will be frequently spoken of, it may here be mentioned that the birth of the two oldest is lost in obscurity. The 45-inch cylinder Carloose, after her first life-time, was renewed in 1775. The little fire-engine is mentioned as working, in Trevithick's account-book of 1765, but there is no intimation of when the 63-inch atmospheric began to work. Watt's 63-inch double engine dates from 1780.

The oldest account-book in Cook's Kitchen Mine shows that under the management of Trevithick, sen., in 1794, Watt was paid 18*l*. a month for the saving of coal. The account-books of Trevithick, sen., as manager of Wheal Treasury Mine, in 1795, show that Bull, Trevithick, jun., and Watt were all rivals, and received payment for the saving of coal by their respective engines; but at this date the powers given to Watt under his patent pressed with a leaden weight on the inventive skill of Cornish engineers.

Trevithick, sen., was, as an engineer, the forerunner and liberal fellow-workman and patronizer of Watt's improvements, but a determined opponent of all illiberal and exclusive acts tending to curtail the growing usefulness of the steam-engine. He has been called a mine manager, because in his time the mine manager was understood to be an engineer. All repairs, and even new work, with the exception of large castings, were made in the mine: thus he was a man of all-work.

One of his Dolcoath account-books is headed "Furnace cost for 1771," showing that in those days the Halvans, or inferior portions of the copper ore, not worth the cost of carriage to the neighbouring copper-smelting works at Copperhouse, were partially smelted in the mine, in the same way as it now comes from South America under the name Regulus. The account is continued for three years, and gives a sample of the labour and wages of the time:—

"Richard Williams, 30 days at 32*s.* per month.
Mary Osbron, 4 days at 5*d.* per day.
Ann Heather, for bucking 319 barrows of calcined work, at 6*s.* for every 15 barrows.
To filling, heaving, and carrying 1190 sacks of Halvans, at 3*s.* per hundred sacks.
To 1 dozen candles from the adventurers, 6*s.* 9*d.*
William John, for a lantern and book, 3*s.*
Richard Woolf, for a watering-pot, 3*s.* 6*d.*
Doctor column, 1*s.* 6*d.*"

Presuming that the labourer's wages at 32*s.* per month meant four weeks, he then worked for 1*s.* 4*d.* a day, while a woman earned 5*d.* a day. Candles at $6\frac{3}{4}$*d.* each seems strange; but there are frequent entries at the same rate. Lanterns and books were supplied by the same man; and the ancestors of Arthur Woolf, the engineer, sold watering-pots. A monthly subscription of 1*s.* 6*d.* to 2*s.* was paid, by the furnace account, to help the sick, under the head Doctor column.

As lord's agent and leading mine manager, he attended the Cornish ticketings. His account-book shows that at the four weekly ticketings at Redruth, in the month of February, 1777, from thirteen mines, 1939 tons of copper ore were sold, worth about 12,748*l.* a month.

The mines selling ore during the two years of 1777 and 1778 are given, that their names may not be lost.[1]

---

[1] SALE OF COPPER ORE AT THE REDRUTH TICKETING FOR THE MONTH OF FEBRUARY, 1777.

From the Account-book of Trevithick, sen.

Date	Mines	Tons of Copper Ore.
Feb. 4, 1777	Dolcoath .. .. Cook's Kitchen ..	414
„ 11 „	Poldice .. .. Wheal Maid .. Wheal Fortune, St. Erth	289
„ 18 „	Roskeare .. .. Wheal Rosewarne „ Lane .. „ Gerry .. Great Close ..	790
„ 25 „	Wheal Virgin .. Carharrack .. .. Great St. George	446
		1939

The average price was 6*l.* 13*s.* per ton, giving a monthly value of 12,748*l.* during the two years. The monthly sales increase slightly.

The names of the various mines selling during the two years are:—

Wheal Burrow.
Burn Coose.
Baldieu.
Cook's Kitchen.
Carharrack.
Chacewater.
Cleggar.
Wheal Cock.
Carzisewood.
Carloose.
Wheal Crenver.
Camborne Vean.
Dolcoath.
Wheal Dinner.
Dudnance.
Wheal Fortune, in St. Erth.
Wheal Fortune, in St. Hillary.
Wheal Fortune, in Redruth.
Wheal Fortune, in Breage.
Wheal Gerry.

[Great

They amount to sixty-four mines, and probably each of them used one or more steam-engines. It was at this period that Watt erected his first Cornish engine; the early use of the steam-engine in Cornwall having caused him to visit it in search of orders.

The mine manager was responsible for the proper carrying out of the underground operations, and the description by Pryce, together with the account-books of Trevithick, sen., establish the fact that, prior to Watt's first working engine in 1776, three steam pumping engines had been at work in Dolcoath under Trevithick, sen., probably for many years, for the Little fire-engine was at work ten or more years before, and during those ten years the 63-inch and the 45-inch cylinder Newcomen atmospherics were erected. Watt saw them on his first visit to Dolcoath Mine in 1777, with the bobs and cranks in the water-engine shafts, spoken of by Trevithick, sen., in 1765, a dozen years before Wasbrough patented the crank as his invention

---

Great Close.
Great St. George.
Wheal Gons.
Gunnis Lake.
Horrowbough.
Hallamannin.
Wheal Hatchet
Herland.
Wheal Jelbis.
Wheal Kitty.
Kestril Adventurers.
Wheal Lane.
Lelant.
Longclose.
Wheal Laity.
Wheal Maid.
North Downs.
North Pool.
Oatfield.
Wheal Orphan.
Poldice.
Wheal Prosper.
Wheal Providence.
Wheal Publick.
Roskeare.
Wheal Rosewarne.
Wheal Raven.
Retallack.
Wheal Rack.
Relubis Vean.
Wheal Sparrow.
South Grambler.
Wheal Seymour.
St. Quick.
Trevenson.
Wheal Truan.
Treskerby.
Wheal Treasure.
Wheal Towan.
Wheal Virgin.
West Wheal Virgin.
West Good Success.
West Grambler.
West Penhellick.

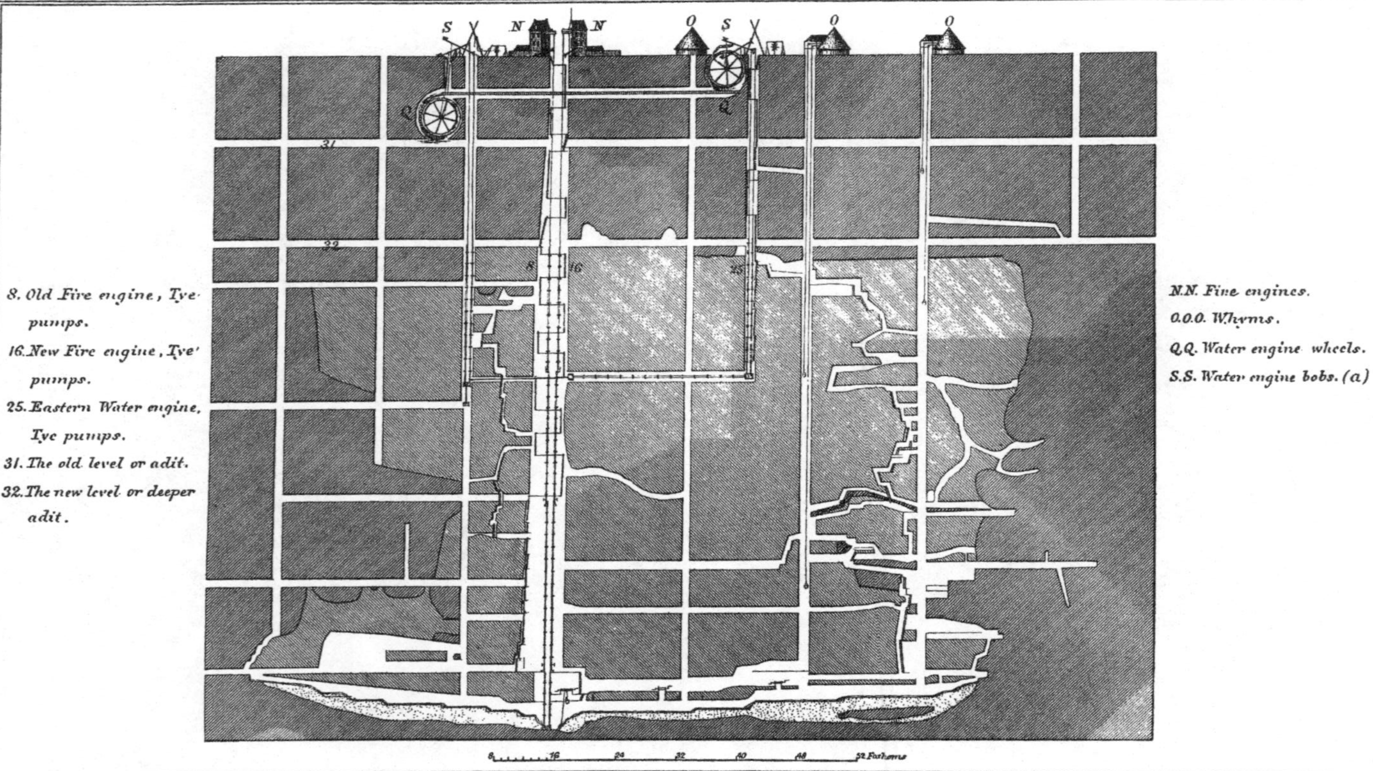

London: E.& F.N. Spon, 48, Charing Cross

Kell. Bros Lith London.

DALCOATH MINE

when applied to the steam-engine. Pryce's drawing, Plate I., showing the new adit by Trevithick, fixes the period of a most important stride in the science of mining in Cornwall. The vertical lines represent shafts, the horizontal lines levels; the large central space, excavated lode, downward from level to level; the bottom line of this excavation shows the slight slope of the ground for drainage toward the bottom of the mine, where the wind-bore or lowermost pipe in the pump-lift is fixed.

Immediately after 1778 such a change was effected by Trevithick, sen., in the principle of mining, that a comparison of modern workings with those given by Pryce makes evident its importance; and from that time the miner, instead of breaking ground under his feet, broke it from over his head, he having first sunk the drainage shaft to the required depth for a lower level, from which the ore may be excavated up to the level above, while by this process the drainage and the railway for removing the broken rock are not disturbed.

The light portions show the extent of the lode worked away, 500 or 600 feet in length, and 400 or 500 feet in depth; the width of the mineral lode varying from a few inches to two or three, or more feet, in many places widening out to several feet. The shafts for raising water and mineral are shown vertical; a cross-section would show many of them in very varying inclines, following the underlay of the lode. The main-pump shaft is generally perpendicular for the convenience of the pump-work, in which case the course of the lode is not followed, the shaft being sunk to meet the underlying lode at an understood depth. The horizontal lines are levels on the course of the lode, 4 or 5 feet wide by 6 or 7 feet high. The levels branch from the shaft 50 or 60 feet one below another.

If the lode is too unproductive for profit, the level is continued in search of better portions or bunches; when such are found, the whole ground is excavated from level to level, of the width of the lode.

The old plan was for each miner to dig or break the ground at his feet, destroying the road over which the broken rock had to be conveyed in barrows, and also preventing the free drainage of water from the miner's work. The new plan is for the miner to work upwards, towards the level above him; by this means the very serious inconvenience of working in a pool of water is avoided, and the roadway kept good. Apply such changes to hundreds of men breaking and removing hard rock, in the cramped space and unhealthy atmosphere of a mine, and its importance is evident.

Mr. Henwood believes that the idea was brought to Dolcoath from Germany, by Mr. Raspe, about 1782;[1] the putting it into practice was the work of Trevithick, sen., the manager,[2] and this system of working is now universal in the Cornish mines.

Dolcoath of the present day is 1800 feet deep, employs 1000 work-people, and yields annually 50,000*l.* worth of tin; during the sixteen years ending 1865, it sold 748,891*l.* worth of ore, giving a profit to the shareholders of 130,655*l.* The invested capital is only nominal, outlay for new machinery being paid for as a working expense. Trevithick's accounts take us back more than a hundred years, but Dolcoath must have

[1] Raspe has been called the author of 'Baron Munchausen.'

[2] 'Address to the Royal Institution of Cornwall,' by William Jory Henwood, F.R.S., F.G.S. 1869.

been a mine many years before that, for two steam pumping engines worked there in 1765.

"In 1796, Watt, writing to Boulton, says of Cornishmen:— The rascals seem to have been going on as if the patent was their own. We have tried every lenient means with them in vain; and since the fear of God has no effect upon them, we must try what the fear of the Devil can do.'

"Legal proceedings were begun accordingly. The two actions on which the issues were tried were those of Boulton and Watt *v.* Hornblower and Maberley; and they were fought on both sides with great determination.

"The proceedings extended over several years, being carried from Court to Court; but the result was decisive in both cases in favour of Boulton and Watt.

"It was not until January, 1799, that the final decision of the judges was given, almost on the very eve of expiring of the patent which had not then a full year to run."[1]

When Watt, in 1796, challenged Trevithick, sen., he could answer for himself, but at the last-mentioned period the son had buckled on his father's armour

An account-book of Richard Trevithick, sen., headed "Wheal Treasury Cost," commencing March, 1795, says:—

"To paid Bull 20*l.* per month for the saving of fuel by his engine.

"To paid Richard Trevithick, jun., for saving of coals by his engine, 18*l.* per month."

In April, 1796, is the following:—

"Note of transfer of a share in the mine, with five engines, &c., subject to pay any demands, which may be hereafter made, by Messrs. Boulton and Watt, for savings in the said mine."

In May of the same year:—

"At a meeting at Praze-an-Beeble, it was unanimously re-

[1] See Smiles' 'Life of Watt.'

solved, That the savings claimed by Boulton and Watt should not be paid until the validity of their patent should be fully proved."

The last twenty years in the life of Trevithick, sen., were passed among events of great engineering importance. Watt's arrival in 1777 broke up the old friendly clique of Cornish engineers and mine managers: practical mine workers on one side, and influential mine shareholders on the other, had different views of the proposals of the new engineers. A depressed metal market added to the difficulty. Mines made little or no profit. Many ceased to work; and among them the famous Dolcoath was closed in 1789. Boulton and Watt not only got their monthly payments with difficulty, but had patent lawsuits with the Cornish engineers and mine proprietors.

In 1796 Wheal Treasury, under the management of Trevithick, sen., which had been paying Watt for the saving by his engine, had also at work a rival engine by Bull, and another by Richard Trevithick, jun., each of the three engineers receiving pay for his improved engine, while a unanimous decision was come to by the shareholders to resist Watt's patent claims.

In April, 1797, the accounts in Wheal Treasury ceased to be entered by the old hand; and in December of that year another made up the account, closing with an entry,—"Paid John Glasson for making up the accounts in this book to December, 1797." Trevithick's hand was cold. He had passed his life in the midst of Cornish mines, having numerous steam-engines working around him—the best engines of the day—on his own particular mines, before Watt had erected an engine. When, in 1777, Watt and his first improved steam-engine appeared in Cornwall, Trevithick cautioned him not to promise too much, or to tempt the mine speculators by offering to supply engines on the promise to pay a monthly hire, based on the saving in the consumption of coal, as compared with the Cornish engines then in use, but Trevithick did not oppose its introduction.

Boulton and Watt's patent lawsuits resulted in a verdict in their favour, on the eve of the expiration of their patent. Richard Trevithick, sen., having lived to see his son an engineer, competing with Watt, and married to the daughter of his old friend, John Harvey, died on the 1st August, 1797, at Penponds, near Camborne, and was buried in Camborne churchyard. For many years he had enjoyed friendship, and frequently companionship, with the great religious reformer, John Wesley, who on his visits to Cornwall made a home of the house of Richard Trevithick, and used it as a place of meeting of the brotherhood, Trevithick being the class-leader in his district. This was a dark time for Cornwall,—death, law proceedings, and poverty were rife; and the numerous and prosperous mines of twenty years before had dwindled down to bare walls and barren mine heaps. In the once busy district of Illogan and Camborne only four steam-engines remained at work out of the forty or fifty which Trevithick, as lord's agent, had helped to set in motion. Stone from the engine-houses on the idle Dolcoath were taken to build the present inn at Camborne; miners' houses were untenanted, and the people without employ.

Trevithick, sen., in his active life of half a century, took part in this rise and fall of Cornish mining and engineering. The prosperous part was before the use of Watt's improved engines; while the fall came with their introduction, and so nearly buried Watt in the ruins, that he contemplated giving up their construction, to lead a less harassing and more profitable life.

A little more than half a century has again passed, and that same district now employs one hundred steam-engines; and in this revival we shall find that Richard Trevithick, jun., took an active part.

No sculptured stone points out the resting-place of Richard Trevithick, sen., who lies not far from the Druids' rocks and ancient castle of Carn Brea.

CHAPTER III.

SMEATON AND WATT.

IN forming an estimate of Richard Trevithick, sen., as the forerunner, and of his son as following Watt, in the improvement of the steam-engine, it is necessary briefly to refer to the engines of Smeaton and of Watt in Cornwall.

The question of rebuilding the Eddystone Lighthouse led Smeaton to visit Plymouth in 1756, when he probably saw the Cornish mines; ten years afterwards he erected several large atmospheric engines, one of which was sent to Russia. In 1775 he supplied the Chacewater Mine, in Cornwall, with an atmospheric engine, having a 72-inch cylinder, 9-feet stroke, working nine strokes a minute. The main beam was of twenty pieces of fir timber, bolted together, and the cylinder was supported on twelve pieces of fir, in place of the beam of oak used by Trevithick, sen. The cylinder bottom was a half sphere, instead of the flat bottom as in Dolcoath. Three hay-stack boilers, each 15 feet in diameter, supplied steam of 1 lb. on the inch above the atmosphere, one of which was placed under the cylinder. The piston bottom was cased with wood to prevent loss of heat. The low steam pressure reduced its power of quick motion, and its hollow cylinder bottom increased the consumption of fuel. It was to do the work of a 62-inch and a 64-inch cylinder that before had drained the mine. Smeaton says that, prior to 1769, one hundred engines had worked in the Newcastle-upon-Tyne collieries; fifteen of them averaged a duty of 5·59 million lbs. raised one foot high by 84 lbs. of coal: such was the use and the economic power of the steam-engine prior to the Watt patent in 1769.[1]

---

[1] "My method of lessening the consumption of steam, and consequently fuel, in fire-engines, consists in the following principles: First, that the vessel in which the powers of steam are to be employed to work the engine, which is called the cylinder in common fire-engines, and which I call the steam-vessel, must, during the whole time the engine is at work, be kept as hot as the steam which enters it; first, by enclosing it in a case of wood, or any other materials that transmit heat slowly; secondly, by surrounding it with steam or other heated bodies; and thirdly, by suffering neither water nor other substance colder than the steam to enter or touch it during that time.

"Secondly, in engines that are to be worked wholly or partially by condensation of steam, the steam is to be condensed in vessels distinct from the steam-vessel or cylinder, though occasionally communicating with them. These vessels I call condensers; and whilst the engines are working, these condensers ought at least to be kept as cold as the air in the neighbourhood of the engines, by application of water or other cold bodies.

"Thirdly, whatever air or other elastic vapour is not condensed by the cold of the condenser, and may impede the working of the engine, is to be drawn out of the steam-vessels or condensers by means of pumps, wrought by the engines themselves or otherwise.

"Fourthly, I intend, in many cases, to employ the expansive force of steam to press on the pistons, or whatever may be used instead of them, in the same manner as the pressure of the atmosphere is now employed in common fire-engines. In cases where cold water cannot be had in plenty, the engines may be wrought by the force of steam only, by discharging the steam into the open air after it has done its office.

"Fifthly, where motions round an axis are required, I make the steam-vessels in form of hollow rings, on circular channels, with proper inlets and outlets for the steam, mounted on horizontal axles like the wheels of a water-mill; within them are placed a number of valves that suffer any body to go round the channel in one direction only. In these steam-vessels are placed weights so fitted to them as entirely to fill up a part or portion of their channels, yet rendered capable of moving freely in them by means hereinafter mentioned or specified.

"When the steam is admitted in these engines between these weights and the valves, it acts equally on both, so as to raise the weight to one side of the wheel, and by the reaction of the valves successively to give a circular motion to the wheel; the valves opening in the directions in which the weights are pressed, but not in the contrary.

"As the steam-vessel moves round, it is supplied with steam from the boiler, and that which has performed its office may either be discharged by means of condensers or into the open air.

"Sixthly, I intend in some cases to apply a degree of cold not capable of reducing the steam to water, but of contracting it considerably, so that the engines shall be worked by the alternate expansion and contraction of the steam.

"Lastly, instead of using water to render the piston or other parts of the engines air or steam tight, I employ oils, wax, resinous bodies, fat of animals, quicksilver, and other metals in their fluid state."

"In 1770 Watt sent his drawings to Soho for his first engine. The castings were made at Coalbrookdale, but were found exceedingly imperfect, and were thrown aside as useless.

"In 1776 the first Watt engine was built at Soho, after much difficulty in finding suitable workmen, and many inquiries for the new engine were coming from Cornwall.

"In 1777 Watt went to Cornwall and erected an engine at Wheal Busy, near Chacewater, and another at Ting Tang, near Redruth. Here he met the Hornblowers, who had been erecting engines in Cornwall for fifty years, they having come from Staffordshire.

"Bonze seems to have been one of the early practical men. Watt says he found five of Bonze's engines in Cornwall with cylinders of from 60 to 70 inches in diameter.

"In 1778 Watt was again in Cornwall, and says, 'The engine at Chacewater goes satisfactorily, making fourteen strokes per minute.'"[1]

Watt's close top to the cylinder, and parallel motion, were great improvements over the open-top cylinder and segment-headed beam before in use; but the condensing away from the cylinder, and use of an air-pump, were the leading features of his claims as an improver of the steam-engine. These changes led to a saving of fuel, and the improved mechanical form gave greater control, but neither the speed nor power of the engine was materially increased. The pressure of steam remained about the same. The vacuum was more perfect in the cylinder, but the labour of working the air-pump reduced the net gain, and the engine was still a low-pressure steam vacuum engine, though greatly improved in usefulness.

"Watt says, in 1761 or 1762:—'I made some experiments on the force of steam in a Pepin's digester, and formed a species of steam-engine. But I soon relinquished the idea of construct-

[1] Smiles' 'Lives of Boulton and Watt.'

ing an engine upon this principle, from being sensible it would be liable to some of the objections against Savery's engine; from the danger of bursting the boiler, and the difficulty of making the joints tight, and also that a great part of the power of the steam would be lost, because no vacuum was formed to assist the descent of the piston.'

"In 1782, Watt patents an *expansive engine*, applicable both to double and single engines. In both these forms of the apparatus the steam acts not only to form the vacuum, but to depress the piston. But still, during the operation of the counterpoise, it produces no effect; and when it was required to move machinery, this suspension of impulse was a great drawback on its utility. This, however, was not objected to its general merit when used as a mover of pumps, and the more so, as it was common to the atmospheric engine.

"The power of the condensing engine is easily known by ascertaining the temperature of the steam which moves the piston, the area of the piston, and the temperature of the vapour which remains in the condenser. It is, however, found most expedient to raise the steam to a somewhat higher temperature than 212°, so as to produce a pressure between 17 and 18 lbs. on each square inch of the piston; yet, in practice, from the imperfect vacuum which is made in the condenser, and after making allowance for the friction of the piston on the sides of the cylinder, and for the friction of the various parts of the intermediate machinery, this pressure of 18 lbs. on each square inch of the piston cannot raise more water per inch than would weigh about 8½ lbs., so that somewhat more than a half of the whole power of the steam is absorbed to give motion to the intermediate mechanism."[1]

In 1778 Smeaton applied to Watt for a licence to attach the patent condenser and air-pump to the atmospheric engine, and received the following reply:—

"By adding condensers to engines that were not in good order, our engines would have been introduced into that county

[1] Stuart's 'History of the Steam-Engine,' pp. 97, 127, 131; published 1824.

(which we look upon as our richest mine) in an unfavourable point of view, and without such profits as would have been satisfactory, either to us, or to the adventurers. Besides, where a new engine is to be erected, and to be equally well executed in point of workmanship and materials, an engine of the same power cannot be constructed materially cheaper on the old plan than on ours. The idea of condensing the steam by injecting into the eduction-pipe, was as early as the other kinds of condensers, and was tried at large by me at Kinneal. We shall have four of our engines at work in Cornwall this summer; two of them are cylinders of 63 inches diameter, and are capable of working with a load of 11 or 12 lbs. on the square inch."[1]

Such was the estimate by those two eminent men of the relative power of the Watt and the Newcomen engine, using steam of a pressure just sufficient to overcome the weight of the atmosphere, trusting to the condensation and vacuum for its effective power. The vacuum was more perfect in the Watt engine than in the earlier atmospheric engine of Newcomen, but this seeming gain was much reduced by the power required to work the air-pump. A gross power, made up of 14 or 15 lbs. from vacuum, and 2 or 3 lbs. steam pressure, was reduced to one-half by the numerous drawbacks in the movements of the engine.

In 1712 the Griff atmospheric engine had an effective force of 10 lbs. on each square inch of the piston. In 1746 the atmospheric engine at Pool Mine, with steam of 2 or 3 lbs. to the inch above the atmosphere, worked fifteen strokes a minute. In 1758 Borlase said that the quickness of movement of the atmospheric engine in Cornwall depended on the steam pressure driving the engine faster or slower. In 1775 the Bullan Garden atmospheric worked sixteen strokes a minute. In 1778

[1] Farey 'On the Steam-Engine,' p. 329.

Watt's Chacewater engine, with steam about 1 lb. on the inch above the pressure of the atmosphere, worked fourteen strokes a minute.

The working therefore of an engine by the elastic force of steam was more advanced a quarter of a century before Watt's patent, than a quarter of a century after it. The Newcomen pumping engine, when at rest, had its piston at the bottom of the cylinder, under which the valve admitted steam of 2 or 3 lbs. on the inch, counterbalancing the weight of the piston, and raising it to the top of the cylinder; a jet of cold water produced a vacuum, and caused the piston to descend with a force of 14 lbs. of atmospheric pressure on each square inch of its surface, together with its own weight of 2 or 3 lbs. on the inch.

The Watt pumping engine had a close-topped cylinder, with its piston at the top when at rest; a valve admitted on it steam of 1 or 2 lbs. on the inch; the equilibrium valve then allowed the steam to pass also into the portion of the cylinder under the piston; an exhaust-valve at the bottom of the cylinder then passed the steam from under the piston to the condenser, where a jet of cold water caused a vacuum, and the piston descended with a force of 14 lbs. from vacuum, and 1 or 2 lbs. of steam pressure on the piston, being together no more than the atmospheric engine.

It is beyond our comprehension that for fifty years the use of the steam-engine was confined to the pumping of water.

In 1780 Mr. Matthew Wasbrough and Mr. James Pickard gave rotary motion from the steam-engine by a crank. Watt objected to the use of his patent engine, if the crank was attached to it, and invented sun-and-planet wheels in place of the crank.

The rivalry of patent claimants, or ignorance, retarded the use of the rotary steam-engine until 1784. This is the more remarkable when bobs with connecting rods attached to cranks on the axles of water-wheels were in common use in Cornwall before 1758, and must have been seen by Watt during his residence there in 1777. They were so used by Trevithick, sen., before the time of Watt, in Bullan Garden and other mines in which Watt erected engines. There is this difference, that in the one case the crank gave motion to the bob, and in the other the bob gave motion to the crank.

"More effectual is the water-wheel and bob: an engine whose power is answerable to the diameter of the wheels, and the length of the bobs, fastened to its axis by large iron cranks; a perpendicular rod of timber to each end of the bobs works a piston in a wooden or (which is far better) a brass hollow cylinder."[1]

The drawing of Bullan Garden in Chapter II. shows the cranks, connecting rods, and beams, in general use in Cornwall for half a century before Watt used them.

"In 1781, Wasbrough having entered into an arrangement with the Commissioners of the Navy to erect a crank engine for grinding flour at the Deptford Victualling Yard, made a formal application to Boulton and Watt to supply their engines for the purpose: Watt protested that he could not bring himself to submit to such an indignity."[2]

"Up to 1780, the use of the steam-engine was confined to the raising of water. The earliest of Mr. Watt's steam-engines, giving a rotatory movement,

---

[1] See Borlase, published 1758.
[2] Smiles' 'Lives of Boulton and Watt.'

were erected in 1784—one for Mr. Whitbread's brewhouse, and one for the Albion Mills."[1]

The crank was also about that time applied to atmospheric engines, but the disputes between Watt and others, on their priority of claim and patent rights, on the new invention of the old crank, retarded its use in the steam-engine for several years. Watt's first rotary engine, in 1784, had his sun-and-planet wheels, and the simple but unfortunate crank, with its patent honours, had to wait still longer before it was allowed to take its place as the most useful limb in the growing steam-engine.

Cook's Kitchen water-wheel, still at work, though well known before the time of Watt, was thus spoken of in 1778:—

"The water-engine at Cook's Kitchen Mine is 48 feet in diameter, and works her tiers of pumps of 9-inch bore, which, being divided into four lifts, draw 80 fathoms under the adit. The water-wheel with bobs, whose power is answerable to the diameter of the wheel, and the sweep of the cranks, fixed in the extremities of the axis; over them the large bobs are hung, upon brass centre gudgeons, supported by a strong frame of timber, and rise and fall according to the diameter of the sweep of the crank.

"To each crank is fixed a straight half-split of balk timber, that communicates with each bob above; at the other end or nose of the bob, over the shaft, a large iron chain is pendent, fastened to a perpendicular rod of timber, that works a piston in an iron or brass hollow cylinder."[2]

This wheel, used by Trevithick, sen., a hundred years ago, was at work before Watt erected his 63-inch cylinder engine in Dolcoath, from which it was distant

[1] Memoir of James Watt, 'Mechanics' Magazine,' August 30, 1823.
[2] See Pryce, published 1778.

but a stone's throw. Hunter renewed it under Trevithick, jun., in 1803.[1] In the Valley smiths' shop near it, Trevithick, jun., constructed his globular tubular boiler of 1800, and also his cylindrical boiler of 1811 for the Dolcoath engine.

---

[1] See chap. vii.

COOK'S KITCHEN WATER-WHEEL AND VALLEY SMITHS' SHOP. [W. J. Welch.]

CHAPTER IV.

EARLY LIFE OF TREVITHICK.

RICHARD TREVITHICK, jun., was born on the 13th of April, 1771, under the parental roof, in the parish of Illogan, county of Cornwall. His birth-place is now a double cottage, around which clouds of mineral sand from the surrounding mine-works float in the wind, depositing layers of pounded rock on everything in house and garden. A hundred years ago it was the manager's residence, delightfully situated at the foot of the north-west slope of Carn Brea Hill, with its ancient castle and Druidical legends. It is in the centre of the famous old and rich mines of Dolcoath, Cook's Kitchen, Pool, Tin Croft, and Roskear, and within a mile of each of them.

He was the first surviving son of five children, and was the mother's pet. Shortly after his birth the family removed a few miles, to a leasehold of Penponds, near Camborne. His mother's wedding-ring, which the writer has just placed on his finger, is seven-eighths of an inch internal diameter, and weighs a quarter of an ounce. On the inside of the ring are the words, "God above, increase our love." The letters are not engraved, but indented with a common chisel, probably the lover's handiwork. The boy's first and only school was in the adjoining small town of Camborne; the master reported him a disobedient, slow, obstinate, spoiled boy, frequently absent, and very inattentive.

Stories are told of his remaining by himself for hours, drawing lines and figures on his slate in place of the school lesson.

His school attainments were limited to reading, writing, and arithmetic. The master once said, "Your sum may be right, but it is not done by the rule." Trevithick replied, "I'll do six sums to your one." His father wished him to sit at the office desk in one of the mines, but he chose to wander through the mines by himself, holding little converse with others, but well able to defend himself in case of attack.

A difficulty among mine agents about some underground levels led young Trevithick to offer to correct the error. Old heads disapproved at first, and then allowed him to try his hand. Genius enabled him to comprehend the rude surveying instruments of that day, and the untrustworthy character of the magnetic needle when near iron tools or machinery; and he laid down a course which was successfully followed, where older heads with more experience had failed.

A letter, written in 1845 by Captain Henry Vivian, describes Trevithick, and his partner in connection with the first high-pressure steam-engines.

"My father was a man of great inventive and arithmetical powers of mind; he has often made up the duty of an engine while we have been walking the road together, multiplying six figures by four figures, and giving the answer without aid of pen or paper, by retaining the figures in memory.

"Mr. Trevithick was a man of still greater powers of mind, but would too often run wild, from want of calculation. They did well together, but badly when separated."

Among the traditions of Trevithick, a favourite one is the story of a mine account, at which the mine agents and adventurers met, and after settling the accounts, partook of a good dinner and mine-account punch, followed as a matter of course by rough jokes, which passed freely in those times. Captain Hodge, a large, strong man of six feet in height, ventured on a little friendly tussle with Trevithick. Hodge was seized by the middle, and turned upside down; the print of his shoes being left in the ceiling of the room. What a jolly roar of laughter burst from those originals! That account-house in Dolcoath was allowed to stand for many years as a memento of his great strength. He threw a sledge-hammer, they say, 14 lbs. weight, over the engine-house chimney, but the writer's informant could not tell him the height; some said it never came down again.

He was famous also as the only one who could throw a ball over the Camborne church-tower, when standing near enough its base to touch it with one foot.

One day the mine athletes, in their trials of strength, attempted to lift from the ground a 9-inch cast-iron pump, such as was used in the mines, weighing seven or eight hundredweight. When Trevithick's turn came, he lifted the pump on to his shoulder and carried it off.

His excusable vanity was gratified by a request from a member of the College of Surgeons that he would show them his strong frame; and their telling him that they had never before seen muscle so finely developed, interested his Cornish friends, who delighted in physical strength.

Captain Joseph Vivian, when a young man, knew Trevithick well. Bowls was a favourite game in Cornwall, and sometimes they used to try who could throw the jack ball to reach the church door before it touched the ground. Captain Dick threw the jack not only to reach the church, but clean over it, and it fell in Jacky Williams' garden. At Crane Mine, the young men standing in the door of the smiths' shop tried to throw a sledge against the wall of the engine-house across the yard; but Captain Dick happening to come by, threw the sledge across the yard and over the roof of the engine-house.

Captain Andrew Vivian's story was, that one day he and Trevithick were walking in London, when a lot of pickpocket fellows hustled them about; Captain Dick laid hold of one in each hand, and knocking their heads together, gave them a swing and scattered the whole lot.

Vivian had often seen him write his name quite easily on a beam about six feet above the floor, with a 56-lb. weight suspended from his thumb, the arm being extended at full length. Captain Andrew thought himself a good hand at figures, and sometimes when he was working away at a calculation, Trevithick, who had not made a figure, would say, "Well, Captain Andrew, it will be about so and so;" and was always very near the truth.

"Mr. Hugh Hunter, aged eighty-seven years, the foreman carpenter in Cook's Kitchen for the last sixty-six years, has often seen Captain Trevithick lift the mandril, and hundreds of people used to come to see him do it. He used to put a bar of iron inside the mandril and fasten another bar to it so that he could get a good hold. A strong stool was placed on each side of the mandril, upon which he would stand with the mandril between his legs, and would lift it off the ground. He was an uncommon quick-spirited man, and the strongest ever known. The weight of the mandril was ten hundredweight."[1]

"John Vivian, the foreman smith, had worked in the mine for fifteen years. The mandril was always known as the one

[1] This mandril is in the Patent Museum at Kensington. The statement was by Hunter at Pool to the writer in 1869.

that Captain Dick used to lift, and that nobody else could. His father worked in Dolcoath under Captain Dick about 1800, and he used to say how Captain Dick would climb up the great shears, or triangles, fifty or sixty feet high, and, standing on the top of the three poles or shear-legs, would swing around a heavy sledge-hammer. He did it for exercise, and to steady his head and his foot."[1]

When the writer was a young man, Kneebone, a strong Cornish wrestler, working in the tin-smelting works at Hayle, lifted him from the ground with one hand, at arm's length; he then raised two blocks of tin, each of them three hundredweight, from the ground, remarking, "Captain Dick Trevithick could lift three blocks as easy as I can two; and he also lifted an old piston-rod, seven or eight hundredweight, with a man sitting on each end of it."

Two rocks rising from the sea, a short distance from the cliffs, at St. Agnes Head, near Redruth, named in charts "The Man and his Man," were for many years known in the neighbourhood under the familiar term, "Captain Dick and Captain Andrew;" one of the rocks is much larger than the other.

The earliest *trace* of steady occupation followed by Richard Trevithick, jun., is in his father's account-book of Stray Park Mine in 1790, where he received 30*s*. a month. His father, as the manager of the mine, stands at the head of the book at 40*s*. a month.

Trevithick was then eighteen years old, and in the rate of wages paid in the mine came next to the manager; he must therefore have had years of experience before that period. Boys then left the village school for the fight of life at ten or twelve years of age.

---

[1] Statement by Vivian, working in Cook's Kitchen Mine, to the writer, 1869.

The name appears in the monthly accounts without intermission for two years, until 1792, when he was employed by the shareholders in Tin Croft Mine to examine and report on the relative duties, or work done with a certain quantity of coal, in the patent engine of Watt, and in that of the double-cylinder engine of Hornblower, who stood high in Cornwall as an engineer before the time of Watt.

At the early age of twenty-one Trevithick was therefore in public and professional contact with Watt, and from that period dates the competition of the great low-pressure engineer and his youthful and vigorous high-pressure rival.

In 1795 Trevithick erected a steam-engine at Wheal Treasury Mine, competing with Watt, and also with Bull, jun., both of whom received pay for the saving of fuel by improvements in the steam-engine in mines under the management of Trevithick, sen. This continued until the end of 1797, when the account ceases.

"*Wheal Treasury Cost, June,* 1795:—

"Richard Trevithick, jun., 13 days at 3*s.* 6*d.*

"Richard Trevithick, sen., for expenses at Helston, at sundry times, and at other places, waiting on the adventurers, 7*s.* 6*d.*

"Richard Trevithick, jun., for moving and erecting the eastern engine, 21*l.*

"Ditto, for one month's attendance on ditto, 1*l.* 11*s.* 6*d.*

"Ditto, for one month's saving respecting the eastern engine, 18*l.*

"To Edward Bull, for his attendance, 2*l.* 2*s.*

"To Mr. Bull's engine, for two months' saving respecting his engine, 43*l.* 12*s.*

"1796, *April*:—

"The said Richard Moyle subjects himself to pay any demands which may be hereafter made by Messrs. Boulton and Watt, for savings in the said mine.

"At a meeting of the old adventurers of Wheal Treasury Mine, held this 27th day of March, 1797, at Praze-an-Beeble, it was unanimously resolved that the earnings claimed by Boulton and Watt should not be paid till the validity of their patent should be fully proved."

Trevithick, jun., had taken down an old engine, probably a Newcomen atmospheric, and re-erected it with his own particular improvements, from which he received about the same rate of "saving money" as Bull. There is no entry of saving money paid to Watt; but in the following year an outgoing shareholder was made liable for any demand Watt might make, and again in the next year the mine determined to refuse Watt's claim for "saving money" Smiles says that Wheal Treasury had been supplied with an engine by Watt; so there were three rival engines and rival engineers under the management of Trevithick, sen., as there had been twenty years before at Dolcoath.

Trevithick, sen., put but a small money value on his labours, for Helston was about ten miles distant; his travelling to it and to other places, "sundry times, waiting on the adventurers," led to a bill of 7*s.* 6*d.* for expenses. His son, who had been entrusted with the improving and re-erecting of an engine, received as daily pay 3*s.* 6*d.*; and at the age of twenty-four was in open and active opposition to the sweeping claims of Watt.

The late Captain Richard Eustace related, as an early incident in Trevithick's life, that "at Wheal Treasury one of Boulton and Watt's pumping engines worked badly, and at last stopped. The engineman in charge could do nothing with her; the water was rising in the mine; when Trevithick, jun., offered his services, and made things right. His father boasted

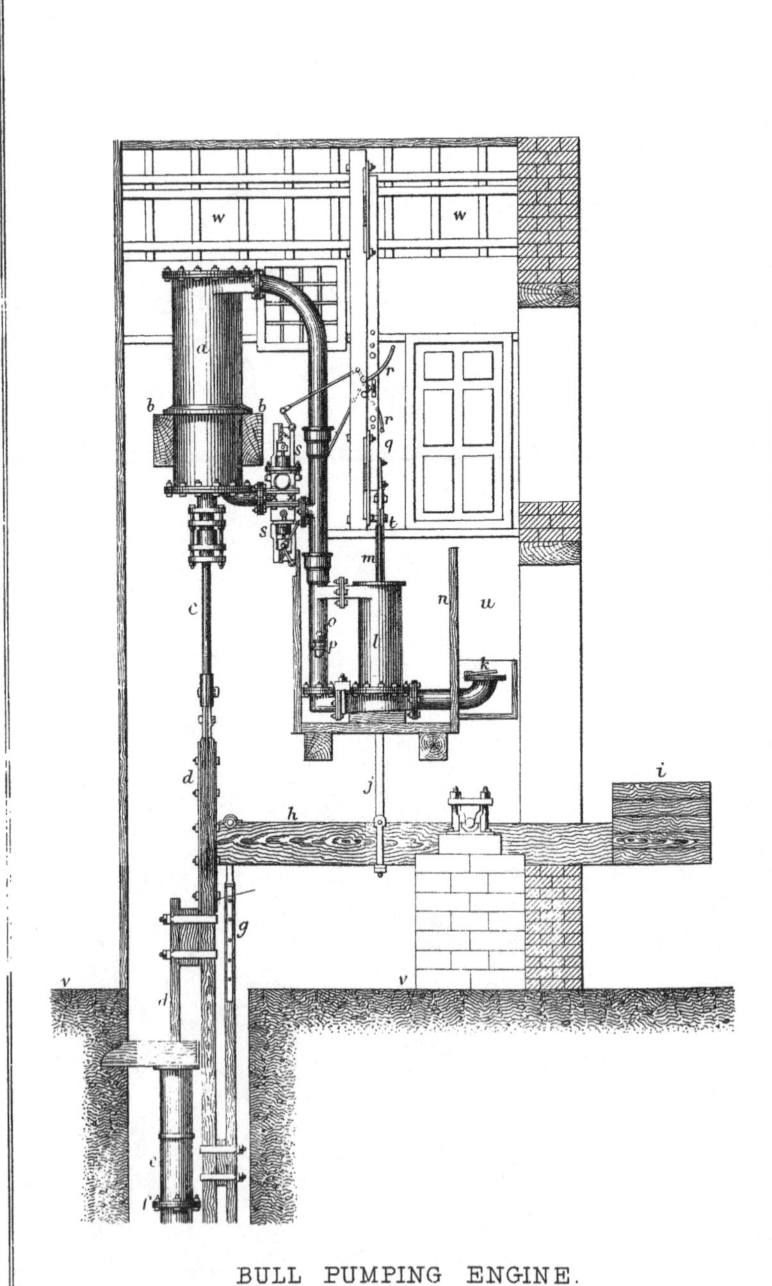

BULL PUMPING ENGINE.

London: E. & F.N. Spon, 48, Charing Cross.

Kell, Bros. Lith. London

that the best man in the mine could not do what his boy had done."

There is no record of the particular kind of engine then erected by Trevithick; he was not, as has been stated, in partnership with Bull, jun., for during the three years the monthly saving in coal by his engine and the entries to Bull are quite distinct, and also their monthly payment as engineers keeping the engines in order. They both strove to avoid Watt's patent claims by combinations and improvements, especially by an engine since known as Bull's. A drawing[1] made by Trevithick probably in those early days represents this kind of engine, which remained in use in Cornwall for many years; several are at work in the London Water-works, and one or two at Battersea at the present moment.

This double-acting engine has never received its fair share of credit, for as compared with its rival, the Boulton and Watt patent engine, it is more simple, and less costly; matters of more importance then than now. The doing away with the heavy beam and parallel motion, the placing the injection-valve in the exhaust-steam pipe, thereby avoiding the larger and separate condensing vessel used by Watt, and the double-acting air-pump, were bold and successful ideas, easy of construction, economical in operation, and widely different from the Watt engine.

Plate II. shows a side elevation of this direct-action or Bull engine, drawn by Richard Trevithick, probably in 1795. *a* is a steam-cylinder; *b*, wood beams supporting cylinder; *c*, piston-rod; *d*, pump-rods;

[1] The drawing was given by Trevithick to the writer, in 1830, as a sample of a properly-made mechanical drawing: it had neither name, date, nor scale attached—a strong evidence of an original Trevithick drawing.

e, pumps; *f*, engine-shaft; *g*, balance-beam connecting rod; *h*, balance-beam; *i*, box for balance-weights; *j*, plug-rod and air-pump bucket connecting rod; *k*, snifting valve; *l*, air-pump; *m*, air-pump bucket-rod; *n*, condensing water cistern; *o*, pipe-condenser; *p*, injection-valve; *q*, plug-rod; *r*, gear-handles; *s*, the valve-nozzles; *t*, floor of engine-house; *u*, condensing chamber; *v*, ground level; *w*, roof of engine-house.

The four walls of the engine-house surround the pump-shaft, over which the steam-cylinder with close top and bottom is placed, supported on strong beams crossing from wall to wall of the house, the piston-rod going direct to the pump-rods, by which the necessity for the large beam and parallel motion is done away. The usual balance-bob attached to the pump-rods gives motion to the air-pump bucket and to the plug-rod for working the valves. The condensation of the steam takes place in the exhaust-pipe, on which the injection-valve is fixed. The valves are worked by tappets and toothed segments. The air-pump bucket was a solid piston, that of Watt a piston with valves. This kind of engine remained in use in Cornwall for thirty years, and is the early type of the modern direct-action engines; it had many new ideas, and clever modifications of those not new.

Trevithick in 1797 was the engineer at Ding Dong Mine, near Penzance, where he had erected an engine on which Boulton and Watt laid an injunction for infringement of their patent; to avoid payment the engine was converted into an open-top cylinder, without beam or parallel motion. About the same time a Bull engine was erected at the Herland Mine, in the spirit of rivalry with Watt, who had placed there one of his best engines.

The late Captain Samuel Grose said:—"My first occupation, after leaving school, was as a young assistant-engineer for Mr. Trevithick. In 1802, an engine worked at the Herland Mine, which had been erected by Bull, jun., and Trevithick about 1798, whose cylinder, 60 inches in diameter, was placed over the shaft; the piston-rod attached to the pump-rods; it had an air-pump, and low-pressure boilers; there were no large high-pressure engines then. Bull, sen., died about 1780, and shortly after Captain Trevithick was the head engineer in the county."[1]

It seems, therefore, that the elder Bull died in the earlier part of the fight between Cornish engineers and Watt. His son Edward Bull followed in his father's steps as an engineer.

Trevithick, jun., and Bull, jun., worked together in the improvement of the steam-engine; but no trace has reached the writer of the form of the Bull engine on the death of William Bull, sen., at which time Trevithick, jun., was nine years old. Trevithick used and improved the Bull engine after the death of its inventor, and in conjunction with Edward Bull, jun., erected engines to compete with Watt during the very height of the acrimonious and long-pending patent lawsuits.

James Bolitho "had worked in Ding Dong for fifty years; his father, Thomas Bolitho, worked all his life in the mine, and frequently spoke about the engine that Captain Trevithick put up; she had not worked long before Boulton and Watt came down with an injunction printed out, and pasted it up on the door of the engine-house, and upon the heaps of mine-stuff, and

---

[1] Statement given in 1858, not long before Captain Grose's death.

nobody dared to touch them. But Captain Trevithick did not care; he and Bull and William West came and turned the cylinder upside down, right over the pump-rods in the shaft; they took off the cylinder top (it was the cylinder bottom before they turned it upside down); water and oil used to be on the top of the piston to keep it tight. Captain Trevithick at that time put a wind-engine in the mine; sometimes it went so fast that they could not stop it; some sailors came from Penzance and made a plan for reefing the sails."[1]

This was one of the many skirmishes with Watt, causing the engine to work as an open-top cylinder atmospheric, rather than pay the patent demands.

Davies Gilbert thus describes his first acquaintance with Trevithick:—

"MY DEAR SIR, "EAST BOURNE, *April* 29, 1839.

"I will give as good an account as I can of Richard Trevithick. His father was the chief manager in Dolcoath Mine, and he bore the reputation of being the best informed and most skilful captain in all western mines; for as broad a line of distinction was then made between the eastern and western mines (the Gwennap and the Camborne Mines) as between those of different nations.

"I knew the father very well, and about the year 1796 I remember hearing from Mr. Jonathan Hornblower that a tall and strong young man had made his appearance among engineers, and that on more than one occasion he had threatened some people who contradicted him to fling them into the engine-shaft. In the latter part of November of that year I was called to London as a witness in a steam-engine cause between Messrs. Boulton and Watt and Maberley. There I believe that I first saw Mr. Richard Trevithick, jun., and certainly there I first became acquainted with him. Our correspondence commenced soon afterwards, and he was very frequently in the habit of

[1] Statement by James Bolitho in 1868.

calling at Tredrea to ask my opinion on various projects that occurred to his mind—some of them very ingenious, and others so wild as not to rest on any foundation at all. I cannot trace the succession in point of time.

"On one occasion Trevithick came to me and inquired with great eagerness as to what I apprehended would be loss of power in working an engine by the force of steam, raised to the pressure of several atmospheres, but instead of condensing to let the steam escape. I of course answered at once that the loss of power would be one atmosphere, diminished power by the saving of an air-pump with its friction, and in many cases with the raising of condensing water. I never saw a man more delighted, and I believe that within a month several puffers were in actual work.

"DAVIES GILBERT.

"J. S. ENYS, Esq."

Probably the meeting with engineers at those legal contests and the shrewd questioning of lawyers led him to ponder on the possibility of working an engine without air-pump or vacuum, and within a few months several steam-puffers were at work; the idea was soon made practically useful, showing not only the energy of Trevithick's resources, but also his practical authority as an acting engineer. The high-pressure steam-engine may be said to date from 1796.

The first account-book in my possession in the handwriting of Richard Trevithick, jun., is dated 1797. The pages are headed Dr. and Cr.; but during the several years the book was in use not one single account was balanced; it commenced as a ledger, but was used as a day-book. Comparatively large amounts of money passed through his hands. He was the engineer supplying machinery, and having mechanics in his pay in the following mines:—Ding Dong, Wheal Bog, Wheal Druid, Hallamanin, Wheal Prosper, Wheal Hope, Wheal Abraham, Dolcoath, Rosewall Mine, Polgrane,

Trenethick Wood, Baldue, Trevenen, Wheal Rose, Wheal Malkin, East Pool, Wheal Seal-hole, Cook's Kitchen, and Camborne Vean.

Many of the entries are for his improved plunger-pole pit-work, and also for his high-pressure portable steam-engines; for from that time he made no more low-pressures.

On the death of his father in 1797 he was elected and employed to fill the vacant position of leading engineer in Cornish mining. Supreme in the district west from Dolcoath, but having an opposing, or Watt party, in the Gwennap, or eastern district.

About this time, while preparing machinery at Hayle Foundry, his friend William West had fallen in love and married Miss Johanna Harvey. Trevithick followed his friend's example, and while on high-pressure engine business at Mr. Harvey's foundry, fell in love with Miss Jane Harvey. The marriage was in 1797. Both families were well known for business ability and good looks. The bride was tall and of fair complexion, with brown hair. The bridegroom 6 feet 2 inches high, broad shouldered, well-shaped massive head, blue eyes, with a winning mouth, somewhat large, but having an undefinable expression of kindness and firmness.

Their first residence was at Moreton House, near Redruth, within a stone's throw of Murdoch's house, and but little farther from Watt's residence at Plane-an-Guarry. Murdoch's house still stands, in Cross Street, Redruth; those still live who saw the gas-pipes conveying gas from the retort in the little yard to near the ceiling of the room, just over the table; a hole for the pipe was made in the window-frame. The old window is now replaced by a new frame. Mrs. Trevithick had seen Murdoch once in her house; he was considered a clever and agreeable person. Mr. Trevithick was friendly with him, but not intimate. Watt lived close by, but never came to their house.

Moreton House was taken for a year, but after nine months' occupation, Camborne became their place of residence. Soon after their removal a demand was made for a second year's rent for Moreton House. Trevithick threatened to shake the applicant; did not everybody know that he had left the house months before? "Perhaps so," said the man, "but you promised to give me the key, and I have not got it yet." The key was found in some cast-off coat-pocket of Trevithick's, and his carelessness cost him a year's rent.

The Williamses, Foxes, and others, men of money and business experience, controlled the Gwennap district, and encouraged Watt to erect his engines in Cornwall. Trevithick, jun., belonging to the western or Camborne district, gave evidence against Watt's claim in 1796, and was then spoken of as the tall and strong young man who cared not for Watt.

The idea of working an engine by the pressure of steam flashed on his mind as a real and important discovery; his scientific friend, who was perfectly conversant with Watt's engine, viewed it in the same light, though failing to trace its full practical bearing. "A puffer-engine would lose the power from vacuum, minus the power required to work the air-pump, and the conveyance of condensing water." This was the methodical answer of the man of learning, describing the difference between the low-pressure steam vacuum engine of Watt, commonly called steam-engine, and the proposed engine of Trevithick, relying on the pressure of steam for its power; but the latter saw its extensive range and value in a practical sense not dreamt of by his friend, and within a month several such engines were at work, some of which have continued in constant operation almost to the present day.

The new principle then discovered is unchanged, and is most prominently illustrated in the locomotive engine, where the vacuum engine of Watt could not be used.

CHAPTER V.

PLUNGER-POLE PUMP AND WATER-PRESSURE ENGINE.

AN account-book, in Trevithick's writing, commences:—

"1797. To Ding Dong, for five weeks' attendance, at a guinea a week.

"Ditto, fixing a 7-inch pole-case with new lifts and wood wind-bore.

"1798. Fixing, in East Pool Mine, a 4½-inch pole, &c.

"Ditto, the same in Prince William Henry."

These and similar entries are frequent in what appears to have been commenced as Trevithick's ledger-account of outgoings and incomings on his taking the post of chief mine engineer, made vacant by the death of his father. Men's wages are charged at numerous mines for putting in pit-work, building engine-houses and boiler-houses; and various entries for wood, leather, pumps, &c., for pit-work.

The increasing depths of the mines, and greater power of the steam-engines for raising water, necessitated a change in the rude, weak pit-work, with pumps made of wood hooped with iron, having buckets with leather-cup packing and valve which only raised water with the *upward* motion of the rods. Pistons without valves, similar to those used in steam-engines, worked in brass or iron pump-barrels, forcing the water upward with the *downward* movement of the pump-rods, reducing the strength and weight of the pump bucket-rods by one-half. Their descending weight about balanced the ascending water, while with the valve-bucket pump the rods had to be strong enough to bear their own weight and also that of the column of water.

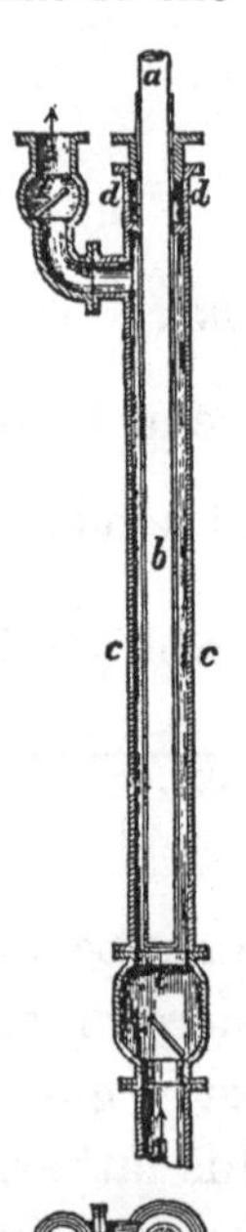

TREVITHICK'S PLUNGER-POLE PUMP, 1797.

a, the pump-rod, of wood, fastened into the hollow cast-iron pole; *b*, the pole or plunger; *c*, the pole-case, allowing space for the passage of water round the pole; *d*, the stuffing box; *e*, the bottom valve, allowing the mine water to ascend into the pole-case on the ascent of the pole; *f*, the top valve, through which the water is forced upward on the descent of the pole.

The practical cause of failure of the solid piston was its liability to jam in the pump-barrel by sand or gravel; while the breakage of a leather-cup bucket, or leakage of its leather valve, was apt to cause serious accident to the costly pump-rods, and even to the steam-engine.

The plunger-pole made useful in the Cornish mines by Trevithick, met all the requirements; raising the water by the descending pump-rods, having great simplicity of structure, and freedom from breakage or liability of jamming.

The great value of this invention or application is proved by its continued use in pump-work, precisely as erected by Trevithick in 1797. Plunger-poles, fitted to the case, the latter having longitudinal grooves for the passage of the water, had been used in France; but Trevithick's plunger-pole worked in an unbored pole-case, the sides of which were not touched by the pole. The upcast pipe shown close to the top of the pole-case allowed the free escape of air; it is now, however, more frequently attached to the bottom of the pole-case. By leaving the packing in the stuffing box a little slack, air escapes; while a water cup on the top of the stuffing-box gland prevents the admission of air through the slack packing.

"With the old atmospheric engine common bucket pumps were employed, the whole depth of the shaft being divided into several lifts; and the rods of each pump, ascending separately to some distance above the topmost lift, were then united to one rod attached by a chain to the engine-beam.

"When Watt's engine was introduced, the pit-work remained pretty nearly the same, except that, instead of the long separate rods to each lift, one main rod was substituted, to which smaller rods were attached by ties, to work the pumps at the various levels where they were required. This arrangement has continued to the present day, and is much superior to the old plan."[1]

"Among other things, Murdoch proposed the use of Sir Samuel Morland's plunger-pump in the pit-work, not as a general substitute for, but as an addition to, the lifting pumps, in order to suit the double-acting engine, by making the pumps double-acting also In 1796, one of those was employed at Ale and Cakes, a mine now forming the eastern part of the United Mines."[2]

Murdoch apparently used, in Ale and Cakes Mine, a plunger-pole to suit the particular requirements of Watt's double-acting engine. Trevithick's account, commencing in the following year, deals with plunger-pole pit-work, not as a makeshift, but as a principle on which pit-work in a mine should be constructed. He removed the old lifting-bucket pit-work and replaced it by the plunger pit-work in several mines, in 1797.

Probably neither Murdoch nor Trevithick knew of its having been patented one hundred years before, which takes from them the claim of invention, though the practically useful introduction rests with Trevithick.

Gregory's 'Mechanics,' a much-esteemed book on pump-work, speaks highly of Trevithick's temporary forcer, and gives a drawing of it.

---

[1] Enys "On the Cornish Engine," 'Trans. Inst. C. E.,' vol. iii.

[2] Ibid. Appendix G to Tredgold. Pole 'On Cornish Engines,' p. 112. The plunger-pump was invented by Sir Samuel Morland, and patented by him in 1675 (p. 115).

"Forcer, temporary, for a pump, is a contrivance to produce a constant stream. A very simple forcer of this kind has been devised by Mr. R. Trevithick; it consists in fixing a barrel with a solid piston along the side of the common pump, in such a manner that the lower space of the additional barrel may communicate with the space between the two valves of the pump, and lastly, by connecting the rods so that they may work together. The effect is, that when the pistons are raised the spaces beneath A and B become filled by the pressure of the atmosphere, at the same time that the upper column flows out at E. But again, when the pistons descend the valve C shuts, and consequently the water driven by the piston in B must ascend through A, and continue to produce an equal discharge through E in the downstroke."[1]

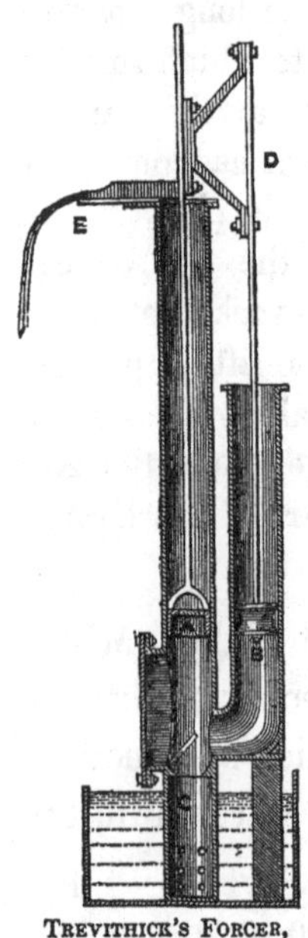

TREVITHICK'S FORCER, TEMPORARY, PUMP, PRIOR TO 1797.

This combined bucket and forcer pump, known under the name "Trevithick's pump," gives him the claim of invention. His account-book, commencing in 1797, does not warrant the supposition that his plunger-pole pit-work was only then used for the first time.

The diffusion of useful knowledge was in those days so remarkably slow, that the managers of neighbouring mines went each his own particular course, in ignorance of changes and improvements that others were benefiting by. Lean tells us that,

"In 1801, in Crenver and Oatfields, in the parish of Crowan, he found the pit-work to consist of leathern buckets, with two

---

[1] 'Nich. Journ.,' No. 7, N.S. Gregory's 'Mechanics,' p. 196, vol. ii., published 1806.

or three pistons, such as were at that time in general use for plungers, in a very bad state; and it may be safely asserted that the engines were idle at least one-third of the time, repairing the pit-work and changing the buckets. And here he first introduced (what is now so generally used, and with so great advantage) the plunger-pole, instead of the common box and piston, wherever he found it practicable."[1]

It is evident that the plunger-pole, before Trevithick took it in hand, was a floating idea struggling to become a reality. Its great value is in its simplicity, amounting to a principle, not generally known to the miners in Cornwall in 1801, as the pump to supersede all others, in deep mines.

Trevithick's first move was to meet the requirements sought for by Murdoch—the lifting of water by the down-stroke of the pump-rods as well as by the up-stroke, exemplified in the temporary forcer, in which the pump-bucket, with its valve, and the solid piston, were placed side by side, one common bottom-valve serving for both pumps. The result of the combination was a constant upward stream of water and the non-necessity for the top valve in the water column; the water from the force-pump barrel passed up through the bucket-valve in the pump-barrel on the down-stroke, while the up-stroke propelled the water from the pump-barrel. Gregory calls it the "Trevithick forcer, temporary," as though it was soon to be replaced by the still more simple plunger-pole.

The pole-case was a plain casting with a box for the hemp packing; the pole was a cast-iron pipe turned on the outside, and fixed like a ferrule on the end of the pump-rods. Two valves formed of two plates

---

[1] Lean's 'Historical Statement of Steam-Engines in Cornwall,' published 1839.

of iron, with a piece of leather between them, the lower plate being less in size than the upper, caused the leather to form the face of the valve, making it water-tight without employing skilled labour in its construction.

The perfect simplicity of each small part, built on ideas partly new and partly seen or heard of before, but scarcely traceable, constituted the invention.

"Mrs. Dennis recollected Mr. Trevithick at Ding Dong about 1797 fixing his new plan of pumps there, and at Wheal Malkin and Wheal Providence, adjoining mines. Her parents lived at Madron, near these mines, and for two or three years Mr. Trevithick came frequently to superintend the mine-work, staying at their house a few days, or a week at a time. He was a great favourite, full of fun and good-humour, and a good story-teller. She had to be up at four in the morning to get Mr. Trevithick's breakfast ready, and he never came to the house again until dark. In the middle of the day a person came from the mines to fetch his dinner; he was never particular what it was. Sometimes, when we were all sitting together talking, he would jump up, and before anyone had time to say a word, he was right away to the mine."[1]

"Henry Clark went to work in Dolcoath smiths' shop in 1799. Captain Trevithick was putting in a plunger-pole lift; everybody said it would never answer, but the same lift is working there to this day (1869). Before that time they used buckets and pistons packed with gasket and a ring screwed on it; they used to jam with sand and gravel."[2]

Trevithick followed up the application of his new plunger-pump with such energy, that during the succeeding four or five years several of the principal mines had removed their old bucket-lifts, to make room for the new plunger-pole lift, and among them Dolcoath Mine.

---

[1] Residing at Penzance, 1869. [2] Residing at Redruth, 1869.

In 1798 Trevithick, by one of those moves common to master-minds, converted the outline of the plunger-pole pump, serving as the agent of the steam-engine, into its rival as the prime mover in positions commanding a stream of water through pipes from elevated ground. In this new position it was called the water-pressure engine, and its first erection was in Prince William Henry Mine,[1] for giving motion to the pump-rods. The pole and case took the place of the cylinder and piston of the steam-engine, and were fixed over the mouth of the shaft. The pole working through a stuffing box on the top of the pole-case, had a cross-head, from which two side rods descended to the pump-rods; water was brought in iron or wood pipes from the neighbouring high ground, and being admitted by a valve to the bottom of the pole, caused it to perform the up-stroke; the release of the water from under the pole allowed the weight of the rods to cause the down-stroke. This first Cornish water-pressure engine continued to work satisfactorily for seventeen years, when it required repairs, and was re-erected.

"Captain Joseph Vivian, the manager of Roskear Mine in 1868, also worked there in 1815, when they re-erected the old water-pressure engine that Trevithick had first put up in 1798. It was spoken of as the first water-pressure engine ever erected with a pole and side rods. The water was brought through pipes for working it."

"Mr. Symons, when a boy, in 1804 or 1805, used to amuse himself by placing a handkerchief on a plug in the wooden water-pipes with iron bands, bringing water down from Carnkie to work Captain Trevithick's water-pressure engine at Wheal Druid. By loosening the plug, it flew from the pipe up into

---

[1] In Watt's time called Wheal Chance; then Prince William Henry; and in the present day Roskear Mine.

the air, carrying the handkerchief up with it, to the great fun of the boys; only sometimes, when the engineman could surprise them, they had the rope's end."[1]

Rees' 'Cyclopædia' thus speaks of Trevithick's Wheal Druid pressure-engine, at the foot of Carn Brea Hill, close to his home of boyhood, and now known as Carn Brea Mine:—

"A very complete pressure-engine was erected by Mr. Trevithick at the Druid copper mine in Illogan, near Truro, in Cornwall; it acted with a double power, that is, the piston pressure was first applied to one side of the piston to force it up, and then on the other to force it down, in the same manner as a double-acting steam-engine. It acted by two slide-valves instead of one, but they were so made that one opened rather before the other shut, and this, though it wasted a small quantity of water by permitting it to escape, prevented the concussion of stopping the column of water, which indeed is always in motion."

"Mr. James Banfield[2] in 1818 was employed in Wheal Clowance Mine; the pumps were worked by Captain Trevithick's water-pressure engine. It had a pole about twelve inches in diameter, working in a pole-case; two side rods went down from the cross-head on the top of the pole, to a cross-head under the pole-case; to this bottom cross-head the pump-rods were joined; the cross-heads worked in wooden guides; there was a balance-bob from which a plug-rod worked the valves, which had gear-handles just like a steam-engine. The engine was fixed underground at the adit level; the head of water came from Clowance Park. Several similar engines were then at work in the Cornish mines; they used to be called Captain Dick's pressure-engines."

The temporary forcer pump was a combination of pump-bucket and solid forcer piston. Before it had time to be practically established in general use, it was

---

[1] Mr. Symons resided at Camborne in 1868.
[2] Resident in Penzance in 1869.

followed by the plunger-pole pump. Then came the pole pressure-engine, for giving motion to the plunger-pole pumps, followed by the Druid Mine double-acting pressure-engine, with solid piston, and water pressure acting both on its up and down stroke like the Watt steam-engine, but using water pressure and vacuum in lieu of steam and vacuum.

The pressure-engine described by Rees was double-acting, having a piston in a cylinder; that by Banfield was single-acting, with a pole, of which several were at work in Cornwall in 1815, and were known as Captain Trevithick's pressure-pole engines.

Gregory's 'Mechanics,' published in 1806, gives a drawing and full particulars of Trevithick's Wheal Druid pressure-engine.

"Pressure-engines for raising water by the pressure and descent of a column enclosed in a pipe, have been lately erected in different parts of this country. The principle now adverted to was adopted in some machinery executed in France about 1731 (see Belidor de Arch., Hydraul., lib. iv., ch. i.), and was likewise adopted in Cornwall about forty years ago. But the pressure-engine of which we are about to give a particular description, is the invention of Mr. R. Trevithick, who probably was not aware that anything at all similar had been attempted before. This engine, a section of which, on a scale of a quarter of an inch to a foot, is shown in Pl. xxiii., was erected about six years ago at the Druid copper mine, in the parish of Illogan, near Truro. A B represents a pipe 6 inches in diameter, through which water descends, from the head, to the place of its delivery, to run off by an adit at S, through a fall of 34 fathoms in the whole; that is to say, in a close pipe down the slope of a hill 200 fathoms long, with 26 fathoms fall; then perpendicularly 6 fathoms, till it arrives at B, and thence through the engine from B to S 2 fathoms. At the turn B the water enters into a chamber C, the lower part of which terminates in two brass cylinders 4 inches in diameter, in which two plugs or pistons of lead, D and E, are capable of moving up and down by their piston-rods, which pass through a close packing above, and are attached to the extremities of a chain leading over, and properly attached to the wheel Q, so that it cannot slip. The leaden pieces D and E are cast in their places, and have no packing whatever. They move very easily, and if at any time they should become loose, they may be spread out by a few blows with a proper instrument, without taking them out of their places. On the sides of the two brass cylinders, in which D and E move, there are square holes, communicating towards F and G, which is a horizontal trunk or square pipe, 4 inches wide and 3 inches deep. All the other pipes, G, G, and R, are 6 inches in diameter, except the principal cylinder, wherein the piston H moves; and this cylinder is 10 inches in diameter, and admits a 9-feet stroke, though it is here delineated as if the stroke were only 3 feet. The piston-rod works through a stuffing box above, and is attached to M N, which is the pit-rod, or a perpendicular piece divided into two, so as to allow its alternate motion up and down, and leave a space between, without touching the fixed apparatus or great cylinder. The pit-rod is prolonged down into the mine, where it is employed to work the pumps, or if the engine were applied to mill-work, or any other use, this rod would form the communication of the first mover. K L is a tumbler, or tumbling bob, capable of being moved on the gudgeons V from its present position to another, in which the weight L shall hang over with

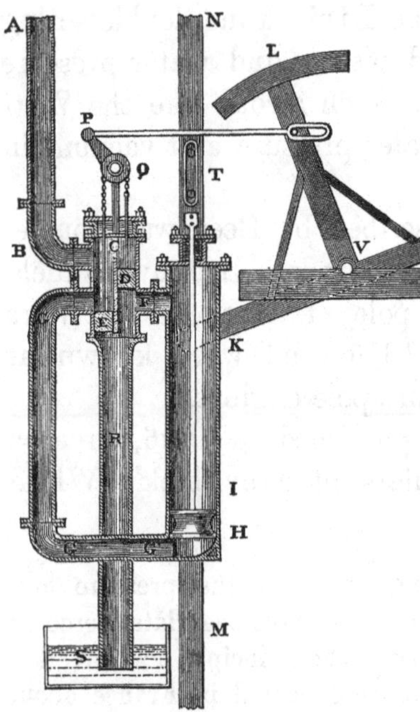

TREVITHICK'S WHEAL DRUID DOUBLE-ACTING WATER-PRESSURE ENGINE, 1800.

the same inclination on the opposite side of the perpendicular, and consequently the end K will then be as much elevated as it is now depressed. The pipe R S has its lower end immersed in a cistern, by which means it delivers its water, without the possibility of the external air introducing itself, so that it constitutes a Torricellian column, or water barometer, and renders the whole column from A to S effectual, as we shall see in our view of the operation.

"Let us suppose the lower bar K V of the tumbler to be horizontal, and the rod P O so situated as that the plugs or leaden pistons D and E shall lie opposite to each other, and stop the water-ways G and F; in this state of the engine, though each of these pistons is pressed by a force equivalent to more than 1000 lbs., they will remain motionless, because these actions being contrary to each other, they are constantly in equilibrio. The great piston H being here shown as at the bottom of its cylinder, the tumbler is to be thrown by hand into the position here delineated.

"Its action upon O P, and consequently upon the wheel Q, draws up the plug D, and depresses E, so that the water-way G becomes open from A B, and that of F to the pipe R; the water consequently descends from A to C, thence to G G G, until it acts beneath the piston H. This pressure raises the piston, and if there be any water above the piston, it causes it to rise, and pass through F into R. During the rise of the piston (which carries the pit-rod M N along with it), a sliding block of wood I, fixed to this rod, is brought into contact with the tail K of the tumbler, and raises it to the horizontal position, beyond which it oversets by the acquired motion of the weight L. The mere rise of the piston, if there were no additional motion in the tumbler, would only bring the two plugs D and E to the position of rest, namely, to close G and F, and then the engine would stop; but the fall of the tumbler carries the plug D downwards, quite clear of the hole F, and the other plug E upwards, quite clear of the hole G. These motions require no consumption of power, because the plugs are in equilibrio, as was just observed. In this new situation the column A B no longer communicates with G, but acts through F upon the upper part of the piston

H, and depresses it; while the contents of the great cylinder beneath that piston are driven out through G G G, and pass through the opening at E into R.

"It may be observed that the column which acts against the piston is assisted by the pressure of the atmosphere, rendered active by the column of water hanging in R, to which that assisting pressure is equivalent, as has already been noticed. When the piston has descended through a certain length, the slide or block at T, upon the pit-rod, applies against the tail K of the tumbler, which it depresses and again oversets; producing once more the position of the plugs D E, here delineated, and the consequent ascent of the great piston H, as before described. The ascent produces its former effect on the tumbler and plugs; and in this manner it is evident that the alternations will go on without limit, or until the manager shall think fit to place the tumbler and plugs D E in the position of rest, namely, so as to stop the passages F and G.

"The length of the stroke may be varied by altering the position of the pieces T and I, which will shorten the stroke the nearer they are together, as in that case they will sooner alternate upon the tail K.

"As the sudden stoppage of the descent of the column A B at the instant when the two plugs were both in the water-way might jar and shake the apparatus, these plugs are made half an inch shorter than the depth of the side holes, so that in that case the water can escape directly through both the small cylinders to R. This gives a moment of time for the generation of the contrary motion in the piston and the water in G G G, and greatly deadens the concussion which might else be produced.

"Some former attempts to make pressure-engines upon the principle of the steam-engine have failed, because the water, not being elastic, could not be made to carry the piston onwards a little, so as completely to shut one set of valves and open another.

"In the present judicious construction the tumbler performs the office of the expansive force of steam at the end of the stroke."

The head of water at Wheal Druid had a fall of 204 feet. The design of the valve-box alone showed genius sufficient to place a man above his fellows. Two cylinders of brass, easily fitted up, two leaden piston-valves, *cast in their places*, without any fittings, which in case of wear could be hammered out to the necessary size without taking them from their places. These two valves were moved by a piece of flat chain passing over a pulley, attached to a wooden lever, with a balance-weight judiciously placed, called a tumbling bob.

The operation of these simple valves was that each of them bearing a pressure of more than 1000 lbs., worked as balance-valves with little weight or friction, or wear and tear, and enabled the engine to change its direction of stroke without jar, and without materially stopping the downward flow of water in the supply-pipe. The plugs, since called by modern inventors cylindrical slide-valves, being made half an inch shorter than the water passages in the brass cylinder, or valve-case, turned the water pressure from the main cylinder into the waste-pipe during the moment of time occupied in changing the direction of the stroke. This one little point in the engine was thus valued by Gregory:—"Some former attempts to make pressure-engines upon the principle of the steam-engine have failed, because the water, not being elastic, could not be made to carry the piston onwards a little, so as completely to shut one set of valves and open another. In the present judicious construction the tumbler performs the office of the expansive force of steam at the end of the stroke." The bottom end of the waste-water pipe emptied into a cistern 12 feet below the engine, to prevent air from going up the pipe. The weight of water, therefore, in this pipe gave a vacuum in the working cylinder of several pounds to the inch. This apparently small engine was of great power, the pressure being about 100 lbs. to the inch on the piston, giving a force of three or four tons, thrown rapidly up and down the long stroke of 9 feet.

Trevithick was for several years the engineer and part proprietor in Trenethick Wood Mine, and introduced his improved plunger-pole pumps, worked by his piston water-pressure engine. The cylinder was 17 inches in diameter, with a stroke of 9 feet, double-acting, worked by a four-way cock, made of brass, in a brass shell. An air-vessel was suggested to remove the shock of the change of movement in the column of water. The head of water was 78 feet, giving a power of nearly four tons.

The drawing of this engine, still in the possession of the Honourable Mrs. Gilbert, of Trelissick, was made by Trevithick, and given to his friend Davies Gilbert (then Giddy), who wrote the following particulars on it:—

"Drawing of the pressure-engine, erected on the mine in Trenethick Wood, near Helston, 1799. Given me by the engineer, Mr. Richard Trevithick. The fall of water is 13 fathoms; the engine works double; a 9-feet stroke. The parts are,—a tube communicating with the head of water; the cylinder or working piece; a tube communicating at the bottom with the cylinder; and a tube acting as a siphon, and thus adding the length of the cylinder to the head of water. When the cock is turned, as here represented, the water from the tube presses on the piston in the cylinder. The water under the piston is forced up the tube, and flows down the siphon. When the cock is turned a quarter round, the water from the tube is pressed down the tube; from thence acts against the under surface of the piston in the cylinder. The water above the piston escapes into the siphon.

"*N.B.*—An air-vessel may be adapted to the tube, and thus

the flow of water rendered almost uniform, notwithstanding the checks of the engine. The apertures through the cock and nozzles are made one quarter of the cylinder."

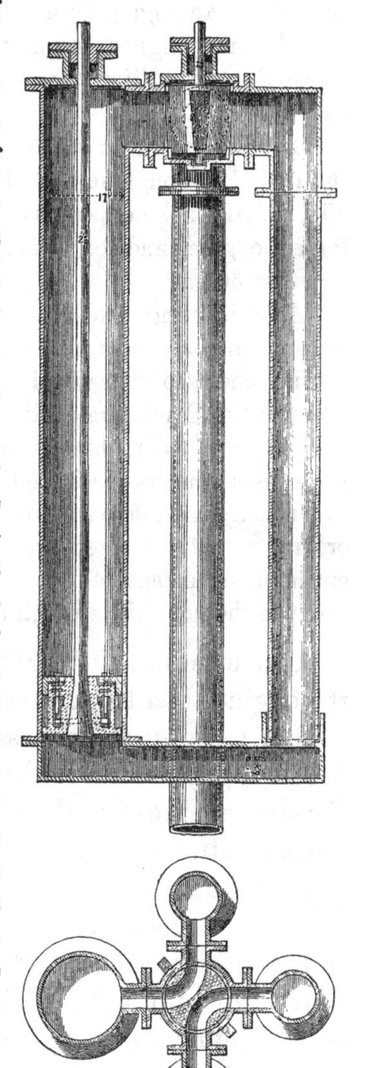

TREVITHICK'S TRENETHICK WOOD DOUBLE-ACTING WATER-PRESSURE ENGINE, WORKED BY A FOUR-WAY COCK; ERECTED 1799.

The balance-valves are here replaced by the four-way cock. The simplicity of form and parts leads one to pass by this machine as merely a few pipes bolted together; but its great power and economy in working led to its use in other than Cornish mines.

"At a meeting of the Hill-Carr Sough proprietors, held June 25th, 1801, it having been represented to the meeting that pressure-engines of the description proposed to be erected near Youlgreave, are now in use in Cornwall, where they give great satisfaction, and Mr. Richard Trevithick, the engineer who erects them, having made an offer to come to view the mines paying composition to this Sough, on condition of the expenses of his journey being paid; and that in case of not contracting for the erection of the engine, he would furnish us with drawings and every information gratis.

"*March 25th*, 1802.—At a meeting this day, the adjourned question respecting the propriety of erecting an engine to lift the water below the level of this Sough was further considered; Mr. Richard Trevithick having given in a drawing and an estimate of an engine for a lift of 8 fathoms.

"The undersigned proprietors agree to accept the above proposals on condition that the whole of the expenses shall not exceed 1800*l.*

"In the reckonings for the quarter ending 27th March, 1802, the Hill-Carr Sough and the Shining Sough partners each pay 15*l.*, as half of the expenses of Mr. Trevithick in making a plan for an engine, and for coming over out of Cornwall with it; together 30*l.*

"At a meeting held March 31st, 1803, Mr. Richard Trevithick having appeared, and produced estimates of the pressure-engine, and also of a wheel calculated to lift the whole of the water in the Black Shale Pit and Gay Vein Gates to the depth of 8 fathoms, from which it appeared that the expense of the pressure-engine was less, and also came recommended by Mr. Trevithick in preference to the wheel; it was unanimously ordered that the former should be adopted, and the whole and entire management of it be entrusted to Mr. Trevithick, pursuant to the conditions which he required."

At a meeting, June 28th, 1804, allusion is made to the engine then being erected.

Half a century had passed since these resolutions of the managers of the Derbyshire lead mines to entrust a Cornish engineer with their drainage, when the following notice was attached to this memorable old water-pressure engine,—"November, 1851, Messrs. John Taylor and Sons advertise for sale the various engines and pumps in those old mines;" and in 1852 they were drawn up and sold, Trevithick's engine going to the scrap-pile after fifty working years.

"ALPORT, near BAKEWELL,
"*March 8th*, 1869.

"DEAR SIR,

"I am glad to learn by your favour of the 4th that the particulars I sent of the pressure-engine will be of service, and that, with some help from Mr. Darlington, you will be enabled to make a drawing of it. There is one thing I believe I have not named, it is that this engine has many times worked for fifteen or sixteen weeks together, without missing a stroke or being stopped for any purpose (a moment once or twice a day, I do not recollect which) but to oil the bearings. The water was very clear and did not wear the valves and gearings of the buckets very much.

"Yours truly,

"F. TREVITHICK, Esq." "SAMUEL BENNETTS.

"MINERA, near WREXHAM,
"1st *March*, 1869

"DEAR SIR,

"The sketch you have sent me of Trevithick's engine is perfectly correct, and I can only answer the questions asked in your letter.

"The pump-work at the Alport mines were two 33-inch drawing lifts, connected above the cylinder to the piston-rod by a cross-head; the bottom of the cylinder was at the adit level. The tumbling beam for working the boxes or valves was fixed about 12 feet above the revolving wheel, with a small rod connected to a short lever fixed to the axle of the revolving wheel. The tumbling beam was worked by two tappets, screwed on the rod that was connected with a powerful balance-beam on the surface. This rod was also connected with the cross-head below, stayed, or worked through guides, as the ordinary Cornish pit-work.

"The water for working this engine descended to the piston below, from the surface, through a 15-inch downfall column, about 150 feet in depth. The regulating valve to this engine was on the surface, and the water was turned on according to the necessary speed of engine. I hope you will see how to form your sketch of this engine from what is stated above. If there is any other question you wish to ask, I will endeavour to answer it.

"Yours very truly,

"F. TREVITHICK, Esq" "JOHN DARLINGTON.

Mr. J. Glynn, in a report on water-pressure engines, says :—

"The first water-pressure engine used in England was erected by Mr. William Westgarth, at a lead mine belonging to Sir Walter Blacket, in the county of Northumberland, in the year 1765. The piston was attached by a chain to the arched head of the beam, working just like the atmospheric steam-engine. Mr. Smeaton suggested to Westgarth the use of a stuffed collar, which he used with success. In 1770 Mr. Smeaton made a small water-pressure engine for supplying Lord Irwin's residence, closing the cylinder-top and using the piston-rod with a stuffed collar.

"After Mr. Smeaton's time the water-pressure engine seems to have remained in abeyance, and I am not aware that any more of them were made until Mr. Trevithick revived their use, in constructing several water-pressure engines, one of which was erected in Derbyshire in the year 1803, and is still, I believe, at work (in 1848) at the Alport mines, near Bakewell, to which place it was removed from its original situation, not far distant."[1]

Trevithick thus speaks of it, in a letter written at the commencement of 1804:—"It has been at work about three months, and never missed one stroke, except when they let a tub swim down the descending column."

In a subsequent letter he complains of non-payment for his foreman's attendance and travelling expenses, and for the wages of the men he sent, adding, "If they found fault with the engine there would be some reason for not paying, but they say it is the best in the world."

The cylinder was 25 inches in diameter, with a 10-feet stroke, in which the water column gave a pres-

[1] 'Report on Water-pressure Engines,' by Joseph Glynn, F.R.S., M. Inst. C. E., &c. Published by the British Association for the Advancement of Science, 1849.

sure of 75 lbs. on the square inch; and was fixed underground at the adit level, 48 feet above the bottom of the mine, giving motion to two large pumps, each 33 inches

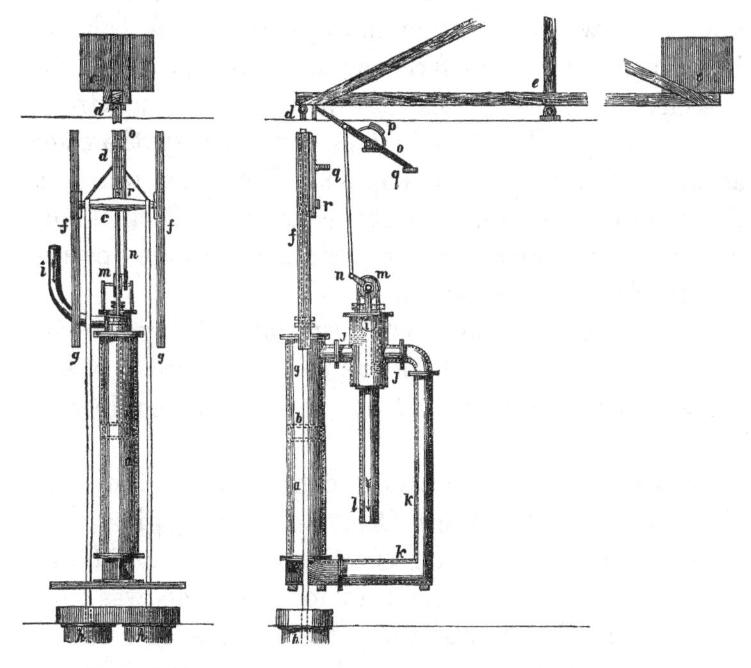

TREVITHICK'S DERBYSHIRE WATER-PRESSURE DOUBLE-ACTION ENGINE OF 1803.

a, the water-pressure cylinder, 25 inches in diameter, 10-feet stroke, working at a pressure of 75 lbs. to the inch; *b*, the piston; *c*, piston-rod cross-head; *d*, rod from cross-head to the balance-bob fixed at the surface: *e*, balance-bob and box for counterpoise; *f*, guides for piston-rod cross-head; *g*, two pump-rods descending from the piston-rod cross-head to the pumps; *h*, the two pumps, each 33 inches in diameter, raising water from the bottom of the mine to the adit level, about 48 feet; *i*, the water-pressure pipe, 15 inches in diameter, bringing the water from the surface down to the engine fixed at the adit level, a depth of 150 feet; *j*, the two piston-valves shown in the drawing as allowing the pressure water above the piston to escape to the waste-pipe, while the other valve allows the pressure water from the column to pass through the pipe *k* to the under side of the piston—the change of position of the valves reverses the action; the pressure water from the column is turned or to the top of the piston, while that below the piston passes back through the pipe *k*, and escapes through the waste-pipe; *k*, the pipe communicating from the pressure-pipe and waste-pipe with bottom of the cylinder; *l*, the waste-water pipe; *m*, a wheel over which a chain passes, giving movement to the rods from the two piston-valves; *n*, lever and rod connecting the valves with the tumbling beam; *o*, tumbling beam, moved by tappets on the main rods, the weighted part accelerating its movement when the weight *p* leans on either side of the centre; *p*, balance-weight, causing the lever to fall quickly just before the finish of the stroke; *q*, rests on which the ends of the tumbling beam fall; *r*, plug-rod attached to the main rods, on which are the two shifting tappets for lifting or depressing the end of the tumbling beam just before the finish of the stroke. The pressure column had a valve at the surface, so that, by contracting or increasing the supply of water, the speed of the engine was decreased or increased. The valves may have been worked wholly by the tappets, or modified by the balance or tumbling bob, or by the use of a cataract,—all three methods having been used in early engines.

in diameter, with a stroke of 10 feet, lifting water a height of 48 feet.

This pressure-engine was double-acting, or of equal power on the up as on the down stroke; the pumps gave their weight only on the up-stroke. A balance-bob placed at the surface equalized the powers; the descent of the weighted box assisting the up-stroke of the pump-rods. The valves were nearly the same as in the Druid pressure-engine. The water raised in the pumps was equal to the combined effect of the descent of the balance-weight together with the upward stroke of the pressure-engine.

This early machine, working night and day for half a century in the bowels of the earth, exerted a power greater than the large and overrated steam-engines of Watt.

Trevithick's water-pressure engine had a force of about 16 tons; the Watt Dolcoath great engine, with its 63-inch cylinder and effective pressure of 10 lbs. on the inch,[1] had a force of 14 tons, but this was only exerted on the down-stroke, the engine exerted no power during the up-stroke. The water-pressure engine to do the same kind of pumping-work gave out power from both the up and the down stroke, a compound stroke giving a power of 32 tons. This simple, durable, and cheap water-pressure engine, with all its disadvantages of position, was thus more than twice as powerful as the great Dolcoath steam-engine of Watt.

The writer has introduced this lengthened account of the plunger-pole pit-work and the water-pressure engine to make good a seeming void in mechanical history.

[1] See calculation in Stuart's 'History of the Steam-Engine,' chap. iii., p. 27, giving 8½ lbs. effective pressure.

The former is still universally used, the latter has to some extent been put aside for the more easily applied steam-engine, though the writer in 1831, at the Hayle Foundry, made drawings for a 10-inch pole water-pressure engine, 7 feet 6 inches stroke; the pole-case 1 inch larger in diameter than the pole; the head of water was 300 feet, giving a pressure of 150 lbs. on the inch; the inlet and outlet valves were 4½ and 5½ inches in diameter, double-beat, worked by a plug-rod and cataract; the general outline was a copy from Trevithick's work of thirty years before.

A thorough knowledge of mechanical movements was one of Trevithick's greatest acquirements, and the correct application of the mechanical powers a matter of comparative ease in his hands. His Wheal Druid engine worked at a pressure of 100 lbs. on the square inch—a force but seldom used in our modern *stationary* engines. It must be considered that, not only was he single-handed in his daring inventions, but, at a time when wooden pipes were still much used in water-works, and iron pipes and machinery of very rude construction, he was compelled to turn manufacturer, oftentimes having to utilize the larger portion of an old machine in the construction of a new one.

The following tradition of Mr. John Harvey, the father of Mrs. Trevithick, and founder of the Hayle Foundry Engine Works one hundred years ago, opens our eyes to the immensely greater difficulties in the path of the engineer in those days as compared with the present time. Trevithick, sen., is spoken of in a former chapter as having gone to Bristol and elsewhere for iron pipes in 1775; for the use of iron had not at that time wholly driven out the wooden pipes from the Cornish mine pit-work.

"On a Sunday morning, in the year 1770, a rich friend of Mr. Harvey's came into his smiths' shop, at Carnhell Green, having lost one of his silver shoe-buckles when taking a run with his beagles before church time, and said, 'How can I go to church with one buckle?' 'Give me one of your old silver spoons, and lend me that buckle, and I'll soon set you up again,' said John Harvey. The rich man was pleased, and asked what his friend would do if he had some of his money. 'Why, go down to Foundry, and make cast-iron pumps for Trevithick's mines in place of the wooden ones.'

"Another story says, that Mr. John Harvey went up the country to see how castings were made. Having been refused admittance, he dressed in rude clothes, feigned to be half-witted, idled about the doors of the foundries, singing songs, and now and then ventured in; offering to carry water, or errands, and so discovered how to make cast-iron pipes."[1]

"Mrs. Edwards, of Perran, in St. Hilary, when a girl, used to go with Betsy Edwards to Hallamanin Mine when she carried dinner to her father and brother, about 1798. They worked in the carpenters' shop, boring out the wood pumps, and we used to carry home the borings for lighting the fire. Captain Trevithick was the engineer. I recollect him very well; he was a tall, thin man. It was about that time that the Brumagem company had to give in, and Captain Trevithick started the mine again."[2]

This was one of the numerous mines in which the Watt low-pressure and Trevithick high-pressure fights were contested, on the first working of high-pressure steam-engines.

It is curious that this battle-ground was within a mile of the heap of stones forming the grave of the old granite boiler at Gwallon Mine, and but a little farther from Ludgvan-lez, where Newcomen had erected one

[1] Banfield's tradition, related in 1869.
[2] Recollections of Mrs. Edwards, living at Perran in 1869.

of his earliest atmospheric engines, which in course of time was replaced by a Boulton and Watt—the old mine-heaps are now called Wheal Boulton—while just across the valley, at Hallamanin, worked a Trevithick high-pressure steam-engine.

CHAPTER VI.

HIGH-PRESSURE WHIM-ENGINE.

TREVITHICK'S account-book gives :—

"1798.—To Prince William Henry, for a boiler, 38*l.*

"1799.—To a boiler for new engine at Rosewall, 21*l.*

"To the same, for Wheal Seal-hole little engine, 21*l.*

"1800.—For carrying the little engine to Wheal Hope, 10*s.* 6*d.*

"To Arthur Woolf going to London as engine fireman with Shelland engine, at 30*l.* a year.

"To received 350 guineas from Mr. Millett for a steam-whim."

A short time ago the writer visited St. Agnes Head in search of evidences of times gone by. A miner,[1] in an almost idle mine, said, "This is Polberrow New Adventure; that engine works in Seal-hole shaft, but old Seal-hole shaft, a few yards to the west, where Captain Dick Trevithick put his puffer-engine, has run together. Them pit-holes and shafts we do call old men's workings for tin before steam-engines were made; they do go down from 10 to 40 fathoms to the old adit in the cliff above the sea."

The models of 1796 grew rapidly into the practical high-pressure steam-puffer engine of 1798, and became active rivals of the low-pressure steam vacuum engine.

In 1800 a high-pressure portable steam-engine was carried in a cart to Wheal Hope, and one was sent to London in charge of Arthur Woolf, the well-known engineer. In the same year the first high-pressure

[1] Capt. Robert Hancock.

steam-whim was made for Mr. Millett, at a comparatively small cost.

This latter application of steam, to supersede the horse-whim in raising mineral from the shafts, was of great importance to the miner. An old account-book, in Cook's Kitchen Mine, has entries bearing on Trevithick's doings at that period.

"1798, *June* 23*rd*.—A survey held for setting the sump, Roger's and Bramblam's whims, on Dunkin's lode, to draw by the hundred kibbals for one year; and if the adventurers shall, during this time, think it proper to erect a steam whim or whims, to draw out of any shaft or shafts, then this contract to be void. Signed by nine working men, and one +, and by Captain Joseph Vivian for adventurers.

"1799.—The above agreement renewed for another year.

"1803. *May*.—To Thomas Tyack for making smith's work for the steam-whim."[1]

Trevithick then believed himself to be the originator of the steam-whim; but a memorandum, written by him in 1830, has the following note:—

"I have since heard that Boulton and Watt had before erected a steam-whim at Wheal Maid, which did not go well, and soon ceased to work."

William Pooly, working in Dolcoath, says:—

"I worked in this mine, in 1816, a whim-engine, which they used to say was first put up in Wheal Maid, in Gwennap, by Boulton and Watt. I never heard what date it was; but people said it was the first steam-whim; that she was sent from Gwennap to Wales, and when Boulton and Watt were at the Herland mine with their engines, she was brought back from Wales and tried at Herland; she was moved from Herland and tried at Dolcoath, about 1816, when I worked her; she worked with the old Boulton and Watt hearse-boiler. Several of Captain Trevithick's high-pressure whims were, before that, working

[1] Extracts from book in Cook's Kitchen Mine in 1870.

close by, some condensing and one or two puffers. Captain Trevithick's boiler was like a cast-iron cylinder, with the fire-door in one end, and the steam-cylinder in the other end; two rods standing out from the cylinder end for guides, and a connecting rod going to the crank."[1]

Mr. Hugh Hunter, the oldest workman in the mine, says:—

"I have been the foreman carpenter in Cook's Kitchen for sixty years. I went to the mine in 1802 or 1803; Captain Trevithick's whims worked there then; one was on Chapell's shaft, another on Bramblam's, and one called the Valley-puffer. Chapell's engine, or Bramblam's, had two cylinders, and a double crank; the whim-cage was fixed high up, near the roof. The engines were fixed on to the boiler; it was something like the present boilers in shape. The piston-rod cross-head worked in guide-rods fixed to the cylinder; connecting rods went from the cross-heads to the cranks. Steam was turned off and on by a four-way cock. I recollect the engine, because the two cranks going one after the other was a new thing; and I recollect the valley engine, because she was a puffer, and you could hear her for miles. I can't recollect the kind of the third engine. They used to be called Trevithick's first steam-whims."[2]

This is the first mention of a double-cylinder high-pressure steam-engine in Cornwall, and it used the blast-pipe. Similar engines were erected in Wales about the same time.[3]

"I am eighty years old, and still work in Dolcoath; in 1815 I worked the valley-puffer engine; the cylinder was horizontal, and fixed in the end of the boiler; the piston-rod worked in guides, not much above the ground, with a connecting rod from the cross-head to the crank on the winding cage. She puffed the waste steam into the air; the two other whims, close by in Cook's Kitchen, condensed; they were at work long before I saw them."[4]

[1] Narrated in 1869.

[2] Mr. Hugh Hunter lived at Pool in 1869.

[3] See Mr. Crawshay's statement, chap. ix.

[4] Narrated by John Smith, 1869.

"James Hosking, working in 1869 one of Trevithick's first lot of steam-whims in Cook's Kitchen, knew the engine in 1838; she is exactly the same now as then. I wouldn't wish to work a handier engine; you need not move from your seat, except to put coals in the boiler. Mr. Samuel Hocking, the engineer, looked in a day or two ago, and said he had known her for fifty years, and could see no difference in her."

"Captain Charles Thomas, the manager of Cook's Kitchen, says, the engine was erected by Trevithick in 1800 or 1801. It works extremely well with 25 lbs. of steam, and promises to do so for many years, and draws six kibbals per hour from the 264-fathom level. The gross weight, including the chain to the bottom of the shaft, is 70 cwt.; the mineral in each kibbal about 6 or 7 cwt."

These facts raise a question of much interest in meting out the just reward of praise to the improvers of the rotary steam-engine.

Did Trevithick construct the first engine using a crank in Cornwall; and had he ever seen an engine with a crank before that of his own construction?

Watt's Wheal Maid failure, a low-pressure steam vacuum engine, erected about 1784, with sun-and-planet wheels, may be taken as the first rotary engine in Cornwall. Trevithick was a boy, and neither saw it nor heard of it; and moreover, it had not a crank, neither did the crank form a part of the steam-engine for some years after that date, and it certainly was not seen in Cornwall until Trevithick constructed his high-pressure steam models, working with a crank, in 1796. He had never been out of the county until about that time, when he went to London to give evidence in the Watt Patent lawsuits. He probably heard of the crank disputes between Watt and Wasbrough, but the crank could not then have been in extensive use as a part of a steam-engine, and may never have been seen by Trevithick.

One of these mementoes of Trevithick's practical genius in 1800, still worked at Cook's Kitchen Mine in 1869.

The accompanying drawing represents the only one of those three Cook's Kitchen engines still working.

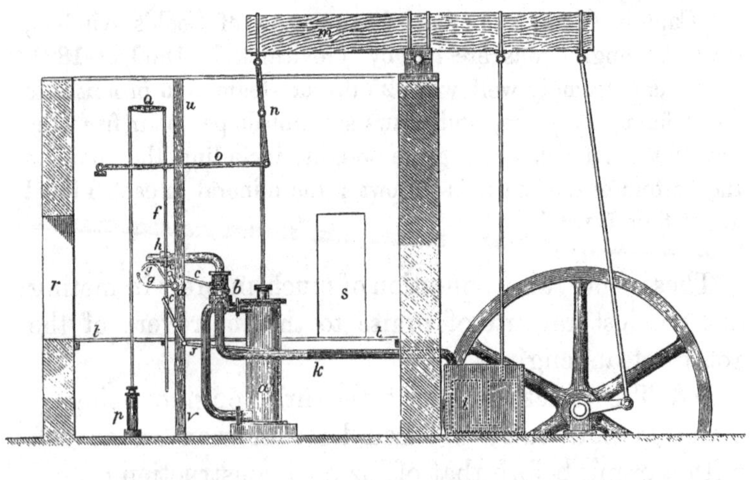

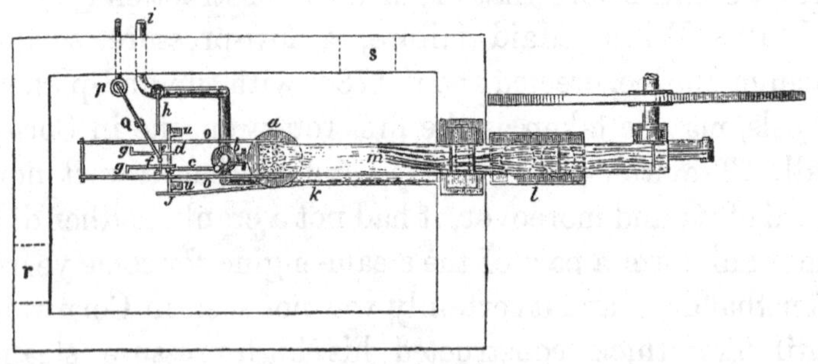

TREVITHICK'S HIGH-PRESSURE EXPANSIVE STEAM CONDENSING WHIM-ENGINE, ERECTED AT COOK'S KITCHEN, 1800.

Description of condensing steam-whim of 1800, taken 1869, from the engine then at work:—*a*, steam-cylinder, 19 inches in diameter, 5-feet stroke; *b*, the four-way cock, by a quarter turn giving the steam in the top of the piston and that under it to the condenser, and *vice versâ* on the return movement; *c*, rod, joining the cock-lever with the Y-shaft lever; *u*, the Y-posts; *d*, the Y-shaft, carrying two gear-handles and the cock-lever; *e*, the two tappets on the plug-rod, for moving the two gear-handles and four-way cock; *f*, plug-rod; *g*, two gear-handles; *h*, steam regulating cock; *i*, steam-pipe to boiler-house; *j*, handle and rod to injection-valve; *k*, exhaust-steam pipe to condenser; *l*, condensing system, containing air-pump and condenser; *m*, main beam of oak; *n*, piston-rod cap; *o*, parallel motion; *p*, feed-pole and rod; *q*, lever-working feed-pole; *r*, door to boiler stoke-hole; *s*, door to top of boiler; *t*, floor of engine-house.

The chief working engineer in the mine for thirty-two years has now replaced parts worn out, by new pieces of precisely the form of the original.

The boiler is of wrought iron, cylindrical, with internal tube and fire-place; the plates have been renewed more than once. Steam is worked at 25 lbs. to the inch above the atmosphere. All the regulating or gear levers are within reach of the engineman without his moving from his seat. It has a Boulton and Watt air-pump and condenser. The parallel motion differs from Watt's patent, and is known as the spider parallel motion. A four-way cock is the only valve used, except the steam shut-off cock from the boiler, and the injection and feed cocks. Its present work is to lift three tons and a half, including the kibbal and chain, when starting from a depth of 1584 feet.

The adjoining mines of Dolcoath and Stray Park made similar use of Trevithick's high-pressure steam whim-engines, erected about the same time, and each of them on a different plan, two or three of which remained at work forty or fifty years.

"Mr. John Vivian, of St. Ives Consols Mine, when a little boy, in 1801 or 1802, carried the pay-money to his uncle, Captain Andrew Vivian, for men working Trevithick's new high-pressure whim-engine in Stray Park Mine. Crowds of people came to see it."

"Henry Vivian worked the first Trevithick high-pressure puffer whim-engine in Stray Park Mine. It worked with a four-way cock, and puffed the waste steam into the air; it was a wonder at that time."

"Captain Joseph Vivian, of Reskadinnick, near Camborne, about 1800 or 1801 saw Trevithick's first puffer whim-engine in Stray Park Mine. The cylinder was in the boiler, and the piston-rod worked horizontally. Trevithick put up another whim-engine in the same mine, working with high-pressure steam, and with the Boulton and Watt air-pump; it had a wooden beam. As a boy, he watched the engine with the beam, and the one without a beam, to understand why they were different. In a year or two several were erected. People who were for Boulton and Watt were against Trevithick. A Boulton and Watt whim was, a year or two after, put up in Dolcoath to beat Trevithick's puffer-whim. The report was in favour of Boulton and Watt. I worked in Dolcoath at that time. Everybody said the agents were told beforehand which way the trial ought to go. Captain Glanville put up the Boulton and Watt low-pressure whim. Captain Trevithick's high-pressure puffer had a horizontal cylinder fixed in the boiler, and the steam puffed up the chimney."

"Henry Clark, of Redruth, was a boy in Dolcoath smiths' shop in the valley, in 1799 or 1800, just after she went to work again. Captain Trevithick was the head engineer; but the adventurers grumbled because he did not give all his time to the mine. Drawings came down from London for the new high-pressure puffer-whim for the valley; Henry Vivian put her up; but Benjamin Glanville, the mine carpenter, was made head man over the engines, because Captain Trevithick was away so much. The mine managers tried the new valley-puffer against a Boulton and Watt engine. When Captain Trevithick heard it, he sent a letter that he would bet Glanville fifty pounds that his puffer should beat Boulton and Watt. Trevithick came down from London, and tried the puffer, and said to his brother-in-law (Captain Henry Vivian), 'Why, how have you got the piston half an inch smaller than the cylinder?' A new brass piston was put in, and she beat Boulton and Watt all to nothing. I was landing the kibbals for the puffer, because we tried Captain Trevithick's new kibbal and pincers made in our shop; we unloaded the kibbals, and sent them down again, almost without stopping the engine. The adventurers fixed upon a trial for three or four days; coals were weighed to each engine, and persons appointed to take account of the kibbals raised and coals burnt. Captain Trevithick's engine did the best; and, after the trial, a little pit was found with coal buried in it, that Glanville meant to use in the Boulton and Watt engine. Trevithick's puffer had a cast-iron cylindrical boiler, about 6 feet in diameter, and 10 or 12 feet long, with cast-iron ends; a wrought-iron tube passed through the boiler, with a fire-place in one end of it, and a wrought-iron chimney at the other. The cylinder was let into the boiler, and the steam puffed off by the side. The trial was about 1804 or 1805, when several of Trevithick's whims, some of which condensed, were working in the adjoining mines."

"James Richards, of Camborne, landed kibbals in 1832, with Trevithick's old whim-engine now in Camborne Vean Mine, but then in Stray Park; she is unaltered, except that the four-way cock is removed to make room for lifting valves. Henry Vivian was the first man to work her; she used to be called the first steam-whim in Cornwall."

"William Eustace, for twenty years foreman smith in Camborne Vean Mine (1869), knew the old Trevithick whim when John West, the engineer, put in valves in place of Trevithick's four-way cock. The cock-pipe was stopped up with lead, and remained on the cylinder for many years, to make sure that the new valves were an improvement, before removing the cock: has often heard the old men talk of this beam condensing engine, and the machine whim with the fly-wheel up in the roof, with the piston-rod working in guides, and no beam, as the two first steam-whims ever put up. They then worked in Stray Park Mine."

"Captain Nicholas Vivian, late of Camborne, was in 1809 appointed manager to re-work Wheal Abraham Mine, which had been idle for some years.

"The old miners spoke of Captain Trevithick's puffer-engine with the cylinder in the boiler, which had worked in the mine just before she stopped.

"Trevithick demanded payment, which the adventurers refused until a longer trial of the engine, or because the mine was going to stop; he therefore took some men into the mine at nightfall, and by the morning the engine had disappeared.

"Captain N. Vivian's first act as manager was to purchase for the mine a second-hand Trevithick high-pressure steam puffer-whim."

"The late Captain Samuel Grose said:—'In 1805 I saw at the Weith Mine a portable engine made by Captain Trevithick, called a puffer; the cylinder was fixed in the boiler, the steam pressure was 30 lbs. on the square inch above the atmosphere. A wood shed sheltered the engine. Their cheapness made them to be much used.'"

In these four or five years we have accounts of four or five distinct kinds of steam-whims—high-pressure condensing engines, with and without beam, having either one or two cylinders, working vertically or horizontally, either as high-pressure puffers or as condensers.

Those various forms were to meet special requirements. Thus an established engine-house with convenient condensing water, had a fixed condensing engine; a temporary shaft had a portable puffer-engine.

Trevithick's aim in all his inventions was to utilize as much as possible the power of steam; so that not only was the steam-engine made to assume various new forms, but many appliances connected with it received his careful attention, and still bear witness to his inventive genius.

Up to about 1800, mineral had been raised by horse-whims from the Cornish mines in buckets, barrels, or kibbals, made of wood, and bound with iron. The watchful care of the driver helped to lessen the breakage, or wear and tear, from the kibbals rubbing on the rough sides of the underlie shafts. The greater speed and power of the steam-whims increased the breakage, and led Trevithick to introduce the use of wrought-iron kibbals, shaped like an egg, with flattened top and bottom. These improved kibbals worked with less friction and wear and tear than the wood kibbal. The method of emptying it was also much improved. The old bucket was turned over by manual labour, and the mineral shovelled into the barrow or tram-waggon; the new plan caused the saving of this labour, by discharging the mineral into the waggon.

"Henry Clarke was striker in the Valley smiths' shop in Dolcoath Mine in 1799 or 1800, and helped to make the first pincers and work for landing Captain Trevithick's new plan of iron kibbals, just then coming into use. William Branch, of Gwinear, was the smith.

"Captain Trevithick was standing by, telling Branch the shape to make them. The first pincers had a spring to throw them open; after a few trials the spring was taken out, not being necessary. The pincers was exactly the same shape as those now used (1869), and so are the iron kibbals, only they may be a bit bigger."

"Captain Joseph Vivian recollects Budge's spiral barrel, put aside by Captain Trevithick in his first steam-whims, because the power of his engine did not require them.

"The weight of the whim-chain, reaching to the bottom of the shaft, was too much for the horses, or the horse-whim, to lift. Budge's spiral was like a watch-barrel. A stone at the end of the chain from the spiral rose or fell in a shaft; this equalized the work for the horses according as the chain and kibbal was near the bottom or the top of the shaft."

"Mr. Tippet[1] helped to erect Captain Trevithick's puffer winding engine in Tin Croft Mine. The new iron kibbals were two or three times as large as the old ones for the horse-whim. The poppet-heads were raised, that the kibbal might go higher. The landing man said, 'I wonder who is going to land them big kibbals; I sha'nt land them.' Captain Trevithick said, 'Can't you wait a bit?' When it was all complete the kibbal didn't want any landing. It turned upside down, and the stuff went into the waggon without any landing or shovelling. The engine worked with very high-pressure steam. Saunders, who

[1] Mr. Tippet had been employed in Cornish mines, and was known to the writer in 1840.

was Captain Dick's head boiler-maker, brought a little plunger-pole pump for the feed-water; it worked by an eccentric on the fly-wheel shaft."

The new plan was to carry the loaded kibbal a certain distance above the waggon; a chain from one side and pincers were attached to a hook in the kibbal bottom; the loaded kibbal was then lowered the required distance; the fixed chain drawing the kibbal off the line of the shaft and over the waggon, and at the same time hanging it bottom upwards, turned its contents into the waggon: a pull on a catch freed the pincers, and the kibbal dropped over the line of the shaft ready for descending. This method, introduced in 1800, is still in use, though square boxes of iron on wheels, sliding in guide-rods in the shaft, are now preferred. The eccentric working the feed-pole is the earliest record the writer has met with of its use.

The foregoing, especially if taken in conjunction with the various applications of Trevithick's high-pressure steam-engines in Wales and elsewhere, fully establish their rapid extension during the first five or six years of the present century: two of them remaining fit for use, after nearly seventy years, proves the original skill of their designer and constructor.

The best test of the relative value of the low-pressure steam vacuum, and high-pressure steam whim engines by the rival engineers, is in a comparison of their performance and durability.[1]

Boulton, writing to Watt, describes one used temporarily for such a purpose at the Consolidated Mines, in 1784:—"There was a full attendance: Jethro

---

[1] See Trevithick's letter, 5th January, 1804, chap. xx.
" " 5th July, 1804, chap. xx.
" " 2nd January, 1805, chap. xx.

looked impudent, but mortified, to see the new little engine drawing kibbals from the two pits exceedingly well, and very manageable; and afterwards it worked six steam-stamps, each $2\frac{1}{2} = 14$ cwt. lifted twice at each revolution, or four times for every stroke of the engine. I suppose there were a thousand people present to see the engine work."[1]

This happened at Wheal Maid, in Gwennap, when Trevithick was thirteen years old.

Jethro was one of the Hornblowers, clever, practical engineers in Cornwall, before Watt brought his engine there, and strongly opposed to the new comer and his plans.

Trevithick's whim in Cook's Kitchen, of about 1800, was still at work in 1870, and lifting 70 cwt. The power of the Watt whim is not mentioned; but when used as a stamping engine it gave motion to but six stamp-heads, each $2\frac{1}{2}$ cwt.

Such a contrivance as Watt's could not continue to work long in a Cornish whim-engine, requiring most exact control, and frequent sudden stoppage and reversing, for which the sun-and-planet tooth-wheels were not suitable. Its failing to work well, and banishment to Wales during Trevithick's boyhood, accounts for his never having heard of it until forty or fifty years after the date of its erection, though he competed with it in Dolcoath in 1806, without knowing its travels or history under its improved face with a crank instead of its original sun-and-planet wheels, shortly after which it finally disappeared.[2]

A steam-engine with a crank had never been seen in Cornwall when he made his first locomotive models

[1] Letter from Boulton, 1784. Smiles' 'Lives of Boulton and Watt,' p 331.
[2] See Trevithick's letter, 18th February, 1806, chap. xx.

and recommended the use of a steam-whim with a crank.

Such was Trevithick's share in making steam useful for raising mineral from mines, and in adapting the kibbals and detail apparatus, suitable for the increased power and speed of the high-pressure steam-engine.

COOK'S KITCHEN MINE AND STEAM-WHIM. [W. J. Welch.]

CHAPTER VII.

CAMBORNE COMMON ROAD LOCOMOTIVE.

The late Mrs. Trevithick frequently spoke of models of her husband's early engines, the first of which worked at her house in Camborne, about the year 1796 or 1797. It was made by Mr. William West, and was to have been shown in the lawsuits between Boulton and Watt and the Cornish engineers.

Lord and Lady Dedunstanville, the large landed proprietors in the mining district—embracing Dolcoath, Cook's Kitchen, Stray Park, and many more of the early Cornish mines—and Mr. Davies Gilbert, a friend of Trevithick's, came to her house to see the model work.

A boiler, something like a strong iron kettle, was placed on the fire; Davies Gilbert was stoker, and blew the bellows; Lady Dedunstanville was engine-man, and turned the cock for the admission of steam to the first high-pressure steam-engine. The model was made of bright brass.

Shortly afterwards another model was made, which ran round the table, or the room. The boiler and the engine of this second model were in one piece; hot water was poured into the boiler, and a red-hot iron put into an interior tube, just like the hot iron in tea-urns.

In a third model the boiler was heated by a spirit-lamp. This one was taken to London by a gentleman who came down for the purpose of seeing it work.

A model of Trevithick's, now in the Kensington Museum, spoken of by Mr. Radford[1] as having come from the engine-works of Messrs. Whitehead and Co., Soho Iron Works, Manchester, is probably one of those spoken of by Mrs. Trevithick as having been made prior to 1800. It is a perfect specimen of a high-pressure steam-engine, with cylindrical boiler, adapted to locomotive purposes. It served as a guide to Messrs. Whitehead and Co , who manufactured engines for Trevithick in 1804.

The drawing (Plate III.) and description of this working model shadow forth the usefulness of the high-pressure steam-engine of the present day in many of its leading features; the non-necessity for condensing water, the cylindrical boiler, the simplest form of crank, the absence of mason-work for the engine or boiler-flues, and its portability and power of locomotion, so nearly met all the requirements, as to entitle it to the designation, "the first high-pressure locomotive." P is the iron-heater (used in the model to avoid smoke); *s*, the safety-valve, kept down by a simple spring; *t*, screw-plug for supplying water; *b*, steam-cylinder let into boiler, with shell for the four-way cock, and pipe for conveying steam to the top and bottom of cylinder cast with it; *o*, a four-way steam-cock, worked by a rod from the cross-head; *c*, the cross-head; *f*, the two side rods; *g*, the crank-pins attached to the two driving wheels; *i*, the steering wheel, fixed by a screw for running in curved or straight lines; *e*, piston-rod; *d*, guides for the piston-rod cross-head; *h*, two driving wheels; *j*, two legs, by screwing out the lower part of which the driving wheels are lifted off the ground, making a stationary engine; *k*, fly-wheel; *m*, pinion

[1] Letter from Mr. Joseph Radford, in 1850, states that it had been in his family since 1810.

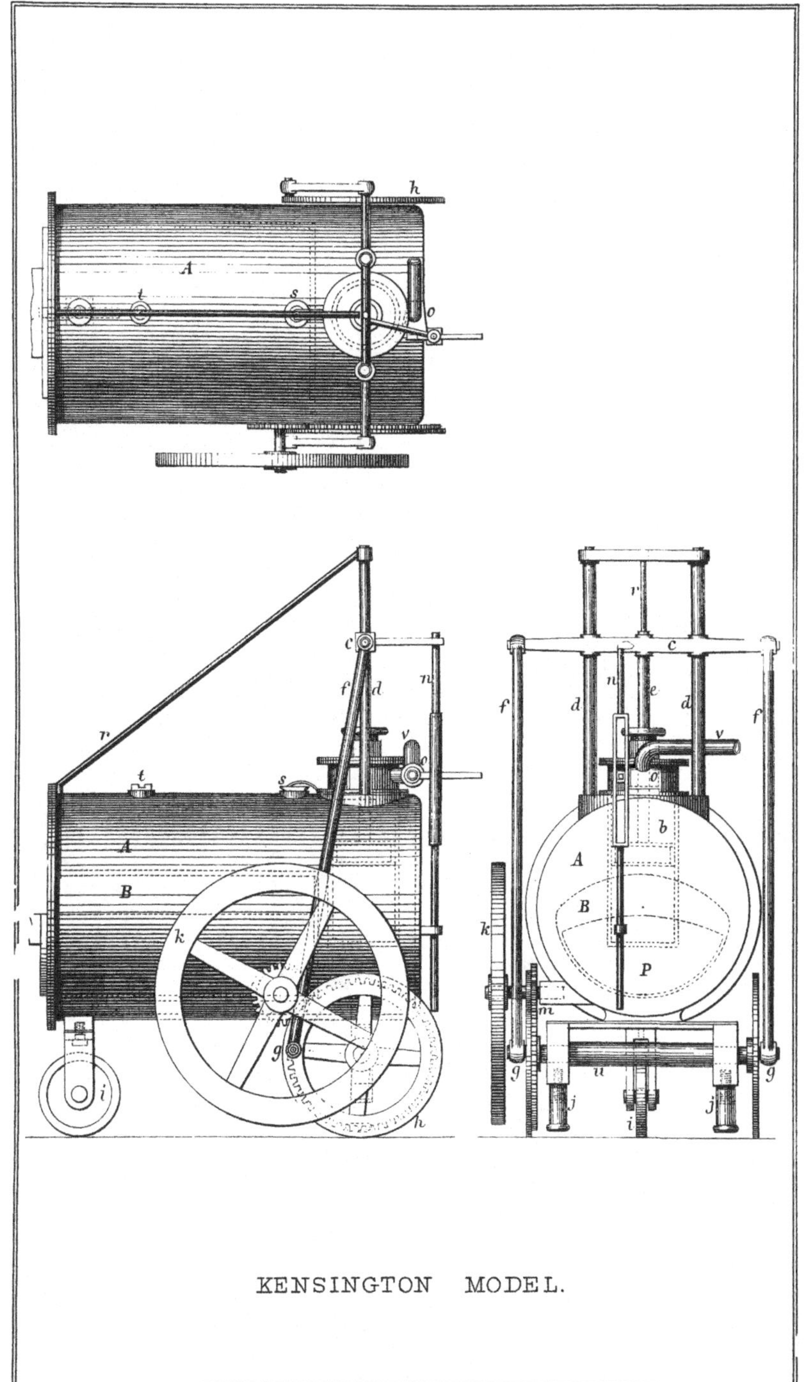

KENSINGTON MODEL.

London: E. & F. N. Spon, 48, Charing Cross.

Kell. Bro.s Lith London.

and gear wheel, connecting the fly-wheel with the driving wheel; *n*, plug-rod attached to cross-head for moving the four-way cock; *r*, stay for the guide-rods; *u*, driving axle; *v*, waste-steam pipe; A, the cylindrical boiler, made of copper; B, the boiler-tube.

Trevithick, after two years spent in numerous working experiments, under very trying circumstances—from the want of sufficient money, from the greatly depressed state of the mining interests in Cornwall, and from the disputes and lawsuits which had led mine adventurers and mine engineers to mistrust one another—had satisfied himself that a steam-engine would work without an air-pump or condensing water; that neither beam nor parallel motion, nor foundations of masonry, were absolutely necessary; and that the boiler, for conveniently supplying high-pressure steam, need not be one quarter of the weight, or cost, of the low-pressure boilers then in use, for producing an equal amount of power. He had conveyed an engine from mine to mine in a common cart, at a cost of 10*s*. 6*d*., and even this expense might have been saved by placing the engine on wheels, and driving them around by the force of the steam. The first commercial application of high-pressure steam was in the portable steam-whims, and then in the steam-carriage.

An old account-book of Trevithick's, dated 1800, gives the detail items in the manufacture of the first steam-carriage. William West for two or three years received pay for constructing models. Other mechanics worked in different mines repairing or improving boilers working the new high-pressure engines; and when they could be spared from the mines, were employed on the new steam-carriage, then being put together in a smiths' shop in Camborne, having in it one small hand-lathe, and one or two smiths' fires.

	£	s.	d.
"Nov. 1800.—To Richard Jeffry, about the fire-carriage, to end of November	10	14	6
To boiler-plates for steam-carriage, 4 cwt. 1 qr. at 38*s*.	7	16	6
To send William West to the Dale for plates.			
To James Oats, about little boiler.			
To James Saunders, about little engine ..	3	13	6
1801.—To Sam. Hambly, for steam-carriage ..	1	2	6
To Francis May, for sundries, do. ..	4	5	0
To little Hocking, do., do. ..	3	0	0
To Arthur Woolf, for steam-gauge and barometer from the Dale	4	7	0
To Sam. Rowe and Wm. Branch, assisting Wm. Jeffry and Sam. Hambly about the carriage.			
To Wm. West	35	10	7
To Lemon, shoemaker, for a bellows for carriage	1	7	0
John Hocking, for brasswork	1	8	7
Files, drink, and oil	0	7	6
Coals from my house	0	15	0 "

In 1801 William West was "at Hayle, about the little boiler," "altering it with wrought-iron plates;" and in the same year "Richard Trevithick was in London about the patent, and in Merthyr Tydfil and Coalbrookdale." The accounts do not enable us to trace the time occupied in building the first steam-carriage, or the necessary alterations during its construction. It was in progress in 1800; and on Christmas-eve, 1801, conveyed the first load of passengers ever moved by the force of steam.

The start was from Tyack's smiths' shop, where the smaller parts had been made.[1] East and west ran the great main coach-road to London, on which the Cornish coach, at that time a van or covered waggon, conveyed the few who travelled on wheels. Northwards, towards

---

[1] The granite remains of the traditional granite steam-boiler at the Weith was within a short distance of Tyack's smiths' shop.

the great house of Lord Dedunstanville, at Tehidy, the road was more hilly. The south road was a rude country lane, in the worst possible order, with a sharp curve at the commencement, and steeper gradients than either of the other roads.

As an indication of the greater difficulty of constructing an engine at that time in Cornwall, thirty years later, when the writer worked in Harvey's engine-factory at Hayle, in the building in which William West had constructed parts of the Camborne locomotive, there were but a few small hand-lathes fixed on wooden benches, a few drilling machines, and but one chuck-lathe. Arthur Woolf was the engineer, and the writer his pupil, and served under the shop foreman, Jeffry, whose father had worked on the Camborne locomotive, and on the Jeffrie and Gribble engine at Dolcoath.

The following statement was given by old Stephen Williams:[1]—

"I knew Captain Dick Trevithick very well; he and I were born in the same year. I was a cooper by trade, and when Captain Dick was making his first steam-carriage I used to go every day into John Tyack's blacksmiths' shop at the Weith, close by here, where they put her together.

"The castings were made down at Hayle, in Mr. Harvey's foundry. There was a deal of trouble in getting all the things to fit together. Most of the smiths' work was made in Tyack's shop.

"In the year 1801, upon Christmas-eve, coming on evening, Captain Dick got up steam, out in the high-road, just outside the shop at the Weith. When we see'd that Captain Dick was agoing to turn on steam, we jumped up as many as could; may be seven or eight of us. 'Twas a stiffish hill going from the Weith up to Camborne Beacon, but she went off like a little bird.

---

[1] Residing at Camborne in 1858.

"When she had gone about a quarter of a mile, there was a roughish piece of road covered with loose stones; she didn't go quite so fast, and as it was a flood of rain, and we were very squeezed together, I jumped off. She was going faster than I could walk, and went on up the hill about a quarter or half a mile farther, when they turned her and came back again to the shop. Captain Dick tried her again the next day; I was not there, but heard say that some of the castings broke. Recollect seeing pieces of the engine in the ditch years afterwards, and suppose she ran against the hedge."

In the same year Mr. Newton[1] informed the writer "that he knew Mr. Trevithick well, and was to have been his pupil in engineering. He rode on the engine the first evening it was tried. It went half a mile up a steep hill, and then returned to the workshop [we stood on the very ground while Mr. Newton told his story]. The fire was blown by a double-acting bellows, worked by the engine. Was well acquainted with Murdoch, and his friends in Cornwall, but never heard he had made a locomotive, or that Trevithick had been his pupil. The engine was called Captain Dick's puffer, from the steam and smoke puffing out of the chimney at each stroke of the engine."

"Captain Nicholas Vivian saw the Camborne steam-carriage, and was familiar with the stories of the early trials, as his friends and relatives were interested in it. It ran part of the way up the Beacon hill when first tried. Something went wrong, and it was taken back to Tyack's smiths' shop; it worked again after that, but he did not know what was then done with it."

The stiffness of the incline on the Beacon hill prevented horse-vehicles from ascending at more than walking speed. In the present day it is straightened, but the old boundary-marks are still to be seen of a sharp curve at the commencement of the journey.

---

[1] Residing at Camborne in 1858.

This southern road from Camborne was the worst of the four that were open to Trevithick's choice for testing his first locomotive, carrying as many passengers as could find standing-room on it—perhaps half a dozen or half a score. A piece of newly-made road with loose stones, just where the incline increased, and when the small boiler had expended its hoarded stock of high-pressure steam, heaped an insurmountable barrier against the small wheels of the engine, and baffled the engineer for the moment. While the road was being smoothed, the steam had increased its elastic force. Another progress was made, and the first half-mile had been travelled on a steam-horse.

The incline was still increasing, almost up to the steepness of the famous Mont Cenis, of 1 in 15 or 20, when the engineer was again beaten, and stuck fast. It was Christmas time: rain drenched the ambitious innovators, and cooled the steam-boiler, and coming darkness added to the gloom. The engine was turned about, and safely conveyed the passengers back again, down this dreadful circuitous hill, to the starting point at Tyack's smiths' shop.

"Captain Joseph Vivian, of Roskadennick, says: When he was a boy about nine years old, he used to go into Tyack's smiths' shop at Camborne to see the steam-carriage Captain Trevithick was having made there.

"Some of the turning work was done in Captain Andrew Vivian's workshop, where there was a small turning lathe. The castings came from Harvey's foundry at Hayle. The first day the new steam-carriage ran about the streets and up the Beacon hill. The next day it went down to Crane, a short mile, that Captain Andrew Vivian's family, who lived there, might see it. Old Mrs. Paul cried out, 'Good gracious, Mr. Vivian! what will be done next? I can't compare un to anything but a walking, puffing devil.'

"It then went nearly three miles to Tehidy Park, that Lord Dedunstanville might have a ride, but something broke just before it reached the house."

The 'Mining Journal,' Oct. 2nd, 1858, has the following:—

"With regard to the Cornwall engine, Mr. John Petherick, writing to his nephew, Mr. Edward Williams, says:—

"'Cardiff, September 11th, 1858. I perfectly remember, when a boy, about the year 1800, seeing Trevithick's first locomotive engine, worked by himself, come through the principal street of Camborne, in Cornwall (Trevithick's native place).

"'This experiment was satisfactory only as long as the steam pressure could be kept up; and during that continuance Trevithick called upon the people to "jump up," so as to create a load on the engine, and it soon became covered with men, who did not seem to make any difference to the power of speed, as long as the steam was kept up. This was sought to be done by the application of a cylindrical horizontal bellows[1] worked by the engine itself, instead of a stack; but the attempt to keep up the power of the steam for any considerable time proved a failure. There were other experiments made which I heard of, but did not witness; but the same objection prevailing, caused the attempt to be laid aside for that time. This was the first public exhibition of the invention.'"

"Henry Clark[2] says he was working in the Valley smiths' shop in Dolcoath, when Captain Dick's locomotive ran. It carried ten or twelve persons from Mr. Harris's gate to Knap's Hotel, about a quarter of a mile, and up hill.

"I worked many years for Captain Dick: he was a nice man. We went to his house in Camborne singing carols; Captain Dick touched his lady with his elbow; she was taking something out of her purse, when Captain Dick said, 'Let me see!' and he

---

[1] The drawing of the Camborne locomotive, made from recollections of the various old men, who, when boys, had seen it, shows an ordinary smith's bellows of the present day. The writer had not then seen the statement of his friend Mr. Petherick, who is probably correct in stating that the bellows was cylindrical.

[2] Living at Redruth, 1869.

turned the purse upside down in my hand, and a pretty lot of silver we had."

"Henry Vivian,[1] a nephew of Captain Andrew Vivian, recollects his father (Johnson Vivian) giving a lecture on the early locomotive in Camborne, and reading a letter he had received from Captain Andrew Vivian describing the trials in the London streets, sometimes going too fast, sometimes not fast enough.

"The omnibus and cab drivers used to throw cabbage stumps, rotten onions, or eggs at them.

"When describing the earlier experiments in Camborne, his father said it was put together in a smiths' shop at Wheal Gerry, near Roskear Mine. He worked about it.

"They started to go to Tehidy House where Lord Dedunstanville lived, about two or three miles off. Captain Dick Trevithick took charge of the engine, and Captain Andrew was steering. They were going very well around the wall of Rosewarne; when they came to the gully (a kind of open water-course across the road) the steering handle was jerked out of Captain Andrew's hand, and over she turned."

These several recollections from eye-witnesses, so long after the event, establish the broad facts, though details may be more or less wanting, showing two distinct experiments: the one up Beacon Hill, the other in the opposite direction, to Tehidy Park. Probably there were several trials, and some repairs and improvements during several weeks or even months.

The blast-pipe gave its puffs of steam and smoke from the chimney. The double-action bellows, for ensuring a good fire in case the steam-blast did not answer, was never used after this first experiment.

Trevithick and Vivian became partners over their Christmas dinner of 1801, and then started for London to secure a patent.

---

[1] Working for Messrs. Harvey and Co., 1869.

"London, No. 1, Southampton Street, Strand,
"Mr. Giddy, "*16th Jan.*, 1802.

"Sir,—No doubt ere this you have been in expectation of hearing from us; but so much time was taken up at Bristol and its environs, in contracting for engines, that we did not arrive here until last Wednesday night.

"We waited on Mr. Sandys, who informed us that the caveat had been secured, and advised us to get the best information we could of persons who were well acquainted with patents and machines, of what title to give the machine, and what was the intended use of it. The next day waited on Mr. Davy with your kind letter, who with the greatest cheerfulness immediately waited on Count Rumford, to whom we had a letter of introduction from Mr. D. to wait on the Count on Friday morning. We found him a very pleasant man, and very conversant about fire-places, and the action of steam for heating rooms, boiling water, dressing meat, &c.; but did not appear to have studied much the action of steam on pistons, &c. The Count has given us a rough draft of a fire-place, which he thinks is best adapted for our carriage, and Trevithick is now making a complete drawing of it.

"We are to wait on the Count when the drawing is completed, and he has promised to give us all the assistance in his power.

"Mr. Davy says that a Mr. Nicholson, he thinks, will be a proper person to assist us in taking out the patent, and we are to be introduced to him to-morrow, and then shall immediately proceed with the business. We shall not specify without your assistance, and all our friends say that if we meet with any difficulty nothing will be so necessary as your presence.

"When we delivered Lord Dedunstanville's letter to Mr. Graham, he said he would give us every assistance in his power gratis, if wanted.

"Mr. Pascoe Grenfell says he can find a way to the Attorney-General, if wanted.

"It is strongly recommended to us to get a carriage made here and to exhibit it, which we also believe must be done.

"Trevithick called on Mr. Clayfield at Bristol, and is to call again on his return from Coalbrookdale to go in the mine. Both Mr. Clayfields beg their most respectful compliments to you. Mr. Davy begins his lectures at the Royal Institution on Thursday next, and has given us tickets of admittance.

"We remain in good health and spirits, Sir,

"Your most obliged, humble servants,

"R. TREVITHICK,
"A. VIVIAN."

On January, 1802, the two partners were busy in London with the intricacies of the patent-laws. The legal men asked, What is the title of the machine? and what its intended use? But these apparently simple questions were not easily answered.

The common road locomotive had been worked. The tramroad and railroad engines were being designed: neither of those, or both combined, would serve as a title describing fully the objects of the patent.

On their way from Cornwall to London they had contracted to erect their patent high-pressure engine for pumping and winding purposes on Mr. Clayfield's colliery.

The drawing to accompany the patent showed three distinct kinds of engines for different uses. How could Trevithick define a use which was to be general and universal?

However, Sir Humphry Davy, Mr. Davies Gilbert, and Count Rumford, all well known for their scientific knowledge (and as active members of the Royal Society, the two former having been Presidents), came to his assistance and helped with their advice.

Count Rumford was not quite up to the idea of the new steam-engine, to be worked solely by the pressure of steam on the piston; still he gave his opinion of a proper fire-place for the boiler of the steam-carriage.

Vivian returned to Camborne, and wrote:—

"DEAR FRIEND, "CAMBORNE, *Tuesday, February 23rd*, 1802.

"I arrived here last evening safe and sound, and missing my wife, was soon informed she was at your house, where I immediately repaired. Your wife and little Nancy are very well, but Richard is not quite well, having had a complaint which many children in the neighbourhood have been afflicted with; they are a little feverish when attacked, but it has soon worn off, as I expect your little son's will also; he is much better this morning, and talked to me very cheerfully.

"Mrs. Trevithick is in pretty good spirits, and requested I would not say a word to you of Richard's illness, as she expected it would be soon over; but as I know you are not a woman, have given you an exact state of the facts. All my family, thank God, I found in perfect health, and all beg their kind remembrances to you, as does everyone that I have met in the village.

"'How do you do?' 'How is Captain Dick?' with a shake by the hand, have been all this morning employed.

"In a day or two you shall have all the particulars of mines, &c. Suffice for the time to say that N. B. Downs is as good as ever. Sold 83 tons, 4th inst., chiefly halvans, at 9*l.* 18*s.* per ton, and have not, as we supposed, postponed the sampling one month, but have sampled again 124 tons.

"In the Falmouth paper are the following lines:—

"'In addition to the many attempts that have been made to construct carriages to run without horses, a method has been lately tried at Camborne, in this county, that seems to promise success.

"'A carriage has been constructed containing a small steam-engine, the force of which was found sufficient, upon trial, to impel the carriage, containing several persons, amounting at least to a ton and a half weight, against a hill of considerable steepness, at the rate of four miles an hour; upon a level road it ran at the rate of eight or nine miles an hour. We have our information from an intelligent and respectable man who was in the carriage at the time, and who entertains a strong persuasion of the success of the project. The proprietors are now in London soliciting a patent to secure the property.'

"The same paper also mentions the increasive population of the parish of Camborne, *viz.* in one week nine women upraised, five pair of banns published on Sunday, and five more delivered to the clerk the Saturday following, eight children christened, and five weddings, *a rare week's work*, which have produced a few lines in verse, which I perused this morning; it describes the parson reprimanding the clerk, sexton, and organist for getting drunk, and himself at the same time reeling against the altar-piece at the Communion table and breaking one of the commandments.

"Pray let me hear from you on the receipt of this. When you go on with the York Water Company be sure to remember in the agreement that the new engine is not to do *more work than the old one unless paid in proportion*, otherwise they may increase their number of tenants, and keep our engine constantly at work.

"With most respectful compliments to Mr. and Mrs. Stamp, Mary, and the little ones,

"I remain, dear Sir,

"Yours most sincerely,

"ANDREW VIVIAN.

"Mrs. T. (your beloved wife) begs her love, and expects to hear from you often.

"Mr. RICHARD TREVITHICK, Steam Engineer,
"1, *Southampton Street, Strand, London.*"

Trevithick returned to Cornwall, to go on with his experiments, and Vivian took his place in London about the patent:—

"SOUTHAMPTON STREET,
"*Tuesday, 23rd March*, 1802.

"MR. TREVITHICK,

"Dear Friend,—Yours of yesterday came to my hand this moment, and Mrs. T. may rely on me to do the needful in procuring the balsam.

"I arrived in town yesterday about three o'clock. I boxed Harry (that is, I went without dinner) that no time might be lost. This I should not have done, you may easily suppose, had the whole of the business been my own; but here you are concerned, which puts all inconveniences to myself out of the question. Of course I immediately repaired to Crane Court, where I was informed that the patent was to be sealed on Wednesday (to-morrow); and that they did not know whether the specification must be lodged on to-day, or a month from the sealing. They recommended me to Mr. Davy, at the Rolls Chapel Office, for information, where I immediately repaired, but could not find the gentleman. From there I shaped my course to Soho Square, and spent two or three hours with Mr. Nicholson; had the necessary alterations made in the rough copy of the specification.

"At the Fantasmagoria Mary and Sarah Stamp and I got about eight o'clock, and returned about ten, well pleased. To-day I have been again at Crane Court, to the Rolls Chapel Office, and to the Patent Office at the Adelphi, and have got the whole business in a proper train.

"Mr. Horton promised me to get the specification engrossed immediately. Mr. H. soon followed me to the Rolls Chapel Office, to request Mr. Davy to engross it, as he knew more of the business. I spent some time with the little gentleman, and have confirmed my opinion of him, that he is a very *clever little fellow.* He will do all in his power to get me out of town on Thursday evening; if so, you will see me on Saturday evening.

"Mr. Davy by some means had heard of me, and says his sisters and self are well acquainted with Mrs. May, before and since she was married; and had I time we certainly should make a round party to some kind of diversion, but time will not permit me the pleasure. To-morrow having no business of the patent until eight in the evening, when I must call at the Patent Office for the great knob of wax; and the former part of the day will be to see Woolf, Watson, &c.

"Mr. and Mrs. Stamp beg their compliments, also Mary, Sarah, and William; and must beg the favour of your best speeches to be employed in my kind remembrance to Mrs. Trevithick, Mrs. Harvey, and all that may happen to be at your house of your good family, and remain, dear friend, yours very sincerely,

"ANDREW VIVIAN."

Davies Gilbert completes this history of the first locomotive by his recollections.

"MY DEAR SIR, "EASTBOURNE, *April 29th*, 1839.

"The travelling engine took its departure from Camborne Church Town for Tehidy on the 28th of December, 1801, where I was waiting to receive it. The carriage, however, broke down, after travelling very well, and up an ascent, in all about three or four hundred yards.

"The carriage was forced under some shelter, and the parties adjourned to the hotel, and comforted their hearts with a roast goose, and proper drinks, when, forgetful of the engine, its water boiled away, the iron became red hot, and nothing that was combustible remained, either of the engine or the house.

"It must have been, I presume, in the preceding summer (that is, of 1801) that Trevithick and myself tried an experiment, on a one-horse chaise, as to the hold of the wheels on the ground for moving it up an ascent.

"We placed the carriage in the middle of one of the roads leading to Camborne Church Town, and we discovered that none of the declivities were sufficient to make the wheels slide in any perceptible degree, as we forced the carriage forward by turning the wheels.

"The success of the experiment induced Trevithick, in concert with Andrew Vivian, either to procure a patent, or to take measures about it.

"After this I have not any distinct recollection of any particular fact, till, being in Oxford with my father and sisters in the winter and spring of 1804, I was earnestly entreated by Trevithick and by Mr. Samuel Homfray, a great ironmaster at Merthyr Tydfil, to come there and assist them in some experiment. I accordingly left Oxford, and reached

Merthyr Tydfil on the 12th of March, and on the 24th the engine which Trevithick had constructed for going on the railway, travelled from Pen-y-darren, Mr. Homfray's, to a works called Plymouth, and back again; but all the weight was accumulated on the same four wheels, with the engine, for none of us once imagined, if the weight was divided, that the wheels of the engine-carriage could possibly hold. In consequence of this great pressure, a large number of rails broke; and on the whole the experiment was considered a failure.

"Within a year or two of this time, Trevithick thought of the iron tanks to be riveted to the timbers of a vessel.

"He took out a patent, and associated himself with some adventurer, who became a bankrupt, and deprived Trevithick of all the advantages to be derived from this excellent adaptation.

"DAVIES GILBERT.

"J. S. ENYS, Esq."

Trevithick had spoken to Davies Gilbert in 1796[1] on the possibility of making an engine to work by the force of steam; and having made models with wheels again requested his friend to come to Camborne, and go with him *in a one-horse chaise* "to test the hold of its wheels on the ground for moving the carriage up an ascent."

This was the same post-chaise that we shall find was specially kept for Watt's use when in the neighbourhood of Redruth some sixteen years before, on his first residence in Cornwall. Mrs. Trevithick has spoken of drives with her husband in this much-envied post-chaise of three-quarters of a century ago, kept for the aristocracy by Mr. Harvey, who lived opposite Newton's Hotel, in Camborne. It was the only comfortable carriage to be let on hire, fit for gentlefolk, in the West

[1] See letter of Davies Gilbert, chap. iv.

of England, to supply the twenty or thirty miles of country from Truro to the Land's End.

Watt engaged the grand carriage for a week or more at a time; Trevithick and his wife only on special holidays; and probably the use of the drag on the steep hills, or other trifling incident, taught Trevithick the amount of friction between a drag or a carriage-wheel, and the road on which it rested.

About 1800 Trevithick requested Mr. Davies Gilbert to come with him and witness the fact that carriages could be propelled by causing their wheels to turn round. The two friends removed the horse from the carriage when on a stiff hill, gave their strength to the spokes of the wheels, and the carriage progressed. This memorable carriage conveyed the two greatest of modern engineers while their thoughts were intent on schemes of national importance, and gave a proof to one of them that if its wheels were forced to turn, the carriage and its contents would progress; and on this hint the building was at once commenced of the first Camborne common road locomotive to be driven by the force of high-pressure steam, which in 1801 was reported in the Falmouth paper as "carrying several persons, amounting to at least a ton and half weight, against a hill of considerable steepness, at the rate of four miles an hour; and when upon a level road, of eight or nine miles an hour."

Stephen Williams was sure it was Christmas-eve when he rode. Davies Gilbert waited at Tehidy for the steam-carriage on the 28th December; but it did not reach the end of its intended journey of three miles, because something broke. The gentlemen adjourned to the hotel for dinner and Cornish punch. Meanwhile the engine, being neglected, was burnt.

Williams heard that the engine ran on the 25th December, the day following that on which he had ridden. These different stories refer to one locomotive tried on various days, which shortly after its destruction was replaced by a new locomotive.

"Mr. Hugh Hunter[1] recollects crowds of people going from Redruth to Tuckingmill to see Captain Trevithick's puffing devil. It worked from Camborne to Tuckingmill and back; the two places are a mile apart. It was a very hilly, bad road, much worse than it is now. 'Tis hilly enough now, but 'tis broad and straight; then it was narrow and very crooked. Some of the crooked parts of the old road may still be seen. I could not leave my work to go to Tuckingmill, but everybody was talking about it; and I thought I should like to work under Captain Dick, who was then engineer in Cook's Kitchen, and, after a bit, I got in there as carpenter about 1802 or 1803, not long after Captain Dick had worked his locomotive; and I have been in the mine ever since, and had to do with many of Captain Dick's engines."

"Mr. Anthony Michell[2] came to live at Redruth in November, 1802, and shortly after, about the spring of 1803, a great many persons went to Tuckingmill to see Captain Dick Trevithick's puffer locomotive that was going to run from Camborne to Redruth, about three or four miles: he was going with Mr. Hocking, a carpenter; Walter Bray, an innkeeper; and John Cole, a carpenter. I could not go. They said, that in going up the Tuckingmill hill toward Redruth, the driving wheels slipped around and sunk into the road, and they could not get her on; it was a very steep and crooked road. Everybody was talking about it, and they said that Captain Dick and Captain Andrew went before the King; and Captain Dick kissed the King's hand, but Captain Andrew was not asked.

"My father, Richard Michell, was one of the old Cornish engineers; he made a model of Boulton and Watt's engine, and took it to London for the trial against Bull, and losing that

---

[1] Living at Pool in 1869. [2] Living at Redruth in 1869.

trial broke Bull's heart. There was no post-chaise in Redruth at that time, but there was one in Camborne that used to be kept for Watt. Watt used to put a piece of his father's roll tobacco to dry on the steam-pipe, and then would roll it in his hands to make a pinch of snuff.

"About twelve years ago my brother John made a working model of a steam-engine that stood upon a fourpenny-piece."

Mr. Hugh Hunter related these early incidents as one reading from a book. He had for more than sixty years worked in Cook's Kitchen Mine, where about 1809 he replaced the famous water-wheel, which fifty years before had, under the management of Trevithick, sen., been spoken of as the largest in the country. Hunter's new wheel was of the same size as the old one. It was of wood, 48 feet in diameter, and 3 feet in the breast.

The locomotive, says Hunter, ran between Camborne and Tuckingmill in 1802 or 1803, on the road to Redruth, from which place crowds of people went to see it, and the talk about it made him solicit employment under Trevithick.

Mr. Michell, who belonged to the Eastern, or Watt Mines, spoke distinctly of the trial of the locomotive between Camborne and Redruth in the autumn of 1802, or spring of 1803, when it stuck fast, because the driving wheels slipped round on going up a steep hill. The road is still, in its altered state, pretty stiff; but thirty or forty years ago the incline was 1 in 15 or 20.

He spoke with great glee of his father's going with Watt in the Camborne post-chaise, and making snuff out of his roll tobacco for the great engineer.

This second trial in Cornwall was a preparation for the greater experiment in the streets of London.[1]

[1] The writer is in possession of a small part of one of these locomotives, said to have been given by Trevithick to the late Mr. Francis Michell.

Stuart thus speaks of the increased power of Trevithick's engine:—

"The steam, however, is capable of a temperature considerably above that required merely to balance the piston, or 14½ lbs. to each square inch; it would, as used in these engines, generally balance four or five times that weight; and, of course, the piston is impelled downwards in the cylinder. When it has made its stroke, the cock is turned by the lever or pin attached to the working beam, or horizontal arm, into the opposite position. This opens a communication between the under side of the piston and the boiler, and between the upper side and the chimney," &c.[1]

Hebert, speaking of the origin of steam locomotion, and of the marked difference between the Watt and the Trevithick engine, says:[2]—

"The merit of the first suggestion of steam-carriages has been attributed to different individuals, but the probability is, that the idea of applying the steam-engine for the purpose of locomotion was coeval with its first invention. Thus Savery, from having considered its possibility, and Dr. Robison, from having suggested it to Watt, have by some been regarded as the inventors; but almost as well might we regard the philosophic poet Darwin to have been the inventor, who prophesied:—

'Soon shall thy arm, unconquered steam, afar,
Drag the slow barge, and *drive the rapid car!*'

"In a note to a late edition of Dr. Robison's 'Mechanical Philosophy,' Mr. Watt states: 'My attention was first directed in the year 1759 to the subject of steam-engines by the late Dr. Robison, then a student in the University of Glasgow, and nearly of my own age. He at that time threw out the idea of applying the power of the steam-engine to the moving of wheel-carriages, and to other purposes; but the scheme was soon *abandoned* on his going abroad.'

"In the patent granted to Mr. Watt in 1784, he gave an account of the adaptation of his mechanism to the propulsion of

---

[1] See Stuart, 'History of the Steam-Engine,' p. 165; published 1824.
[2] See Luke Hebert 'On Railways,' p. 18; published 1837.

land-carriages. The boiler of this apparatus he proposed should be made of *wooden staves*, joined together, and fastened with iron hoops like a cask. The furnace to be of iron, and placed in the inside of the boiler, so as to be surrounded on every side with water. The boiler was to be placed on a carriage, the wheels of which were to receive their motion from a piston working in a cylinder, the reciprocating motion being converted into a rotary one by toothed wheels, revolving with a sun-and-planet motion, and producing the required velocity by a common series of wheels and pinions.

"By means of two systems of wheel-work, differing in their proportion, he proposed to adapt the power of the machine to the varied resistance it might have to overcome from the state of the road. A carriage for two persons *might*, he thought, be moved with a cylinder of 7 inches in diameter, when the piston had a stroke of 1 foot, and make sixty strokes a minute. Mr. Watt, however, never built a steam-carriage.

"It is well known that Mr. Watt retained up to the period of his death, the most rooted prejudices against the use of high steam; indeed, he says himself: 'I soon relinquished the idea of constructing an engine on this principle, from being sensible it would be liable to some of the objections against Savery's engine, *viz.* the danger of bursting the boiler; and also that a great part of the power of the steam would be lost, because no vacuum was formed to assist the descent of the piston.'—*Watt's Narrative.*

"The specification of the patent granted to Messrs. Trevithick and Vivian is descriptive of a high-pressure engine, the most simple and effective ever known, which has thus been characterized by the eloquent Mickleham:—'It exhibits in construction the most beautiful simplicity of parts, the most sagacious selection of appropriate forms, their most convenient and effective arrangement and connection: uniting strength with elegance, the necessary solidity with the greatest portability; possessing unlimited power, with a wonderful pliancy to accommodate it to a varying resistance; it may indeed be called *The Steam-Engine!*'"

Stuart, writing fifty years after the date of the

Watt patent, clearly defined the difference, in principle and in practice, of the rival engineers. Trevithick increased the steam pressure from one atmosphere, or 14½ lbs. on the square inch, to 50 or 60 lbs., and by it impelled the piston with four times the force of a Watt low-pressure steam vacuum engine. Hebert, who wrote thirteen years later, still illustrates the marked difference in the two men by pointing out that, in 1784, Watt gave his views of a steam-carriage, and Murdoch tried his hand at one. Watt proposed a wooden boiler, a cylinder 7 inches in diameter, with a stroke of 1 foot, and sun-and-planet wheels. It is not said that it was to carry condensing water, but such may reasonably be inferred. It was to convey two persons, but on further consideration, Watt said: "I soon relinquished the idea of constructing an engine on this principle (high-pressure steam principle), because of the danger of bursting, and also because no vacuum was formed to assist the descent of the piston."

Trevithick's high-pressure engine, which was worked by the force of steam 60 lbs. or more on the square inch, wholly discarded the vacuum; and certainly without this radical change there could have been no locomotion.

No original drawing remains of this first passenger-carrying locomotive engine, but from the description of those who saw it, there is little doubt of the general correctness of the drawing made by the writer.

One of the gains from the Trevithick high-pressure engine was its portability and cheapness, when compared with the Newcomen or the Watt engine, neither of which could work without condensing water—which locomotive engines cannot procure.

This engine had little or no resemblance to any steam-engine that had preceded it. No stone foundations, air-pump, condenser, condensing water, beam, or parallel motion, all the improvements of fifty years thrown aside; and the great and costly boiler of his predecessors was replaced by one so small that it seemed merely a part of the engine, as a stand-point on which to fix the mechanism.

The boiler was cylindrical, of cast iron, in which the steam-cylinder was fixed; the tubular-boiler flues were of wrought iron, and entirely within the boiler.

The exhausted steam having done its work in the cylinder, at a pressure of 60 lbs. to the inch, passed into the chimney as a steam-blast, causing an intensely hot fire, and in its passage heated the feed-water.

The boiler-flues, together with the steam-cylinder, filled up a large portion of the interior of the boiler, reducing the weight of water carried to its lowest possible requirement, while a feed-pump, steam-gauge, safety-valve, and soft metal plug in the fire-tube, gave accurate supply of water, or escape of steam.

The feed-pole was constructed precisely like the plunger-poles he had fixed in the mine-shafts, and is unchanged to the present day.

Low-pressure boilers had no feed-pumps, being supplied by gravity of the water from a cistern above the boiler.

The heating the feed-water by the waste steam was also an application originating with him. The steam-cylinder had its steam-pipe and bottom cast as a part of it. The inlet and outlet of steam was performed by one simple piece called Trevithick's four-way cock. The piston-rod, cross-head, guide-rods, and connecting rods, were all of simple form. The boiler served as a framing for keeping the four wheels in their proper places, and also the cylinder and working parts. The

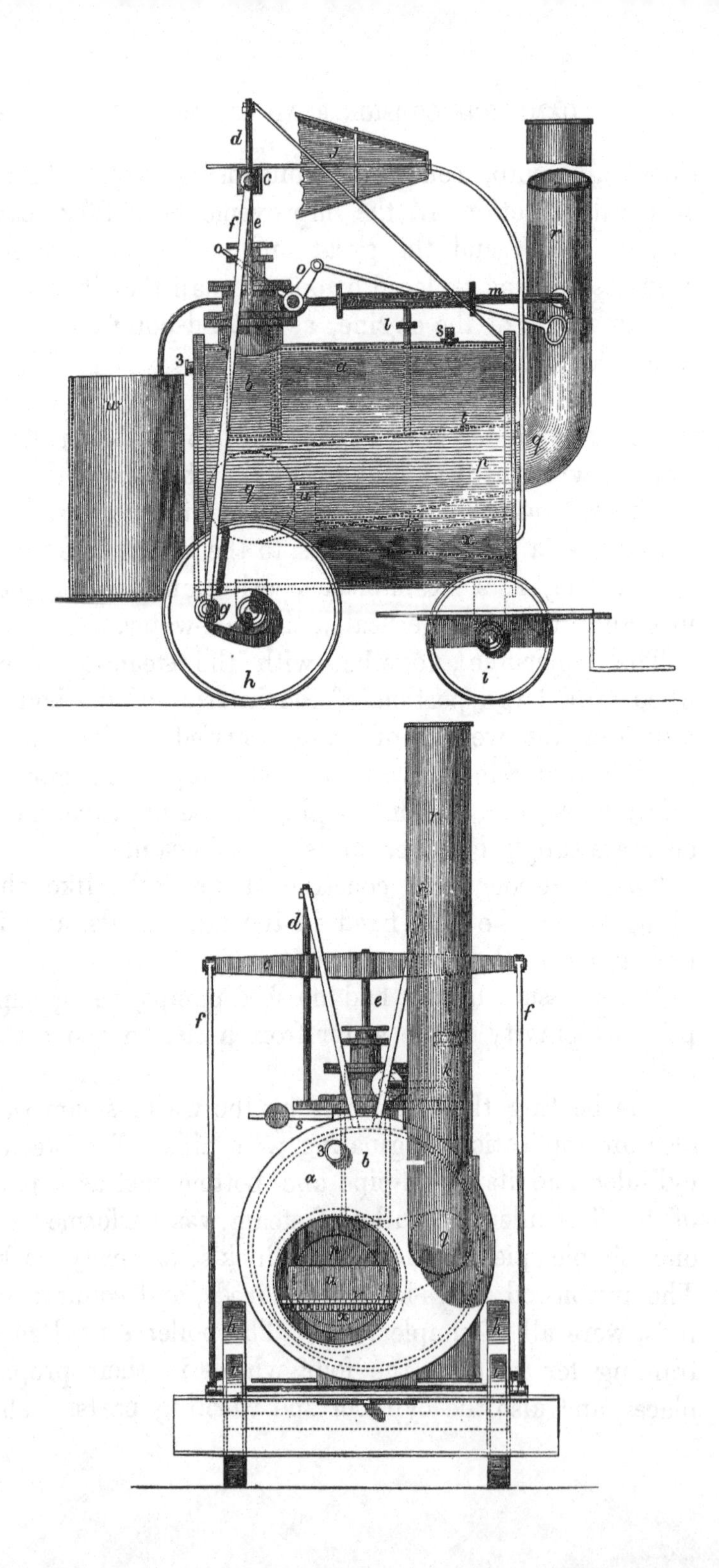

d
c
f
e
o
o
m
l
s
3
a
b
w
r
q
p
q
u
x
g
h
i
d
e
f
f
s
3
b
a
k
q
p
u
x
h
h

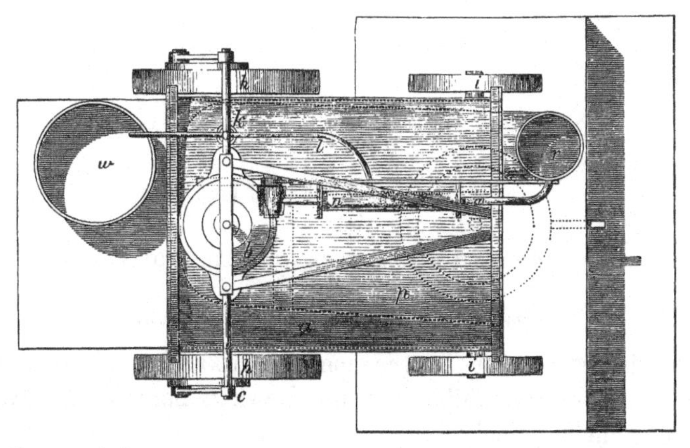

TREVITHICK'S FIRST PASSENGER-CARRYING COMMON ROAD LOCOMOTIVE, CAMBORNE, 1801.

a, cylindrical boiler with wrought-iron ends, having inside it a wrought-iron tube bent as the letter U; *p*, the fire-place, in one end of the tube; *v*, fire-bars; *u*, fire-bridge; *x*, the ash-pit; *q*, the return flue, leading to *r*, the chimney—the fire-door is not shown, as it would confuse the drawing; *z*, the steam-gauge; *s*, safety-valve; *t*, soft metal safety-plug in top of fire-tube; *j*, the bellows, blowing air into the close ash-pit, fixed to the guide-stays, and worked by the arm of its movable middle division connected with the piston-rod cross-head; *b*, steam-cylinder let into the boiler, having a close top and bottom, with pipe for conveying steam to and from the bottom, and also the shell for the four-way steam-cock, and the steam-way from the boiler, all cast with the cylinder; *o*, a four-way steam-cock, worked by a rod from the cross-head, with two tappets striking the lever, *o*, up and down, and having a handle, *o*, suitable for the engineman; *k*, the feed pole-pump, worked from the cross-head; *l*, the feed-pipe; *w*, feed-water cistern; *n*, case for heating feed-water by the passage of the waste steam through *m*, the waste-steam pipe, from the cylinder to the chimney; *c*, the cross-head; *f*, the two side rods; *g*, the two cranks; *h*, two driving wheels; *i*, two steering wheels; *e*, piston-rod; *d*, guides for the piston-rod cross-head.

two front or steering wheels were turned by a rod conveniently placed close to the engineman attending at the fire-door.

One result of these experiments was the immediate application for a patent, granted on the 24th March, 1802, to Richard Trevithick and Andrew Vivian, for steam-engines for propelling carriages, &c., which may be read and studied by the young engineer with pleasure and profit even in "this" age of greatly-improved steam mechanism.

CHAPTER VIII.

PATENT OF 1802, AND LONDON LOCOMOTIVE.

Trevithick and Vivian's Specification.

"OUR improvements in the construction and application of steam-engines are exhibited in the drawings hereunto annexed and explained, namely:—

"In Plate IV., Figure 1 represents the vertical section of a steam-engine with the said improvements; and Figure 2 represents another vertical section of the same engine at right angles to the plane of Figure 1. The dark shaded parts represent iron, and the red parts represent brickwork, and the yellow parts engine brass, excepting only the wooden supporters of the great frame in Figures 4 and 5, and the carriage-wheels in 6 and 7. A represents the boiler made of a round figure, to bear the expansive action of strong steam. The boiler is fixed in a case D, luted inside with fire-clay, the lower part of which constitutes the fire-place B, and the upper cavity affords a space round the boiler, in which the flame or heated vapour circulates round till it comes to the chimney E. The case D and the chimney are fixed upon a platform F, the case being supported upon four legs. C represents the cylinder enclosed for the most part in the boiler, having its nozzle, steam-pipe, and bottom cast all in one piece, in order to resist the strong steam, and with sockets in which the iron uprights of the external frame are firmly fixed. G represents a cock for conducting the steam, as may be more clearly seen by observing Figure 3, which is a plan of the top of the cylinder, and the same parts in Figure 2.

"*b*, Figures 2 and 3, represents the passage from the boiler to the cock G. This passage has a throttle-valve or shut, adjustable by the handle *m*, Figure 2, so as to wiredraw the steam,

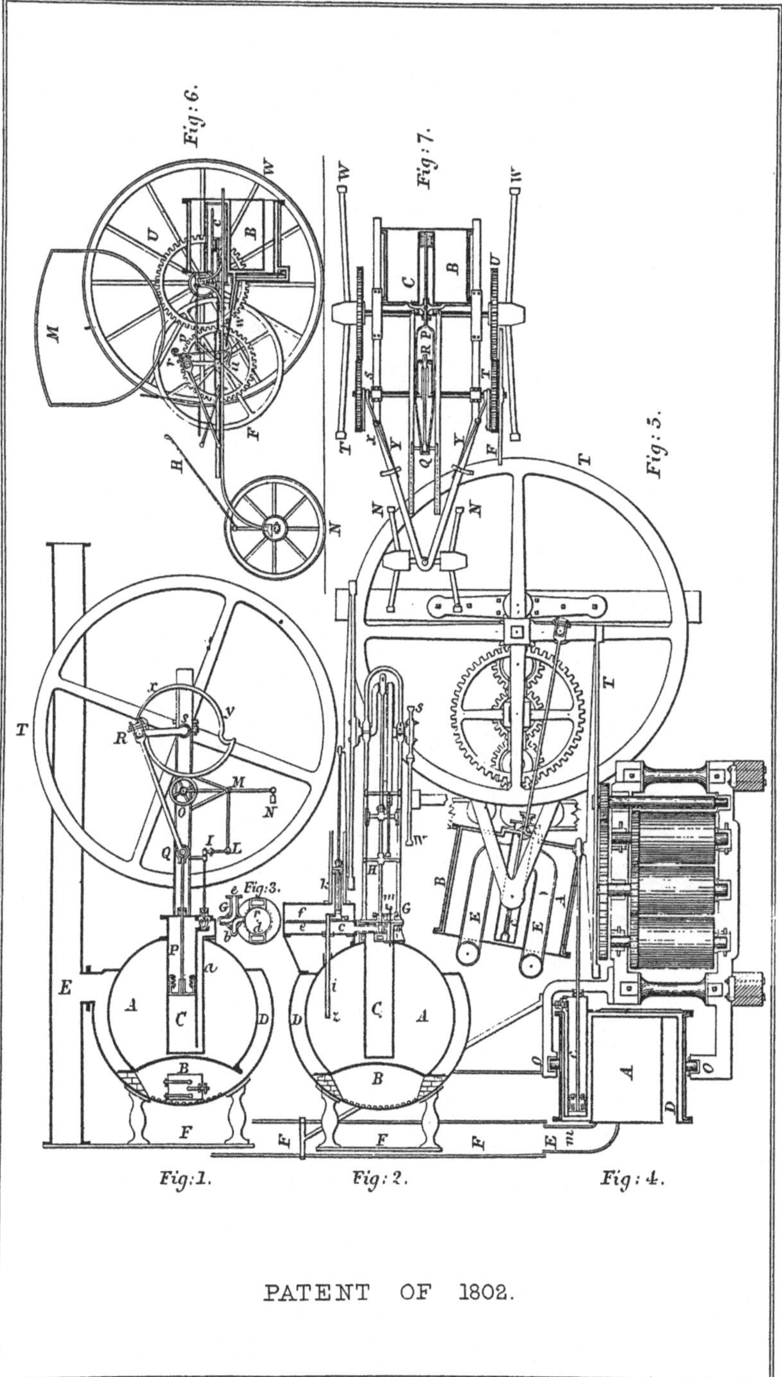

PATENT OF 1802.

London: E. & F. N Spon, 48, Charing Cross.
Kell, Bros. Lith. London

and suffer the supply to be quicker or slower. The position of the cock represented in Figure 3 by the yellow circle is such, that the communication from the boiler through *b*, by a channel in the cock, is made good to *d*, which denotes the upper space of the cylinder above the piston, at the same time that the steam-pipe *a* (more fully represented in Figure 1) is made to afford a passage from the lower space in the cylinder beneath the piston to the channel C, through which the steam may escape into the outer air, or be directed and applied to heating fluids or other useful purposes. It will be obvious that if the cock be turned one quarter of a turn in either direction, it will make a communication (Figure 3) from the boiler passage *b* to the lower part of the cylinder, by or through *a*, at the same time that the passage *r* from the upper part of the cylinder will communicate with *c*, the passage for conveying off the same steam. P, Q, is the piston-rod moving between guides, and driving the crank R, S, by means of the rod Q, R, the axis of which crank carries the fly T, and is the first mover to be applied to drive machinery, as at S and W, Figure 2. The alternations of action are made by the successive pressure of the steam above and below the piston, and these are effected by turning the cock a quarter turn at the end of each stroke by means of the following apparatus, most fully delineated in Figure 1. *x*, *y*, is a double snail, which in its rotation presses down the small wheel O, and raises the weight N by a motion on the joint M of the lever O, N, from which proceeds downwards an arm, M, L, and consequently the extremity L is at the same time urged outwards. This action draws the horizontal bar L, I, and carries the lever or handle H, I, which moves upon the axis of the cock G through one-fourth of a circle. It must be understood that H, I, is foreshortened (the extremity I being more remote from the observer than the extremity H), and also that there is a click and ratchet-wheel in the part H, which gathers up during the time that L is passing outwards, and does not then move the cock G. But that when the part *x* of the snail opposite O, that is to say, when the piston is about the top of its stroke, then the wheel O suddenly falls into the concavity of the snail, and the extremity L by its return at once pushes I, H, through the quarter circle, and carries with it the cock G, and turns the steam upon the top of the piston, and also affords a passage for the steam to escape from beneath the piston; every stroke, whether up or down, produces this effect by the half turn of the snail, and reverses the steam-ways as before described. Or, otherwise, the cock may be turned by various well-known methods, such as the plug with pins or clamps striking on a lever in the usual way, and the effect will be the same, whether the quarter turns be made back or forward, or by a direct circular motion, as is produced by the machinery here delineated, but the wear of the cock will be more uniform and regular if the turns be all made the same way. In the steam-engines constructed and applied according to our said Invention, the steam is usually let off or conducted out of the engine, and in this case no vacuum is formed in the engine, but the steam, after the operation, is or may be usefully applied as before mentioned. But whenever it is found convenient or necessary to condense the steam by injection water, we use a new method of condensing by an injection above the bucket of the air-pump; and by this Invention we render the condenser or space which is usually constituted or left between the said bucket and the foot-valve entirely unnecessary, and we perfectly exclude the admission of any elastic fluid from the injection water into the internal working spaces of the engine.

"In Figure 2 is represented a method of heating the water for feeding the boiler, by the admission of steam after its escape through *c* into the cistern *f*. The steam passes under a false bottom *e*, perforated with small holes, and heats the water therein, a portion of which water is driven at every revolution of the fly by the small pump *k*, through *l*, *z*, into the boiler A. We also on some occasions produce a more equable rotary motion in the several parts of the revolution of any axis moved by steam-engines, by causing the piston-rods of two cylinders to work on the said axis by cranks at one quarter turn asunder. By this means the strongest part of the action of one crank is made to assist the weakest or most unfavourable part of the action in the other, and it becomes unnecessary to load the work with a fly.

"Figure 4 is an upright section, and Figure 5 is a plan of the engine, with rollers for pressing or crushing sugar-canes, moved by a steam-engine improved and applied according to our said new Invention. B is a case, in the form of a drum or cylinder, suspended upon two strong trunnions or pivots at O and O, its flat ends standing upright; within the iron case is fixed a boiler A, not much smaller in its dimensions, but so as to leave a vacant space between itself and the case, and within the boiler is fixed a fire-place, having its grate above the ash-hole D; the heated vapour and smoke rises at the inner extremity, and passes through two flues E, E, Figure 5, which join above at E, *m*, Figure 4, in the chimney E, which is there loosely applied, and is slung between centres in a ring at F. The working cylinder C, with its piston, steam-pipe, nozzle, and cock, are inserted in the boiler as here delineated. The piston-rod drives the fly T, T, upon the arbor of which is fixed a small wheel, which drives a great wheel upon the axis of the middle roller. The guides are rendered unnecessary in this application of the steam-engine, because the piston-rod is capable, by a horizontal vibratory motion of the whole engine upon its pivots O, to adapt itself to all the required positions, and while the lower portion of the chimney E, *m*, Figure 4, partakes of this vibratory motion, the upper tube E, F, is enabled to follow it by its play upon the two centres or pivots in the ring F. In such cases or constructions as may render it more desirable to fix the boiler with its chimney and other apparatus, and to place the cylinder out of the boiler, the cylinder itself may be suspended for the same purpose upon trunnions or pivots in the same manner, one or both of which trunnions or pivots may be perforated so as to admit the introduction and escape of the steam, or its condensation as before mentioned. And in such cases, when it may be found necessary or expedient to allow of no vibratory motion of the boiler or cylinder, the same may be fixed, and the method of guides be made use of, as in Figure 1 or 2. The manner in which the cock is turned is not represented in these two drawings, but every competent workman will, without difficulty, understand that this effect may be produced by the same means as in Figure 1, or otherwise by the stroke of pins duly placed in the circumference of the fly, and made to act upon a cross fixed on the axis of the cock, or otherwise by the method used in the carriage, Figure 6, and hereinafter described. The steam which escapes in this engine is made to circulate in the case round the boiler, where it prevents the external atmosphere from affecting the temperature of the included water, and affords, by its partial condensation, a supply for the boiler itself, and is or may be afterwards directed to useful purposes as aforesaid.

"Figure 6 is a vertical section, and Figure 7 the plan of the application of the improved steam-engine to give motion to wheel-carriages of every description. B represents the case, having therein the boiler with its fire-place and cylinder, as have been already described in Figure 4.

"The piston-rod P, Q, Figure 7, is divided or forked, so as to leave room for the motion of the extremity of the crank R. The said rod drives a cross-piece at Q backward and forward between guides, and this cross-piece, by means of the bar Q, R, gives motion to the crank with its fly F, and to two wheels T, T, upon the crank axis, which lock into two correspondent wheels U upon the naves of the large wheels of the carriage itself. The wheels T are fitted upon round sockets, and receive their motion from a striking box or bar S, X, which acts upon a pin in each wheel. S, Y, are two handles, by means of which either of the striking boxes S, X, can be thrown out of gear, and the correspondent wheel W by that means disconnected with the first mover, for the purpose of turning short or admitting a backward motion of that wheel when required. But either of the wheels W, in case of turning, can be allowed considerably to overrun the other without throwing S, X, out of gear, because the pin can go very nearly round in the forward motion before it will meet with any obstruction. The wheels U are most commonly fixed upon the naves of the carriage-wheels W, by which means a revolution of the axis itself becomes unnecessary, and the outer ends of the said axis may consequently be set to any obliquity, and the other part fixed or bended, as the objects of taste or utility may demand. The fore wheels are applied to direct the carriage by means of a lever H, and there is a check-lever which can be applied to the fly, in order to moderate the velocity of progression while going down hill. In the vertical section, *r, u,* denotes a springing lever, having a tendency to fly forward. Two levers of this kind are duly and similarly placed near the middle of the carriage, and each of them is alternately thrown back by a short bearing lever S, *t,* upon the crank axis, which sends it home into the catch *u,* and afterwards disengages it when the bearing lever comes to press upon V, in which case the springing lever flies back. A cross-bar or double handle *o, p,* is fixed upon the upright axis of the cock, from each end of which said cross-bar proceeds a rod *p, q,* which is attached to a stud *q,* that forms part of the springing lever *r, u.* This stud has a certain length of play, by means of a long hole or groove in the bar, so that when the springing lever *r, u,* is pressed up, the stud slides in the groove without giving motion to *p.* When the other springing lever is disengaged, it draws the opposite end of *p, o,* by which means *p* draws the long hole at *q,* up to its bearing against the stud, ready for the letting off of that first-mentioned springing lever. When this last-mentioned lever comes to be disengaged, it suddenly draws *p* back, and turns the cock one quarter turn, and performs the like office of placing the horizontal rod of the other extremity of *p, o,* ready for action by its own springing lever. These alternations perform the opening and shutting of the cock, and to one of the spring levers *r, u,* is fixed a small force-pump *w,* which draws hot water from the case by the quick back-stroke, and forces it into the boiler by the stronger and more gradual pressure of S, *t.* It is also to be noticed, that we do occasionally, or in certain cases, make the external periphery of the wheels W uneven, by projecting heads of nails or bolts, or cross-grooves, or fittings to railroads when required; and that in cases of hard pull we cause a lever, bolt, or claw, to project through the rim of one or both of the said wheels, so as to take hold of the ground; but that in general the ordinary structure or figure of the external surface of these wheels will be found to answer the intended purpose. And, moreover, we do observe and declare, that the power of the engine, with regard to its convenient application to the carriage, may be varied, by changing the relative velocity of rotation of the wheels W, compared with that of the axis S, by shifting the gear or toothed wheels for others of different sizes properly adapted to each other in various ways, which will readily be adopted by any person of competent skill in machinery. The body of the carriage M may be made of any convenient size or figure, according to its intended uses.

"And lastly, we do occasionally use bellows to excite the fire, and the said bellows is worked by the piston-rod or crank, and may be fixed in any situation or part of the several engines herein described, as may be found most convenient."

A little repetition may be necessary in drawing the reader's attention to the number and variety of valuable practical inventions in this notable patent specification of 1802. Figures 1, 2, 3, show a portable high-pressure steam-engine, the boiler of which is a hollow globe of cast iron, having the under side indented for the fire-place. A casing of iron, coated with fire-clay, forms the flues surrounding the boiler. The chimney is of iron, attached to the flue. The cylinder, of most simple form, is let partly into, and fixed in the boiler. From its top, two upright bars serve as supports for the crank-shaft, and also as guides for the piston-rod. A roller on the top of the piston-rod causes it to work easily in the guides. A four-way cock is moved by a compound eccentric, called a snail. If the action of the eccentric, which was nearly, or quite, a new thing, was unsatisfactory, the cock might be worked by the well-known method of tappets. A throttle-valve regulates the supply of steam. An air-pump may be used, condensing in the exhaust-steam pipe in preference to the larger condensing vessel used in the Watt engine; or the engine may be worked as a puffer, without condensing; in which case the waste steam warms the feed-water, which is forced into the boiler by a plunger-pole.

Figures 4 and 5 show his high-pressure portable steam-engine applied to a crushing machine. The boiler in that engine is a cylinder of wrought iron with flat ends, having three internal wrought-iron tubes. The fire-place is in the lower of the tubes, which tube extends the length of the boiler, and returns to the fire-door end as a double tube, where it joins the wrought-iron chimney. The boiler is enclosed in a case slightly larger, just allowing space for the passage of the used steam, thus sheltering the boiler from external cold, and supplying hot feed-water, or reserve steam for other purposes.

The boiler is supported on two centres or trunnions, on which it moves or oscillates. The cylinder is fixed in the steam space of the boiler. The piston-rod attaches direct to the crank. The trunnions serve as steam-ways. The chimney is wrought iron, suspended on a centre to allow for the oscillation of the boiler; or the boiler may be fixed, the cylinder alone oscillating. Two cylinders are recommended in preference to one, with cranks at right angles.

The crank-axle is fixed on a cast-iron frame, which also supports the boiler, chimney, and crushing rollers; the whole apparatus being complete, without masonry or woodwork.

How very distinct is this description of engines, patented seventy years ago: and these were not mere floating ideas, but the result of true and practical knowledge, tested by prior experience.

This particular combination of parts is similar, in a practical sense, to the double-cylinder, oscillating marine engines of the present day. The tubular boiler of small bulk, so small that it might reasonably oscillate. The boiler-flues wholly internal. The cylinder and the boiler protected from the cold, and the feed-water heated by the used steam on its way to the chimney: everything small, and portable, and complete, and attached to the frame that supported the machine to be put in motion. Trial engines had been constructed before the date of the patent to prove their efficiency.

Figure 6 is of a locomotive and carriage for common roads. The boiler is cylindrical, made of wrought iron, with two internal tubes, for fire-place and flue, called the U, or return tube. The cylinder is fixed in the boiler, and placed in a horizontal position, close behind the driving axle. The end of the piston-rod is split into four rods, working in a guide, that the crank-axle may be brought near to the cylinder. Changeable tooth-wheels with couplings connect the crank-axle with the driving wheels. A fly-wheel equalizes the movement and serves as a brake. The four-way cock is worked by spring levers, resting against projections on the crank-shaft, being an eccentric simplified. A spring lever also works the feed-pole. The two driving wheels are of wood, about 10 feet in diameter. The two guide-wheels about 4 feet. Nearly the whole weight of machinery and passengers comes on the driving wheels. The gear-wheel on the crank-axle is 3 feet in diameter, working into a 4-feet on the driving axle, thus increasing the power of the cylinder. Those gear-wheels could be shifted, to vary the power or speed. The fire-door and chimney were both at the back end, but are not shown. The framing is of wrought iron. The whole engine is new in design, and totally different from the first Camborne engine. The horizontal cylinder takes the place of the earlier vertical cylinder. The forked piston-rod is the mechanical novelty that has since become widely useful, being what is now known as the side rod, or side beam and cross-head movement, largely used in steamboats.

Neither of the drawings shows the passage of the waste steam into the chimney. The steam-blast and the adhesion of the wheels were stumbling-blocks to a generation following Trevithick. He saw both of them plainly, though the best form of the blast-pipe was not quite decided on, and therefore not drawn. In the specification he says of the first drawing,—"The steam may escape into the outer air, or be directed and applied to heating fluids, *or other useful purposes.*" And again, "but the steam, after the operation, is or may be *usefully applied as before mentioned.*" In describing the second engine,—"The steam which escapes in this engine is made to circulate in the case round the boiler, where it prevents the external atmosphere from affecting the temperature of the included water, and affords, by its partial condensation, a supply for the boiler itself, and is or may be *afterwards directed to useful purposes as aforesaid.*" In the third or locomotive engine, having a boiler very similar to that just described, he says, "and lastly, we do *occasionally* use bellows to excite the fire."

The bellows is not drawn any more than the blast-pipe, though both were used in the first Camborne engine prior to the patent, but only the blast-pipe was used after the patent. Years of experience have taught locomotive engineers that the best effect of the blast-pipe depends on its particular shape and position in the chimney. This difficulty Trevithick felt, as shown by the wording of the specification, and absence of a drawing of the pipe; but the idea was distinct, and the proof is in his constant use of the blast-pipe, from the first trial.

On the grip or adhesion of the wheels, he says:— "We do occasionally, or in certain cases, make the external periphery of the wheels uneven, by projecting heads of nails or bolts, or cross-grooves, or fittings to railroads when required; and that in cases of hard pull we cause a lever, bolt, or claw, to project through the rim of one or both of the said wheels, so as to take hold of the ground; but that in *general the ordinary structure or figure of the external surface of these wheels will be found to answer the intended purpose.*"

Mr. Symons[1] recollects distinctly Captain Andrew Vivian explaining to him and others that his partner (Trevithick) must be wrong about his proposed tramway engine. "If the wheels slipped on the rough ground, they would slip much more on the smooth iron." This was said when the patent was being taken out, in which Trevithick and Vivian state "the grip will be found sufficient." "Trevithick was one of those men who would have his own way." On this vital question of adhesion his partner was dead against him, and in his view was backed by general opinion, which remained more or less in force for fifty or more years; while, on the other hand, his early practical inventions were clogged by the hesitating and never-carried-out patent claim of Watt. "I intend in many cases to employ the expansive force of steam, to press on the piston, in the same manner as the pressure

[1] Living at Camborne in 1869.

of the atmosphere is now employed in common fire-engines. In cases where cold water cannot be had in plenty, the engine may be wrought by the force of steam only."

The failing to see or to understand what Trevithick's working experiments clearly pointed out, has caused millions of money to be wasted in avoiding inclines and curves on railways that could easily have been passed by the locomotive engine and its train of carriages.

The patent having been secured, Trevithick was hard at work in Cornwall, on his portable high-pressure engines, to be adapted to any purposes where power from coal was cheaper than from men or horses.

Another common road locomotive was in progress, the trials of which are spoken of in a former chapter as the Tuckingmill locomotive. Its form was not sufficiently described by the old men who recollected its running when they were boys, to enable the writer to give a drawing. It performed longer journeys than the first one, and probably those trials suggested further changes. The engine portion was sent to London, there to be supplied with larger wheels, and a suitable carriage for passengers.

A letter[1] of May 2nd, 1803, speaks of a high-pressure engine working very satisfactorily in London, turning and boring brass cannon for the Government, working with steam of 40 to 45 lbs. on the inch.

The Tuckingmill locomotive engine had at that time just arrived in London. The cylinder was 5½ inches in diameter, with a stroke of 2½ feet. With 30 lbs. of steam it worked fifty strokes a minute.

---

[1] See letter, May 2nd, 1803, chap. viii.

The following extract from Andrew Vivian's account-book authenticates the dates:—

	£	s.	d.
"1801.—Expenses to Wm. West about first carriage.			
1803, January.—Wm. West, expenses at Messrs. Harvey and Co., preparing a new cylinder.			
Feb. and March.—Ditto, preparing a boiler.			
Ditto, five months' expenses in London.			
Jan., 1803.—Andrew Vivian, expenses in London, and quay dues at Falmouth.			
July, 1803.—To Felton, for building the coach	83	5	0
Aug., 1803.—To paid Messrs. Foxes, shippers, Falmouth, for carriage of the engine to London ..	20	14	11 "

A rough draft of a letter from Captain Andrew Vivian[1] to Sir Richard Vyvyan, says:—"At the time it was on the road in London, my partner was not with me; it ran well; however, there was some defect, and our finances being low, we were compelled to abandon it."

A draft of a letter by the late Captain Henry Vivian,[2] a son of Captain A. Vivian, says:—

"They carried their engine to London about the middle of 1803.

"I well recollect its going up and down our town several times; and the rise for about 300 feet is 1 in 20.

"The women used to call it the devil. Not making it answer, they sold it to a company in Merthyr Tydfil, in Wales, where it ran on a tramroad.

"My father went to London in 1801, and again in 1802. He himself worked the engine when it ran from Leather Lane, from the shop of Mr. Felton (who built the carriage, and he and his sons were with the engine all the first day it ran), through Liquorpond Street, into Gray's Inn Lane, by Lord's Cricket Ground, to Paddington and Islington, and back to Leather Lane.

"Mr. Gurney never saw the engine that was built here, for it was destroyed by fire[3] before he came to Camborne with me, as stated in my father's letter."

---

[1] Written in July, 1834.

[2] Written in 1845.

[3] See letter of Davies Gilbert, chap. vii.

This second locomotive, tried in Camborne in the latter part of 1802 or commencement of 1803, was sent to London in January, 1803. William West was then at Harvey's foundry in Cornwall, preparing a new cylinder; and still in February and March he was there preparing a new boiler, after which he was for five months in London, about the steam-carriage; and in August, Felton was paid for building the coach.

The London locomotive of 1803 was a great improvement on the former ones: it was not so heavy; and the horizontal cylinder, instead of the vertical, added very much to its steadiness of motion; while wheels of a larger diameter enabled it the more easily to pass over bits of bad road, which had brought the Camborne one to a standstill. The boiler was wholly of wrought iron, and, with the engine attached to it, was put together at or near Felton's carriage shop in Leather Lane; Trevithick, Andrew Vivian, and William West were with it; and Arthur Woolf (then in Trevithick's pay with the first high-pressure sent to London) came to see what was going on.

Andrew Vivian ran it, one day, from Leather Lane, Gray's Inn Lane, on to Lord's Cricket Ground, to Paddington, and home again by way of Islington—a journey of half a score miles through the streets of London. Trevithick was not on the engine on that occasion. Andrew Vivian was not an engineer, and would not have ventured on so long a run had there not been prior proof of what the engine could do; and the fact of an altered cylinder and boiler having been under construction in Cornwall after the locomotive had been sent to London, proves that several trials had been made, and changes found necessary.

"Captain Joseph Vivian recollects that about 1803, his father, then a captain of a vessel, on his return from London

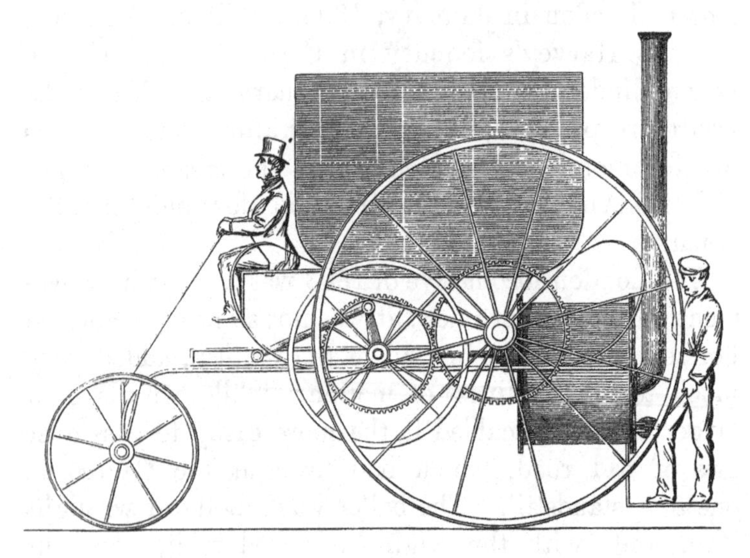

TREVITHICK'S COMMON ROAD PASSENGER LOCOMOTIVE, LONDON, 1803.

told them that he and his nephew, John Vivian, had been invited to take a bit of a drive with Captain Trevithick and Captain Andrew Vivian on their steam-carriage: they went along pretty well through a good many streets, and were invited again for the next day; but Captain Vivian thought he was more likely to suffer shipwreck on the steam-carriage than on board his vessel, and did not go a second time."

"Captain John Vivian, H.M.P.S., was, about the middle of 1803, on board his uncle's vessel in London, and often went to see the steam-carriage putting together at a coach-builder's shop in Leather Lane. Captain Trevithick and Captain Andrew Vivian were there, and Mr. William West was the principal man in putting the engine together. Mr. Arthur Woolf frequently came in, he being engaged close by as an engineer in Meux's brewery.

"Thinks the engine had one cylinder, and three wheels; the two driving wheels behind were about 8 feet in diameter. The boiler and engine were fixed just between those wheels.

The steering wheel was smaller, and placed in front. There were some gear-wheels to connect the engine with the driving wheels. The carriage for the passengers would hold eight or ten persons, and was placed between the wheels, over the engine, on springs. One or two trips were made in Tottenham Court Road, and in Euston Square. One day they started about four o'clock in the morning, and went along Tottenham Court Road, and the New Road, or City Road: there was a canal by the side of the road at one place, for he was thinking how deep it was if they should run into it. They kept going on for four or five miles, and sometimes at the rate of eight or nine miles an hour. I was steering, and Captain Trevithick and some one else were attending to the engine. Captain Dick came alongside of me and said, 'She is going all right.' 'Yes,' said I, 'I think we had better go on to Cornwall.' She was going along five or six miles an hour, and Captain Dick called out, 'Put the helm down, John!' and before I could tell what was up, Captain Dick's foot was upon the steering-wheel handle, and we were tearing down six or seven yards of railing from a garden wall. A person put his head from a window, and called out, 'What the devil are you doing there! What the devil is that thing!'

"They got her back to the coach factory. A great cause of difficulty was the fire-bars shaking loose, and letting the fire fall through into the ash-pan.

"The waste steam was turned into the chimney, and puffed but with the smoke at each stroke of the engine. When the steam was up, she went capitally well, but when the fire-bars dropped, and the fire got out of order, she did not go well.

"I heard afterwards that the framing of the engine got a twist, and she was used to drive a mill for rolling hoop-iron; and also that she ran on a tramroad laid down in Regent's Park."[1]

"In 1860, Mrs. Humblestone recollected Mr. Trevithick's steam-carriage go through Oxford Street; the shops were closed, and numbers of persons were waving handkerchiefs from the houses; no horses or carriages were allowed in the street

[1] Captain John Vivian died January, 1871, at Hayle, aged eighty-seven years.

during the trial. The carriage moved along very quickly, and there was great cheering. At that time she kept a shop next door to the Pantheon, and it, like the others, was closed. Her husband was employed with Mr. Trevithick at the Blackwall dredgers or the Tunnel."[1]

The reader must not imagine that these few records collected of events witnessed by but very few who still live, nor that the slight notice to be found of them in written history are indications of their want of importance, or of their having been mere passing experiments. They were the first and firm steps of the young locomotive, which Trevithick laboured for years to strengthen and make useful, and which would have given to our fathers some of the benefits we receive from steam locomotion had they had the good sense to comprehend Trevithick. Two years of locomotive experiments in Cornwall were followed by six or eight months of trials in the streets of London, commencing with January, 1803, when the locomotive was forwarded from Cornwall.

The day on which Captain John Vivian steered, they started at four o'clock in the morning, going through Tottenham Court Road and the City Road, sometimes at a speed of eight or nine miles an hour, and so successfully that—in joke—they talked of keeping on until they reached Cornwall, when bad steering, rather than engine defects, brought them to a standstill. It passed through Oxford Street at a good pace, amid cheers and waving of handkerchiefs. Undoubtedly other trips were run, but we have evidence sufficient to prove that the streets of London were safely passed through by the steam-horse at about the speed of the living horse.

---

[1] Statement by Mrs. Humblestone, 51, Earl Street, Islington, 1860.

Felton's carriage, to carry six or eight persons, placed over the engine and boiler between the large driving wheels, was supported on springs from the engine framing, which was of wrought iron. The boiler was of wrought iron, with internal return fire-tube. The fire-door and chimney were at one end. The cylinder was fixed in the boiler horizontally, and the crank was just under the carriage: coal and water were carried on the engine platform. The steam-blast in the chimney caused the small boiler to produce the necessary quantity of steam, and the increased size of the driving wheels to 8 or 10 feet in diameter gave a better grip, and enabled the engine and its load to pass more easily over road inequalities.

These continued trials and their consequent cost drained to the bottom the pockets of the inventors, and put an end for a time to locomotive experiments. The steam-carriage was sold for what it would bring, and the engine portion became a hoop-iron rolling-mill engine.

All the partners, except Trevithick, were disheartened at this result of steam travelling. He and Vivian at that time held each two-fifths of the patent, and West one-fifth; none of them received a penny, and they paid their losses as best they could.

Trevithick had learnt, however, from the shocks on the rough pavements, that a smooth road of iron was the thing to enable the steam-horse to run well, and from that time gave his thoughts to the railway form of locomotion.

It has been said that Trevithick was a pupil of Murdoch's, and learned from him the knowledge of the steam-carriage.[1] Murdoch's model was made in 1784,

[1] Smiles' 'Life of George Stephenson.'

when Trevithick was a boy of thirteen. Murdoch's experiments, which never bore fruit, were made in the dark and in secret. His model had a cylinder $\frac{3}{4}$ inch in diameter, with a 2-inch stroke; its boiler was a square box of copper, 4 inches on the side, heated by a lamp. Its smallness did not admit even of a liliputian engineman, and, having no guide, it could not run.

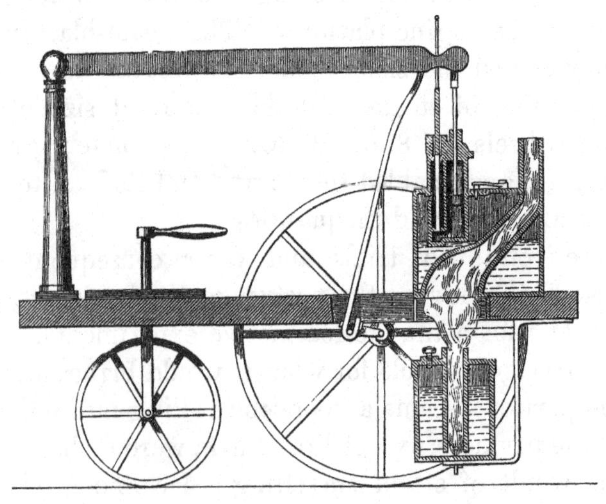

MURDOCH'S LOCOMOTIVE MODEL.

The following particulars were read at the Institute of Mechanical Engineers in 1850, and from them the writer first learnt of Murdoch's attempt, though he had lived many years among engineers within ten miles of Murdoch's residence at Redruth:

"Watt's friend and assistant, William Murdoch, took up the idea of a steam-carriage, broached by his patron, and constructed a non-condensing steam-locomotive, of liliputian dimensions, in the year 1784, the date of Watt's second patent. This locomotive, placed on three wheels, is shown in Fig. 1.

"The boiler is of copper; the flue passes obliquely through it, and is heated by a spirit-lamp. The cylinder is $\frac{3}{4}$ inch diameter, and has 2 inches stroke; it is fixed on the top of the boiler, and the piston-rod is connected to one end of a vibrating beam, to which also is attached the connecting rod for working the crank of the driving axle.

"The slide-valve is double, cylindrical, and worked directly by the beam, which strikes the shoulders of the valve-spindle; and the exhaust steam passes through the hollow of the spindle, going out near the top. One of the wheels only is fixed on the crank-axle, and a single wheel is placed in front, working in a swivel frame, to allow the carriage to run in a small circle. The driving wheels are 9½ inches diameter, and the leading wheel 4¾ inches. Notwithstanding these diminutive dimensions, this little gentleman managed to outrun the inventor on one occasion. 'One night, after returning from his duties at the mine,' in Redruth, Cornwall, where he resided for some time, in charge of the mining engines, 'he wished to put to the test the power of his engine; and, as railroads were then unknown, he had recourse to the walk leading to the church, situated about a mile from the town. This was very narrow, but kept rolled like a garden-walk, and bounded on each side by high hedges. The night was dark, and he alone sallied out with his engine, lighted the fire, a lamp under the boiler, and off started the locomotive, with the inventor in full chase after it. Shortly after he heard distant despair-like shouting; it was too dark to perceive objects, but he soon found that the cries for assistance proceeded from the worthy pastor, who, going into town on business, was met on this lonely road by the fiery monster, whom he subsequently declared he took for the Evil One *in propriâ personâ.*'"[1]

The foregoing led the writer to make inquiries among old men at Redruth who had been intimate with Murdoch. Nothing could be learned, except an indistinct recollection that they had heard that something had been tried. As a last resource, the daughter of the "worthy" but frightened pastor was sought for, and

[1] Biographical notice of William Murdoch, by Mr. Buckle. D. K. Clark's 'Railway Machinery,' published 1855.

though but a child when her parents told the story, gave the following with all the freshness of a modern event:—

"On a dark evening her parents, returning from Redruth to the vicarage, were somewhat startled by a fizzing sound, and

REDRUTH CHURCH. [W. J. Welch.]

saw a little thing on the road moving in a zigzag way. Murdoch was with it; her parents knew him well. They understood that Murdoch wished the experiment to be kept secret, and she does not recollect ever hearing of it afterwards, though she frequently saw Murdoch and heard of his engineering occupations. Mr. Wilson, the agent for Boulton and Watt, was a frequent visitor at her father's house."[1]

[1] Mrs. Rogers' recollections, Penzance, 1869.

Only one of the driving wheels was fixed on the driving axle, which fully accounts for its travelling in zigzags.

The following letter by Trevithick, in 1803, shows that he lost no time in applying his locomotive engine to tramways:—

"Boulton and Watt sent a letter to a gentleman of this place, who is about to erect some of these engines, saying that they knew the effect of strong steam long since, and should have erected them, but knew the risk was too great to be left to careless enginemen. That it was an invention of Mr. Watt, and the patent was not worth anything. This letter has much encouraged the gentlemen of this neighbourhood respecting its utility. As to the risk of bursting, they say it can be made quite secure.

"I believe that Messrs. Boulton and Watt are about to do me every injury in their power, for they have done their utmost to report the explosion, both in the newspapers and in private letters, very different to what it really was. They also state that driving a carriage was their invention; that their agent, Murdoch, had made one in Cornwall, and had shown it to Captain Andrew Vivian, from which I have been enabled to do what I have done. I would thank you for any information that you might have collected from Boulton and Watt, or from any of their agents, respecting their ever working with strong steam, and if Mr. Watt has ever stated in any of his publications the effects of it; because if he condemns it in any of his writings, it will clearly show that he did not know the use of it.

"There will be a railroad-engine at work here in a fortnight; it will go on rails not exceeding an elevation of one-fiftieth part of a perpendicular, and of considerable length. I have desired Captain A. Vivian to wait on you to give you every information respecting Murdoch's carriage—whether the large one at Mr. Budge's foundry was to be a condensing engine or not." [1]

---

[1] See Trevithick's letter, 1st October, 1803, chap. xx.

The simplicity of Trevithick's question brings out the truth: "If Watt condemns the use of high steam in any of his writings, it will clearly show that he did not know the use of it."

Dr. Robison,[1] the scientific and intimate friend of Watt, says: "It is well known that Mr. Watt retained up to the period of his death, the most rooted prejudices against the use of high steam." He also speaks of a design for a locomotive by Watt, in 1784, with a wooden boiler, and sun-and-planet wheels. Those schemes of Watt and of Murdoch may have some relation to the story that Captain Andrew Vivian was to speak of to Davies Gilbert, "whether the large one at Mr. Budge's foundry was to be a condensing engine or not."

Budge[2] was a clever Cornish mechanic, and had small workshops and a foundry at Tuckingmill; working frequently for Trevithick, sen., prior to Watt's appearance in Cornwall. If Watt never worked with high steam, his contemplated locomotive must have been a condensing engine.

This opens up a curious question in the history of the steam-engine. It is evident that Murdoch attempted a working model of a locomotive in 1784, but failed, and Watt blamed his agent for so throwing away his time. It is also evident that Watt had his plans for a o comotive as described by Dr. Robison, and it now seems more than probable that to put his ideas into form he employed Mr. Budge, whom we have met with as the mechanic employed by Trevithick, sen., in erecting his Carloose engine of 1775, and in which an increased pressure of steam was probably used, from the improved boiler.

---

[1] See Dr. Robison's statement, chap. vii.
[2] See Budge erecting Trevithick's engine in 1775, chap. ii.

We may never know what passed between Budge, Watt, and Murdoch. Murdoch's model, with one driving wheel, wobbled like a lame duck, and disappeared after one brief airing on a dark night; his boiler retained the weak flat sides of the low-pressure system, and Watt talked of a still weaker boiler to be made of wood. Fifty years ago the writer wondered what was carried on inside the dirty walls and windows of what had been Budge's foundry, but never dreamed that perhaps the construction had been attempted of a low-pressure steam locomotive vacuum engine, with wooden boiler and sun-and-planet wheels.

There was no similarity between those proposed locomotives and Trevithick's realities.

CHAPTER IX.

TRAM AND RAILWAY LOCOMOTIVES.

THE energy with which Trevithick followed up and made useful his inventions is, if possible, more wonderful than his never-failing faculty of bringing together old things under new and improved forms. It was a saying in Cornwall that Captain Dick would make a capital working steam-engine from the scrap heap or castaways of ordinary engineers.

While busy with the work mentioned in former chapters in Cornwall and London, he carried on other and new designs in South Wales and Shropshire; and in August, 1802, the Coalbrookdale Company were building for him a railway carriage or locomotive, though they feared to attempt its construction, or disbelieved Trevithick's promise that such an engine would work well, without the Watt air-pump and condenser.

He therefore designed an engine suitable for locomotion, and at the same time adapted it to force water a certain height, so that its power might be seen and accurately measured by the unbelievers. A 10-inch pump having a 4-feet stroke was attached to the engine, which forced water up a column of pipes 35 feet high.

This engine had all the leading features of the modern locomotive. The boiler was cylindrical, 4 feet in diameter, and gave steam of any pressure from 60 lbs. to 145 lbs. on the square inch.

A blast-pipe, with a regulating cock, caused the

boiler to give a full supply of steam, and also heated the feed-water.

The cylinder was 7 inches in diameter, with a stroke of 3 feet, and, with its gear-work, was attached to the boiler.

The term blast-pipe was then unknown to describe the thing just brought into use. But the expressions, "the steam continued to rise the whole of the time it worked," until it became unmanageable, and "I was obliged to stop, and put a cock in the mouth of the discharging pipe," and a small hole in the discharge-pipe to allow some of the blast-steam to escape into the feed-warmer, through which it passed, prove the invention of the blast-pipe, with its regulating cock and transmission of heat from surplus steam into the feed-water, to have been as fully understood and practically applied by Trevithick in 1802 as it is by us in the present age, when innumerable blast-pipes are in use.

"COALBROOKDALE, SHROPSHIRE,
"*Aug.* 22*nd*, 1802.

"DAVIES GIDDY, ESQ.,

"Sir,—I should have written to you some time since, but not having made sufficient trial of the engine, have deferred it until it is in my power to give you agreeable information of its progress.

"The boiler is 4 feet diameter; the cylinder 7 inches diameter, 3-feet stroke. The water-piston is 10 inches in diameter, drawing and forcing 35 feet perpendicular, equal beam. I first set it off with about 50 lbs. on the inch pressure against the steam-valve, without its load, before the pumps were ready, and have since worked it several times with the pumps, for the inspection of the engineers about this neighbourhood.

"The steam will get up to 80 lbs. or 90 lbs. to the inch in about one hour after the fire is lighted; the engine will set off when the steam is about 60 lbs. to the inch, about thirty strokes per minute, with its load. There being a great deal of friction on such small engines, the steam continued to rise the whole of the time it worked; it went from 50 lbs. to 145 lbs. to the inch in fair working, forty strokes per minute; it became so unmanageable as the steam increased, that I was obliged to stop and put a cock in the mouth of the discharging pipe, and leave only a small hole open of $\frac{1}{4}$ by $\frac{3}{8}$ of an inch for the steam to make its escape into the water. The engine will work forty strokes per minute with a pressure of 145 lbs. to the inch against the steam-valve, and keep it constantly swimming when burning 3 cwt. of coal every four hours. I have now a valve making to put on the top of the pumps, to load with a steelyard, to try how many pounds to the inch it will do of real duty when the steam on the valve is 145 lbs. to the inch.

"I cannot put many more pumps, as they are very lofty already. The packing stands the heat and pressure without the least injury whatever; enclosed you have some for inspection that has stood the whole of the working. As there is a cock in the discharging pipe which stops the steam after it has done its office on the piston, I judge that it is almost as fair a trial as if the pump-load was equal to the power of the engine. Had the steam been wire-drawn between the boiler and the cylinder it would not have been a fair trial, but being stopped after it has passed the engine, it tells much in its favour for bearing a greater load than it has now on.

"The boiler will hold its steam a considerable time after the fire is taken out. We worked the engine three-quarters of an hour after all the fire was taken out from under the boiler. It is also slow in getting up, for after the steam is atmosphere strong it will take half an hour to get it to 80 lbs. or 90 lbs. to the inch. It is very accommodating to the fireman, for, fire or not, it is not soon felt.

"The engineers at this place all said that it was impossible for so small a cylinder to lift water to the top of the pumps, and degraded the principle, though at the same time they spoke highly in favour of the simple and well-contrived engine. They say it is a supernatural engine, for it will work without either fire or water, and swore that all the engineers hitherto are the biggest fools in creation. They are constantly calling on me, for they all say they would never believe it unless they saw it, and no person here will take his neighbour's word even if he swears to it. They all say it is an impossibility, and they will never believe it unless they see it. After they had seen the water at the pump-head, they said that it was possible, but that the boiler would not maintain its steam at that pressure for five minutes; but after a short time they went off, with a solid countenance and a silent tongue.

"The boiler is $1\frac{1}{2}$ inch thick, and I think there will be no danger in putting it still higher. I shall not stop loading the engine until the packing burns or blows out under its pressure.

"I will write you again as soon as I have made further trial. If I had fifty engines I could sell them all here in a day, at any price I would ask for them. They are so highly pleased with it that no other engine will pass with them.

"The Dale Company have begun a carriage at their own cost for the railroads, and are forcing it with all expedition.

"I remain, Sir,

"Your humble servant,

"RICHARD TREVITHICK."

Wire-drawing steam was an early term for expansive working, and was familiarly spoken of by Trevithick in 1802, though the engineers who came to see the new principle at work, some of them perhaps from Soho, not many miles distant, where none but low-pressure steam vacuum engines were made, said that this new high-pressure steam-engine, without condenser, "degraded the principle," because it did not use condensing water, though at the same time they spoke highly in favour of its simplicity, and of the smallness of the fire-place and boiler resulting from the use of the blast-pipe, which produced a white heat and rapidly generated steam of great expansive force.

Within two years from the expiry of Watt's still-born patent, claiming the invention of an engine working by the pressure of steam, Trevithick had carried his vigorous high-pressure puffers from Cornwall to the outskirts of Soho.

Watt disbelieved, after years of deliberation, the propriety of constructing high-pressure steam-engines; and so did other engineers living near him, who thought an engine deriving its main power from vacuum better than one that was in no way dependent on it.

They all said it was an impossibility, and they would never believe it unless they saw it; but having seen the water at the pump-head, they said that the boiler would not maintain its steam at the pressure for five minutes; and so the sceptics passed from error to error, and were now in the mystery of the blast-pipe, a cause of confusion to many; and among them was Watt, who with a full knowledge of the working of Trevithick's high-pressure steam-puffer engines, and the all-powerful authority of the Soho Works, never, during the remaining seventeen years of his life, from the time of this public proof of the power and compactness of the non-condensing engine, constructed an engine with a blast-pipe.

The boiler bearing this high-pressure steam was a cast-iron cylinder, 1½ inch in thickness, having a wrought-iron internal return tube, similar to those he had during four or five years supplied, first to the Cornish mines as portable engines, and then in the Camborne common road locomotive.

It required an hour after lighting the fire to raise the steam to 80 or 90 lbs. to the inch; but when the engine was at work, and the blast came into play, the steam pressure rapidly increased to 145 lbs. This was a shock of annihilation to the Wattites, whose tongues were silent and countenances changed. Instead of the stock of steam in the small boiler dwindling to nothing in five minutes, it, strange to say, increased more and more in pressure, while the engine at the same time so wonderfully increased in speed and power, that even Trevithick had to stop the supply of steam, until some check could be placed on its startling power, by a vent-hole in the side of the blast-pipe.

The dimensions of this engine correspond very nearly with the locomotive that followed it in Wales, and the Dale Company were so satisfied with its performance, that they at once commenced "a carriage for the railroads."

Unfortunately, there is no further trace of this contemplated railroad of 1802 among his papers.

Trevithick was obliged to leave Coalbrookdale for a time to follow up the common road locomotive trials in Cornwall, and the construction of the locomotive for London, where he and his partners were during the early months of 1803; but this did not prevent his finding time for other applications of the new steam-power. In the beginning of April, 1803, he had put to work high-pressure steam-engines in London working with steam of 40 lbs. to the inch; which were approved of by the Admiralty, and burnt less coal than the Boulton and Watt low-pressure steam vacuum engines, that had before driven the same machinery.

In Derbyshire "they were going on with spirit with their engine." It is not clear if this latter was also a locomotive, for about that time railways or tramroads were used or proposed at both Coalbrookdale and in Derbyshire.

His neighbours in Cornwall, who scarcely noticed

him at home, gave him in London an order for an engine, and the autumn of 1802 promised golden fruit, for London, Derbyshire, Shropshire, Wales, Cornwall, and other places used the high-pressure steam-engines, whose applicability to steam locomotion had been proved, and some one really talked of giving 10,000*l.* for one-quarter of the patent.

"MR. GIDDY, "BRISTOL, *May 2nd*, 1803.

"Sir,—I set going an engine in London about four weeks since, for boring and turning brass cannon. The cylinder is 11 inches diameter and 3½-feet stroke; it does the work of ten horses; it consumed five chaldrons of coal in twenty-one days, and works exceedingly well. It works constantly from six in the morning until eight at night, and keeps in the fire over night. It turns and bores four brass cannon at a time, and turns a mill to grind clay at the same time. It requires the steam at a pressure of 40 to 45 lbs. to the inch to do its work well, working about twenty-six or twenty-seven strokes per minute. It is much admired by everyone that has seen it, and saves a considerable quantity of coal, when compared with a Boulton and Watt. We use but thirty gallons of water in the hour.

"I was sent for to explain the engine at the Admiralty Office. They sent to inspect it, and say they are about to erect several for their purposes, and that no other shall be used in the Government service. An engine worked in the place in which this now stands; it was taken down for this to be put up in its place. The proprietor of the gun factory has one other now at work, and has given an order for one more from me, and will take down the other. He was so satisfied with the engine that he paid for it the second day it was at work. There has been no further trial at the Dale. They are going on with spirit at Derbyshire about their engine. The coach-engine did not arrive in London until last Wednesday. The coach is ready to fix to the engine. I expect we shall be ready to start in about a fortnight. We worked the engine before we sent it from home; it was perfectly tight, and gave double as much steam as we wanted, without blowing, and the chimney was but 2 feet high. It worked fifty strokes per minute with 30 lbs. of steam to the inch. The cylinder was 5½ inches diameter, 2½-feet stroke; much more power than we shall want. I will write again soon. I have a prospect before me of doing exceedingly well. I tell you, as a friend, that I have sold to a gentleman of this place, one-quarter part of the patent for 10,000*l.*; but this must remain a secret. Mr. Williams, Mr. Robert Fox, Mr. Gould, and Captain William Davey were here, and much liked the engine; they gave me an order for one for Cornwall as a specimen. I was at Treadrea the day before I left home, but did not find you.

"I remain, Sir,

"Your humble servant,

"RICHARD TREVITHICK."

The engine of the steam-carriage of 1803 was made in Cornwall; and it used the steam-blast, for without it a chimney of only 2½ feet high would not have caused sufficient draught; and, moreover, the bellows tried in the Camborne locomotive was no longer used.

Having given a few months to the successful experiments with the London common road engine, and to the erection of high-pressure engines, in rivalry with Watt's low-pressure steam vacuum engines,[1] he was, in October, 1803, busily engaged in constructing at Penydarran, in South Wales, a tramway locomotive, to run on rails not exceeding an elevation of 1 in 50, and of considerable length.

"MR. GIDDY, "PENYDARRAN, 15*th February*, 1804.

"Sir,—Last Saturday we lighted the fire in the tram-waggon, and worked it without the wheels to try the engine. On Monday we put it on the tramroad. It worked very well, and ran up hill and down with great ease, and was very manage-

---

[1] Letter, October 1st, 1803, chap. xx.

able. We had plenty of steam and power. I expect to work it again to-morrow. Mr. Homfray and the gentleman I mentioned in my last, will be home to-morrow. The bet will not be determined until the middle of next week, at which time I shall be very happy to see you.

"I am, Sir,

"Your humble servant,

"RICHARD TREVITHICK."

The first tramroad locomotive in Wales worked in the month of February, 1804, running with facility up and down inclines of 1 in 50, and having a full supply of steam and power.

"MR. GIDDY, "PENYDARRAN, *February 20th*, 1804.

"Sir,—The tram-waggon has been at work several times. It works exceedingly well, and is much more manageable than horses. We have not tried to draw more than 10 tons at a time, but I doubt not we could draw 40 tons at a time very well; 10 tons stand no chance at all with it. We have been but two miles on the road and back again, and shall not go farther until Mr. Homfray comes home. He is to dine at home to-day, and the engine goes down to meet him. The engineer from the Government is with him.

"The engine, with water included, is about 5 tons. It runs up the tramroad of 2 inches in a yard forty strokes per minute with the empty waggons. The engine moves forward 9 feet at every stroke. The public are much taken up with it. The bet of 500 guineas will be decided about the end of this week. Your presence would give me more satisfaction than you can conceive, and I doubt not you will be fully repaid for the toil of the journey by a sight of the engine.

"The steam that is discharged from the engine is turned up the chimney about 3 feet above the fire, and when the engine works forty strokes per minute, 4½-feet stroke, 8¼ inches diameter of cylinder, not the smallest particle of steam appears out of the top of the chimney, though it is but 8 feet above where the steam is delivered into it, neither at a distance from it is steam or water found. I think it is made a fixed air by the heat of the chimney. The fire burns much better when the steam goes up the chimney than when the engine is idle. I intend to make a smaller engine for the road, as this has much more power than is wanted here. This engine is to work a hammer.

"The engineer from London will try a great many experiments with these engines, as that is his sole business here, and that is my reason for so much wishing you here. He intends to try the strength of the boiler by a force-pump, and has sent down orders to get long steam-gauges and force-pumps ready for that purpose.

"We shall continue our journey on the road to-day with the engine, until we meet Mr. Homfray and the London engineer, and intend to take the horses out of the coach, fasten it to the engine, and draw them home. The other end of the road is 9¾ miles from here. The coach-axles are the same length as the engine-axles, so the coach will run very easily on the tram-road.

"There have been several experiments made by Mr. Homfray and this engineer in London, lately, on these engines. I am very much obliged to you for your offer to assist in making out a publication of the duty and advantages of those engines. As soon as I can get proper specimens at work, and you as an eye-witness of their performance, I shall value your kind offer and assistance far beyond any other to be got, as you have been consulted, and have assisted me from the beginning.

"I am, Sir,

"Your very humble servant,

"RICHARD TREVITHICK."

"MR. GIDDY, "PENYDARRAN, *February* 22*nd*, 1804.

"Sir,—Yesterday we proceeded on our journey with the engine; we carried 10 tons of iron, five waggons, and seventy men riding on them the whole of the journey. It is above nine miles, which we performed in four hours and five minutes. We had to cut down some trees and remove some large rocks out of the road. The engine, while working, went nearly five miles per hour; no water was put into the boiler from the time we started until we arrived at our journey's end. The coal consumed was 2 cwt. On our return home, about four miles from the shipping-place of the iron, one of the small bolts that fastened the axle to the boiler broke, and all the water ran out of the boiler, which prevented the return of the engine until this evening. The gentleman that bet 500 guineas against it rode the whole of the journey with us, and is satisfied that he has lost the bet. We shall continue to work on the road, and shall take 40 tons the next journey.

"The public until now called me a scheming fellow, but now their tone is much altered. An engine is ordered for the West India Docks, to travel itself from ship to ship, to unload and to take up the goods to the upper floors of the storehouses by the crane, and in case of fire to force water on the storehouses. The fire is to be kept constantly burning in the engine, so as to be ready at all times.

"Boulton and Watt have strained every nerve to get a Bill in the House to stop these engines, saying the lives of the public are endangered by them, and I have no doubt they would have carried their point if Mr. Homfray had not gone to London to prevent it; in consequence of which an engineer from Woolwich was ordered down, and one from the Admiralty Office, to inspect and make trial of the strength of the materials, and to prove that the steam-gauges will admit steam through them in case the steam-valve should be fastened down. They are not to come until everything is complete for those experiments. You shall know of our future experiments as fast as we get on with them.

"Your humble servant,

"Richard Trevithick."

Before a week had passed, from the first getting-up of steam, this pioneer of railway-engines had run several times, drawing a load of 10 tons, and was more controllable than horses. Only two miles of road were to be run over during the first trials, but within the week the engine ran a distance of $9\frac{3}{4}$ miles. The horses were removed from Mr. Homfray's common road coach, the wheels of which were of the same gauge as the tramway, and it was drawn by the engine, together with Mr. Homfray and his companion, a Government engineer, brought down for the purpose of examining and testing Trevithick's engines. This was a practical proof of how street tramway engines could be made to draw by steam ordinary road carriages.

The engine in working order weighed about 5 tons; its cylinder was $8\frac{1}{4}$ inches in diameter, with a stroke of $4\frac{1}{2}$ feet. It took empty waggons up an incline of 2 inches in a yard, at forty strokes a minute, progressing 9 feet at each stroke; in other words, it took its load up an incline of 1 in 18 at the rate of four miles an hour.

Trevithick's own words solve the blast-pipe riddle:—"The steam that is discharged from the engine is turned up the chimney, about three feet above the fire. The fire burns much better when the steam goes up the chimney than when the engine is idle."

It was something like engineering, that within a week of first putting fire in a newly-designed engine for a novel purpose, it took in the morning its luggage-train work up an incline believed by many to be dangerously unmanageable even now, after seventy years of experience, and was prepared for its evening passenger-train work, to bring home the ironmaster and his friend, snugly riding in an ordinary horse-carriage.

Trevithick was anxious that Mr. Davies Gilbert should be near him to observe the drift of events, for he knew the London engineer had been experimenting with his high-pressure engines then at work in London, in competition with Watt; and he also hoped that Davies

Gilbert would make public a report on the advantages of the new principle of working without a vacuum, and state what he had seen at Penydarran.

Immediately after writing a long letter, giving clear and important facts, during the din and hurry of preparing the tramroad-engine for her first long run, a start was made from Penydarran for the Basin, 9¾ miles distant. Five waggons[1] were attached to the engine, loaded with 10 tons of iron and seventy men. Four hours and five minutes were required for the journey, though the engine went at the rate of five miles an hour, but was obliged to stop frequently that trees might be cut down and rocks removed that blocked the roadway. How different to the well-prepared level railway on the Liverpool and Manchester line, on which the much-talked-of locomotives ran a quarter of a century later!

The gentleman who bet 500 guineas that the engine would not draw 10 tons of iron over the distance, rode with Trevithick, and admitted the fact. The seventy men and tools taken at 7 tons, made a net load of 17 tons; and, including the five waggons, a gross load of 25 tons. During the journey, 2 cwt. of coal were consumed, and Trevithick believed, from that day's experience, that he could take a load of 40 tons.

On the return journey, when four miles had been passed, one of the bolts fastening the axle-boxes to the boiler broke, the water and steam escaped through the hole, and caused the engine to stop for repairs.

This roughly-constructed locomotive performed a comparatively long journey with a heavy load, over the worst possible road, with sharp curves and inclines, and

[1] See Trevithick's letter, 26th April, 1812, chap. xviii.

frequently breaking tram-plates; yet such, to other men, insurmountable difficulties are not so much as mentioned in Trevithick's letters.

The writer worked locomotives on those same tramways thirty-three years afterwards, in 1837, and it was then necessary to send platelayers to jump from the waggon-trains to replace broken plates.

He at the same time renewed a long-idle early locomotive on Trevithick's plan, that had been built at Neath Abbey. The double or breeches fire-tube in the boiler was removed to make room for thirty small tubes, such as were then coming into use for locomotive boilers. That all-important part of a locomotive, the blast-pipe, brightening the fire, and increasing the supply of steam with greater or less intensity in proportion to the greater or less amount of work performed, and steam puffs accelerating the draught, was retained in its original form and position as used by Trevithick in 1804.

Boulton and Watt still opposed Trevithick's high-pressure steam plans on the plea that "the lives of the public were endangered," and tried to get an Act of Parliament to prevent the construction of any more engines on the high-pressure principle. The Government sent engineers from Woolwich and the Admiralty to test those engines, both as to the fact of being able to work a steam-engine without using the vacuum from injection, and also as to the pressure that the boilers would safely bear, and the kind of safety-valves or safeguards against explosion.

At that time he was constructing a travelling steam-crane and fire-engine for the West India Docks in London, to unload ships, convey the merchandise to the storehouses, and then lift it to the required floors;

the engine was arranged with pump and hose, and to be in steam night and day.

It seems like a dream that such things were done nearly three-quarters of a century ago; for though we now have steam-cranes, and steam fire-engines, and railways into docks, we have not yet so mastered the detail as to combine the three operations in one engine.

"MR. GIDDY, "PENYDARRAN, *March 4th*, 1804.

"Sir,—We have tried the carriage with 25 tons of iron, and found that we were more than a match for that weight. We are now preparing to get the materials ready for the experiments by the London engineers, who are to be here on Sunday next. We have fixed up 28 feet of 18-inch pumps for the engine to lift water; these engineers particularly requested that they might have a given weight lifted, so as to be able to calculate the real duty done by a bushel of coal.

"The waggon-engine is to lift this water, then go by itself from the pump and work a hammer, then to wind coal, and lastly to go the journey on the road with iron. We shall have all the work ready for them by the end of the week. They intend to stay here about seven or eight days, and as the report that they will make on their return will be the standing or the condemning those engines, it is my reason for so anxiously requesting your presence. As they intend to make trial of the duty performed by the coal consumed, they will state it as against the duty performed by Boulton's great engines, which did upwards of twenty-five millions, when their 20-inch cylinders, after being put in the best order possible, did not exceed ten millions.

"As you were consulted on all those trials of Boulton's engines your presence would have great weight with those gents, otherwise I shall not have fair play. Let me meet them on fair grounds, and I will soon convince them of the superiority of the '*pressure-of-steam engines.*'

"The steam is delivered into the chimney above the damper; when the damper is shut the steam makes its appearance at the top of the chimney; but when open none can be seen. It makes the draught much stronger by going up the chimney; no flame appears. The coal here has but very little bitumen in it, therefore but very little smoke comes from it. We never tried a torch at the top of the chimney.

"Perhaps there may never be such an opportunity when your assistance in those experiments will be of so great a benefit to me as at this time, therefore I hope you will forgive me for again requesting your attendance on a business that may be of such consequence to me.

"I remain, Sir,

"Your very humble servant,

"RICHARD TREVITHICK."

The chimney-damper formed a fixed part of this locomotive. There was also a cock for regulating the blast and for heating the feed-water by the waste steam. These three things have exercised the genius of engineers to the present day, and they have not improved on Trevithick's ignored plans of 1804. The engine-driver of to-day removes a temporary damper-plate from the top of the chimney, while Trevithick had it under mechanical control; the same may be said of regulating blast and feed warming.

Within three weeks from lighting the first fire in the boiler, the Welsh tram-engine had drawn 25 tons of iron, net load. An upright column, 28 feet high, of 18-inch pumps, had been erected to test the power exerted by the engine and the coal consumed in raising a known quantity of water, or rather to discover how much water could be lifted a fixed height by the consumption of a bushel of coal. The reader must bear in mind that a bushel of coal was a misunderstood term: Trevithick's bushel was 84 lbs., Watt's bushel seems to have been 112 lbs.

Trevithick, who was not apt to fear, had an instinctive dread of those London gentlemen. His was a small engine, and small engines never did such good duty as larger engines. "Let me meet them on fair grounds, and I will soon convince them of the superiority of the *pressure-of-steam engines.*"

The unreasonableness of the requirements of the Government engineers proves their ignorance of practical engineering, and almost shows an influence from Soho, judging from the following report by Trevithick:—"The waggon-engine is to lift this water in the pipes, then go by itself from the pump and work a hammer, then to wind coal, and lastly to go the journey on the road with a load of iron." By the end of the week everything would be in readiness, and they thought of keeping the engine under those various tests for seven or eight days.

The blast-pipe continued to act well. "The steam is delivered into the chimney above the damper; it makes the draught much stronger by going up the chimney."

Mr. Davies Gilbert, who was prepared to stand by the man of genius in this contest with men in authority, left Oxford for Penydarran to be present at the experiments, but meanwhile his friend had written that an accident had made it impossible for Mr. Homfray to receive his friends from London, and the official tests were therefore deferred.

"Mr. Giddy, "Penydarran, *March 9th,* 1804.

"Sir,—I am sorry to inform you that the experiments that were to be exhibited before the London gents are put off on account of an accident which happened to Mr. Homfray on Tuesday last. His horse ran away with the gig, and threw him out, hurt his face, sprained his ankle, and dislocated his arm at the elbow. The experiments will go on as usual; everything is nearly ready. If you come now you will not have Mr. Homfray's company or that of the London gentlemen. Perhaps it will be a month before they come down, and they are the persons I wish you to see.

"I hope you will come this way before you go to Cornwall, as we can go through the experiments at any time. I find myself much disappointed on account of the accident, for I was very desirous to make the engine go through its different work, that its effect might be published as early as possible. I should be very happy to see any gentleman you may recommend this way for information, as the more public it is made the sooner the engines will circulate.

"We have not made any experiments since I last wrote. I received a letter from home this morning saying they had seen the steam-carriage in the newspapers, but did not believe it to be truth.

"Wheal Prosper is condemned; Binner Downs is under water, but was never so good as when it stopped. We are removing Wheal Treasury great engine there.

"I cannot see any release for me from this place soon; and intend to go down almost immediately to Cornwall, and bring up my family to spend the next summer here.

"I am, Sir,

"Your very humble servant,

"RICHARD TREVITHICK."

"This letter did not reach Oxford till after I had proceeded to Penydarran. I believe it was never opened till January, 1824.—D. G."

Had Trevithick carried out his wish to remove his family from Cornwall to Wales, that he might give more attention to the locomotive, this country might have had the benefit of railways and locomotives twenty years earlier.

Mr. Homfray thought of becoming a partner in Trevithick's patent, and both of them were at that time full of the wish to extend the general application of the high-pressure steam-engine, causing the partial neglect of the locomotive.

"DEAR SIR, "PENYDARRAN PLACE, 10*th July*, 1804.

"I dare say you will think me remiss in not answering your first letter, enclosing one for Mr. Trevithick. In a day or two after I received it I went up to Staffordshire (and likewise Trevithick), and have only returned yesterday, and find your favour of 3rd; both letters I thank you for, and shall at all times be happy to show you or your friends civility at this place.

"Our engine is at work at the rolls, and goes on very well. It has rolled this last fortnight upwards of 200 tons of iron, from the balls to the rough bars, which is as much as it was first expected, and I hope it will continue improving. It does much more than our old engine of 32-inch diameter cylinder and this is only 28-inch, same stroke. Trevithick went down the tramroad twice since you left us, with 10 tons each time, and though he took his load down, Mr. Hill does not yet allow the 500 guineas, because he did not return again with the empty trams in the same time the horses usually do, and this was owing to the little forcing pump not being quite right to feed the boiler, and he was obliged to wait and fill with cold water; but this little defect is easily cured, and no doubt but Mr. Hill will be satisfied; but Trevithick would not stay here for the present to make another journey, so that it stands over till his return. He is now, I believe, at the Dale, and will not expect to be here this month again.

"Lord Dudley's engine for winding coal is got to work, as well as one at Worcester, for a glover there, which will be applied to various purposes, and show what the engine can do. I beg leave to congratulate you on the honour of a seat in Parliament. Mrs. Homfray and my niece Eliza join me in respectful compliments.

"I am, dear Sir,

"Your obedient servant,

"SAML. HOMFRAY.

"DAVIES GIDDY, Esq."

This letter of the 10th July draws attention to one written by Davies Gilbert on the 3rd, thanking Mr. Homfray for his hospitality; and during the interval the locomotive had gone two journeys over the tramway, with its net load of 10 tons of iron. It commenced work in the early part of February, and continued at work therefore at least five months, and was then in good working order, with the exception of a small bolt to be repaired.

After putting to work a high-pressure whim-engine for Lord Dudley, in Staffordshire, a small engine in Worcester, and a winding engine at Stourbridge, he paid a flying visit to Coalbrookdale, and then to Newcastle-upon-Tyne, to work his locomotive, and while on the journey wrote to Davies Gilbert.

"The tram-engine has carried two loads of 10 tons of iron to the shipping place since you left this. Mr. Hill says he will not pay the bet, because there were some of the tram-plates in the tunnel removed, so as to get the road into the middle of the arch.

"The first objection he started was that one man should go with the engine without any assistance, which I performed myself without help; and now his objection is that the road is not in the same place as when the bet was made. I expect Mr. Homfray will be forced to take steps that will oblige him to pay.

"As soon as I return from here there will be another trial, and some person will be called to testify its effects, and then I expect there will be a lawsuit immediately. The travelling engine is now working a hammer."[1]

Trevithick was not like other men. He was alone at various places erecting and working differently-designed engines suitable to the numerous requirements

---

[1] See letter, 5th July, 1804, chap. xx.

he readily engaged to fulfil. Who constructed the engines? who made the necessary drawings? who kept newly-designed machines in daily work, with steam of 100 lbs. to the inch, more or less? are questions the answer to which is lost in the lapse of a comparatively few years. It is evident that numerous manufacturers and engineers, from Watt downwards, were fully acquainted with those events, and that the latter not only disapproved of the new engine, but actively opposed its use. It may be almost said that every engineer of the time knew of them; yet no one seems to have recorded them, except Trevithick, who never dreamt of being an historian.

While at Coalbrookdale, in 1804, he wrote to his friend:[1]—

"At Newcastle I found four engines at work, and four more nearly ready. Six of these were for winding coal, one for lifting water, and one grinding corn; the latter, an 11-inch cylinder, driving two pair of 5-feet stones. Below I send you a copy of Mr. Homfray's and Mr. Wood's letters."

Some of these engines were probably for the Gateshead Iron Works of Messrs. Hawks and Crawshay. The latter, I believe, was a brother of the Mr. Crawshay in the Welsh Iron Works. Every move of Trevithick and his engines in Wales was therefore known to the coal-works proprietors and engineers of Newcastle.

Mr. Wood says:—"An engine of this kind was sent to the North, for Mr. Blackett, of Wylam, but was, for some cause or other, never used upon his railroad, but applied to blow a cupola, at an iron foundry in Newcastle."[2]

---

[1] See letter, 23rd September, 1804, chap. xvii., p. 385.

[2] Wood 'On Railroads,' p. 130; published 1825.

On the 10th January, 1805,[1] having left Newcastle, he wrote from Soho Foundry, Manchester:—

"I shall go to Newcastle-on-Tyne in about four weeks. By that time the little engine will be sent to London for the coal-ships. A great number of my engines are now making in different parts of the kingdom. There are three foundries here making them. I expect there are some of the travelling engines at work at Newcastle. As soon as I get there I will write to you. This day I received a letter from Jane—sad lamentation on account of my absence. I am obliged to promise to return immediately, but shall not be able to fulfil it at this moment. I should be wrong to quit this business, as there are now seventeen or eighteen foundries going on with those engines, and unless I am among them the business will fall to the ground, and after such pains as I have taken I am very sorry to quit it until I get it established."

Having been at Newcastle in September, 1804, to arrange, amongst many other things, for supplying the Wylam Railway with a locomotive, he counted on being there again in February or March, 1805, to see "*some* of the travelling engines at work."

Numerous small high-pressure engines were being constructed in seventeen or eighteen different manufactories; and in May, 1805, one of the Newcastle travelling engines was ready for its work on the railway.

Mr. Wilson, well known at Newcastle-on-Tyne, has well described it in the following Memorandum, to be seen at the South Kensington Patent Museum:—

"*Memorandum, May 1st,* 1805. (*Copy from R. W.'s Memorandums on Steam-Engines.*)—I saw an engine this day upon a new plan: it is to draw three waggons of coal upon the Wylam waggon-way; the road is nearly level. The engine is to travel

[1] Letter of Trevithick, 10th January, 1805, chap. xv., p. 325.

with the waggons. Each waggon with the coal weighs about 3½ tons, and the engine weighs 4½ tons. The engine is to work without a vacuum. The cylinder is 7 inches in diameter, 3-feet stroke, and is placed inside the boiler, and the fire is inside also. The speed they expect to travel at is four miles per hour.—ROBERT WILSON."

Mr. Wilson, an experienced mechanic, called this high-pressure engine in 1805 an engine on a new plan, working without a vacuum; not knowing that similar engines had been at work for six or seven years.

Having traced the construction of common road, tram, and railway engines up to the year 1805, mainly from Trevithick's letters, it may be necessary to substantiate the facts by disinterested evidence.

"F. TREVITHICK, ESQ., "TREDEGAR WORKS, *May 15th*, 1854.

"Dear Sir,—I received your letter of last month, and should have answered it sooner, but as yet have not been able to go over to Merthyr to see the men that worked for your late father. I had found out the engine-driver; but my son has found out one of the men who put the locomotive engine together, and he believes he shall be able to make a sketch from their description. It was my father that was at Penydarran when the engine was made and tried. It made three journeys to the Basin, nine miles below Merthyr, and in coming up the third journey it broke both axles; the road being only common tramroad-plates 3 feet long. The men say it would have done its work well on a good road.

"The engine had one cylinder 8 inches in diameter and 2-feet stroke; a wrought-iron boiler, with a tube thus, U; the side rods connected to the wheels; and as side coupling rods were not thought of then, it had cog-wheels connecting front and hind axles. To pass the centres it had a fly-wheel. Weight of engine about 7 tons, on four wheels. You shall hear from me as soon as I can obtain all the information I require.

"The engines you saw at Tredegar were made by me. We had the original from Stephenson, Newcastle, about 1829; but it had the old tube, the same as marked on the other side. I believe I am the first that turned the steam into the chimney, never thinking it would give me any advantage in getting up steam. It was done to prevent frightening the horses, as we had a turnpike-road fourteen miles by the side of our railway. You may rely on my getting all the information I possibly can, as my informants are, one seventy and the other seventy-six years old. No time should be lost.

"Dear Sir, yours truly,

"THOS. ELLIS, Sen."

"TREDEGAR WORKS, 22*nd June*, 1854.

"DEAR SIR,

"We have not succeeded well yet with the locomotive. The poor old fellow is very feeble; but we have a part of that done also. From what I can understand the engine was made in Cornwall, and put together in Merthyr; and if we can understand the old man, your father adopted the pipe inside the other to warm the water, and also took the steam up the chimney. If so, that was done before I saw the light. Trusting I shall send you the required information,

"Dear Sir, yours truly,

"THOS. ELLIS.

"F. TREVITHICK, Esq."

Mr. Ellis then published a drawing of the Welsh locomotive, with the following particulars:—

"Trevithick's high-pressure tram-engine, so designated in the original plan, dated 1803, was constructed partly in Cornwall and partly at Penydarran Works, by Richard Trevithick, Esq., engineer, for Samuel Homfray, Esq., proprietor of the Penydarran Iron Works, Merthyr Tydfil, who, while discussing the principles and feasibility of locomotive steam-engine power with Richard Crawshay, Esq., of the Cyfartha Iron Works, made a bet of 1000 guineas that he would convey by steam-power a load of iron from his works to the Navigation House (nine miles distant) along the Basin tramroad, which he effected by means of this engine, and won his wager, although the heavy gradients, sharp curves, and frangible nature of the cast-iron trackway operated against the return of this ingenious though rudely-constructed machine with the empty trains—hence its discontinuance. As may be perceived, the exhausted steam discharged into the stack and the wheels combined; thus to *Trevithick* is the credit due for the application of those two principles to locomotive engines.

"Rees Jones, who aided in the fitting, and William Richards, its driver, are still alive; the former, when shown the plan, instantly identified it; and the latter, now in his eighty-fifth year, has worked no other than *Trevithick's high-pressure engine.* To this day portions of the old engine exist in the one he now works at Penydarran, and during a period extending far beyond half a century, never having had an accident with his boiler.

"THOS. ELLIS, Engineer,
"*Tymaur Ponty Pridd, Glamorganshire.*"

Mr. Menelaus also supplied a drawing of the Welsh locomotive, made by Mr. Llewellyn.

"DEAR SIR, "GLANWERN, PONTYPOOL, *24th November*, 1855.

"I regret that I should have forgotten to send you the drawing of the old locomotive engine which you asked me for some time ago. I have had a pretty long search for it, but succeeded in finding it at last. I observe that nothing is mentioned upon it of its being the engine constructed by Trevithick; but inasmuch as I am aware that my uncle was engaged with him at Merthyr in the trial made there on the old navigation tramway, and have an impression of hearing my uncle say that it was a rough draft of the original engine, I think there is very little doubt upon the subject. It was drawn by my uncle (very roughly you will see) in the year 1803, which was about the time Trevithick was at Merthyr. You will perhaps be kind enough to get the parties you send the sketch to, to return it when done with. I should tell you that I shall send the drawing in another enclosure.

"Dear Sir, yours very truly,

"WILLIAM LLEWELLYN.

"W. MENELAUS, Esq.

"N.B.—The cylinder was 4¾ inches diameter, and 3-feet stroke."

'Engineering' for the 27th March, 1868, has the following:—

"Trevithick was the real inventor of the locomotive. He was the first to prove the sufficiency of the adhesion of the wheels to the rails for all purposes of traction on lines of ordinary gradient, the first to make the return flue boiler, the first to use the steam-jet in the chimney, and the first to couple all the wheels of the engine.

"One of his pupils, John Roe, was not long ago living, and he supplied a correspondent to our pages with many of the particulars of the old engine, as did also Rees Jones, who, at the age of twenty-one, worked on it. The engine made one or more trips from Penydarran to Navigation, but broke a large number of the cast-iron tramway-plates with which the line was laid. After this it worked for two or three years between the blast-furnaces and forges at Homfray's works at Penydarran."

The 'Mining Journal,' Oct. 2nd, 1858:—

"Rees Jones, engine-fitter, Penydarran, says:—Dowlais, Sept. 9, 1858.—I am now eighty-two years old. I came to Penydarran on April 1st, 1794. I was then eighteen years of age. I have been in the employ of the Penydarran Iron Company ever since. I am still able to do a little. I am in the works every day. About the year 1800 Mr. Trevithick came to Penydarran to erect a forge-engine for the company. I was at this time overlooking the engines at Penydarran. I assisted Mr. Trevithick in the erection of the forge-engine. When this engine was finished Mr. Trevithick commenced the construction of a locomotive. Most, if not all, the work of this locomotive was made at Penydarran. Richard Brown made the boiler and the smith-work. I did the most of the fitting, and put the engine together. When the engine was finished she was used for bringing down metal from the furnaces to the old forge. She worked very well; but frequently from her weight broke the tram-plates, and also the hooks between the trams. After working for some time in this way, she took a journey of iron from Penydarran down the Basin Road, upon which road she was intended to work. On the journey she broke a great many of the tram-plates; and before reaching the Basin she ran off the road, and was brought back to Penydarran by horses. The engine was never used as a locomotive after this; she was used as a stationary engine, and worked in this way several years.

"I understood the reason for discontinuing using her as a locomotive was the weakness of the road. The boiler was made of wrought iron, having a breeches tube also of wrought iron, in which was the fire. The pressure of steam used was about 40 lbs. to the inch. The cylinder was horizontal; it was fixed in the end of the boiler. The diameter of the cylinder was about $4\frac{3}{4}$ inches. The three-way cock was used as a valve. The engine had four wheels. These wheels were smooth; they were coupled by cog-wheels. There was no rack-work on the road; the engine progressed simply by the adhesion of the wheels. The steam from the cylinder was discharged into the stack. This statement was made in the presence of the undersigned.

"W. MENELAUS.
"G. MARTIN.
"W. JENKINS."

Mr. Menelaus says:—

"I would call particular attention to the fact that Trevithick in 1803 had satisfied himself that smooth wheels would have sufficient adhesion to propel a load; that he had hit upon the plan of coupling the wheels; and that he discharged the waste steam into the stack. Boilers of the same type as that used by Trevithick in this engine were used successfully for locomotives twenty years after his invention."

A boiler-plate in the Kensington Patent Museum has the following written on it:—

"Piece of S. Homfray's engine that took the load of iron to Navigation House for the 1000*l.* bet—Trevithick, builder, and with two cylinders. The history of our getting it:—After taking the iron down it was made into a small planishing-hammer engine at Penydarran; brought into the Forest by Mr. Protheroe thirty years ago to sink pits at Castle-rag Colliery; from these to Protheroe's lower pits, and sank them; then to Link's Delight for our old company, and we pulled it down and cut a piece out for you. This is as I copied it from old Broad, our manager.

"HENRY CRAWSHAY, *March* 14*th*, 1850."

Rees' 'Cyclopædia' of 1819 says:—

"The application of steam-engines to driving of carriages.—These are now called locomotive engines, and we may date their introduction with the patent of Messrs. Trevithick and Vivian in 1802.

"Mr. Trevithick made a locomotive engine in South Wales in 1804, which was tried upon the railroads at Merthyr Tydfil. The engine was the same as that of which we have given an account of its work in speaking of the high-pressure engine, having an 8-inch cylinder and a 4 feet 6 inch stroke. It drew after it upon the railroad as many carriages as carried 10 tons of bar-iron, for a distance of nine miles; and it performed all that distance without any further supply of water than that contained in the boiler at setting out, travelling at the rate of five miles per hour.

"The boiler of cast iron, of a cylindrical form, 6 feet long, and 4 feet 3 inches in diameter, the fire-place being withinside. The cylindrical boiler was mounted horizontally upon four wheels, and the cylinder of the engine was placed vertically in the end of the boiler, having two connecting rods descending from the cross-bar of its piston-rods to two cranks upon an axis extending beneath the boiler and cylinder, and communicating its motion, by means of wheel-work, to the two fore wheels upon which the engine runs, and by this means the alternate ascending and descending motions of the piston-rods act to turn round the crank and wheels, and draw the carriages forward. In this way no fly-wheel was necessary, because the momentum of the carriage to advance itself forward on the road continued the motion of the wheels and cranks sufficiently to make the cranks pass the lines of the centre."

Mr. Crawshay, who had good means of learning the facts, says the boiler was of wrought iron, and there were two cylinders.

Rees' 'Cyclopædia' makes the boiler of cast iron and the cylinder vertical. The old Welshmen speak of a wrought-iron boiler, though their drawing rather indicates a cast-iron outer casing, with wrought-iron tube, and the steam-cylinder to have been horizontal.

They all agree on the blast-pipe, which the writer has not shown in the drawing, neither the feed-pump, or water-heating apparatus, because they are not given in the original Welsh drawing. The position of the feed-pump, feed-heating pipe, and blast-pipe, may be readily discovered by a reference to the patent drawing of 1802.

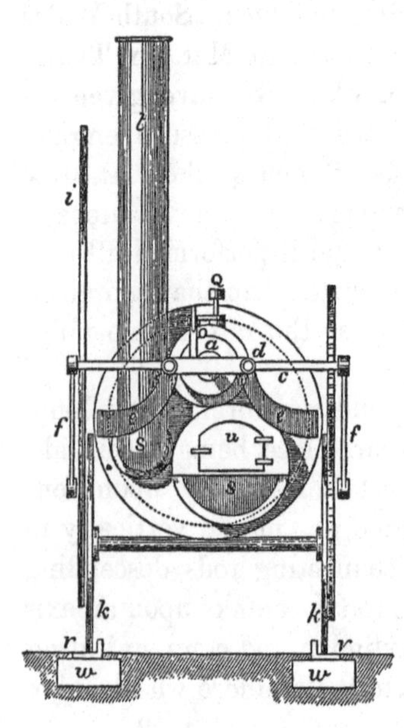

The following particulars are taken from Trevithick's letters where other evidences are conflicting. It is probable that more than one tramroad-engine was constructed in Wales at that time.

Eight years after these

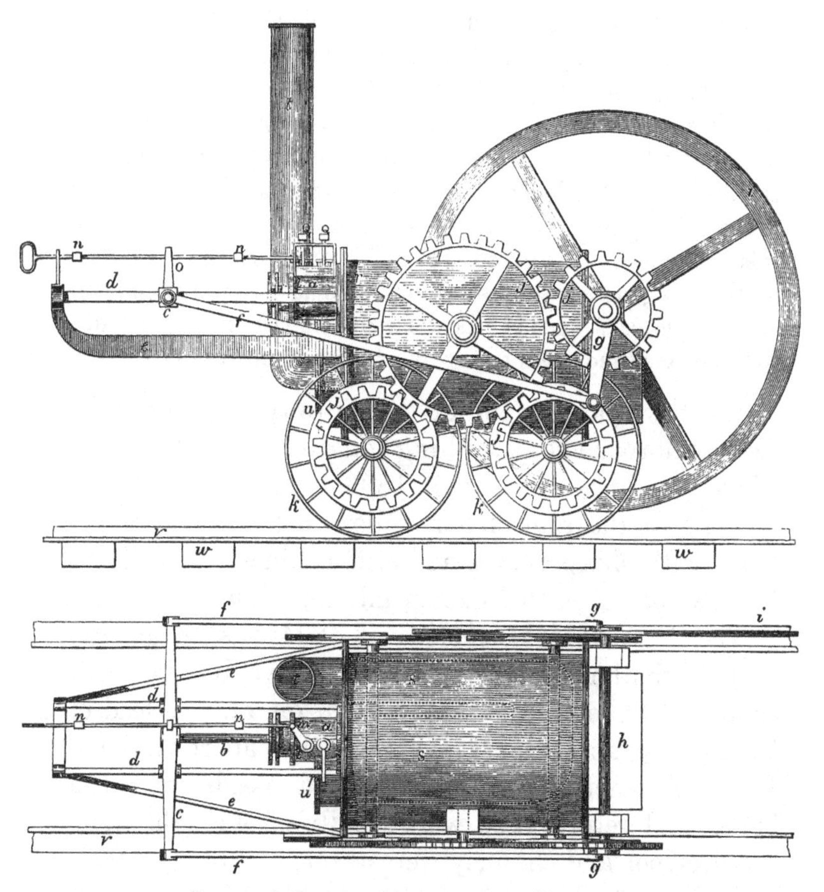

TREVITHICK'S TRAMROAD LOCOMOTIVE, SOUTH WALES, 1803.

a, the cylinder, 8¼ in. in diameter, 4 ft. 6 in. stroke, fixed in the boiler; *b*, the piston-rod, fastened to the cross-head; *c*, the cross-head; *d*, guides for cross-head; *e*, stays for the guides; *f*, the connecting rods from ends of cross-head to cranks; *g*, the two cranks; *h*, the driving axle; *i*, the fly-wheel; *j*, the gear-wheels, connecting the driving axle with the driving wheels; *k*, the four driving wheels, 3 ft. 9 in. in diameter, 4 ft. 1 in. from centre to centre; *l*, the four-way cock; *m*, the cock lever; *n*, the rod and tappets for working the four-way cock; *o*, lever from cross-head for striking the tappets; *p*, regulating cock and handle; *q*, stay and screws for tightening the three-way cock and the regulator cock; *r*, the cylindrical boiler of wrought or cast iron, 4 ft. 3 in. in diameter, 6 ft. long; *s*, the internal fire-place and wrought-iron return tube; *t*, the chimney; *u*, the fire-door; *v*, the tram-road-plates; *w*, the tramroad-sleepers. Weight in working order, 5 tons.

events Trevithick wrote the following recollections of them :—

"About six years since I turned my thoughts to this subject, and made a travelling steam-engine at my own expense, to try the experiment. I chained four waggons to the engine, each loaded with 2½ tons of iron, besides seventy men riding in the waggons, making altogether about 25 tons, and drew it on the road from Merthyr to the Quaker's Yard, in South Wales, a distance of 9¾ miles, at the rate of four miles per hour, without the assistance of either man or beast; and then without the load drove the engine on the road sixteen miles per hour. I thought this experiment showed to the public quite enough to recommend it to general use; but though a thing that promised to be of so much consequence, has so far remained buried, which discourages me from again trying its practice at my own expense."[1]

We have no account of the railway work done by the Coalbrookdale travelling engine of 1802. The Welsh tramroad-engine of 1803 took a gross load of 25 tons, at the rate of four miles an hour, over a bad road, with sharp curves and stiff inclines, and without load ran at a speed of sixteen miles an hour.

The Newcastle locomotive of 1804 was, in general outline, similar to the Welsh locomotive, but in detail superior. The wheels were to run on rails instead of tram-plates, and were 9 inches farther apart than the Welsh locomotive, giving increased steadiness. The boiler and return tube were wholly of wrought iron; the fire-door and chimney were at one end of the boiler, and the cylinder and guide-rods at the other end, giving more room to the engineman than on the Welsh locomotive, which had all those things at one end of the boiler.

The cylinder of the Newcastle locomotive was of the same size as the Coalbrookdale engine of 1802, being 7 inches in diameter, with a 3-feet stroke, and therefore was probably made at Coalbrookdale, from Trevithick's drawings and patterns of 1802, with its regulating blast-pipe and steam of from 60 to 145 lbs. on the square inch.

---

[1] See Trevithick's letter, 26th April, 1812, chap. xviii.

The first Newcastle locomotive is thus mentioned:—

"The death of Mr. John Whinfield, who was concerned in the manufacture of some of the earlier locomotives, has revived the question as to who was the first to apply the power of steam to locomotive purposes; and it has, we think, been well established that the honour is due to Mr. Richard Trevithick of Camborne, who, more than fifty years since, ran a locomotive through the High Street of his native town, and employed a locomotive to draw a load of bar-iron from the Penydarran Iron Works to the Basin. An attempt having been made by the North-countrymen to prove that John Steel, an employé of Mr. Whinfield, was the first inventor of a locomotive, a correspondent of the 'Gateshead Observer' (in continuation of correspondence on the subject) states that—

"'Trevithick's steam-engine was a patent; and Mr. John Whinfield, who had a small foundry in Pipewellgate, and whom I well knew, was a sort of agent of his. The plans to which you refer, and probably also some portions of the engine, may have come from Trevithick; Steel, an old acquaintance of mine, being the local engineer. Steel came, I believe, from somewhere about Colliery Dikes. He was at one time in Wales, where (at Merthyr Tydfil) Trevithick tried his locomotive engine on a tramroad. Whether it was before or after "our first engine" was made that Steel went to Wales I cannot say. These were the primitive days of steam. The appliances and facilities of the present day were then unknown. Everything was comparatively rude; the boiler was of cast iron. Well do I remember "Tommy Waters," who had a small foundry on the Pipewellgate slope, behind the "Blue Bell," in Bridge Street, and who was employed, under Trevithick's patent, to make a locomotive engine for Mr. Blackett, between forty and fifty years ago. Tommy's steam-engines wouldn't always go; and when they were obstinate, he would take hold of the lever of the safety-valve, and declare, in his desperation, "that either *she* or *he* should go!" Poor Steel! both he and the engine went in France—as you stated in your columns of last week.'"[1]

[1] From the Mining Journal,' Saturday, October 2nd, 1858.

Trevithick's letters show how readily he taught untrained men in numerous small mechanical workshops to construct his engines; for certainly Tommy Waters' locomotive with a cast-iron boiler was a copy, though a bad one, of that made by Trevithick in Wales.

"LOCOMOTIVE AND CARRIAGE DEPARTMENT,
"ENGINEER'S OFFICE, SWINDON,
"19*th April*, 1870.

"MY DEAR SIR,

"I am in receipt of your letter as to the engine sent to Wilam in 1804, and have, as your son has informed you, a drawing of it, from which I will have a tracing made and sent to you. I got this drawing made from a tracing which was lent to me by one of my brothers, who resides in Sunderland, and I understood from him that the engine itself was in existence up to a few years ago, and that it was driving a fan at a foundry either in Newcastle or Gateshead; but in order that I may be able to give you more accurate information, I will endeavour to get the particulars of the engine, dates, &c., and will then write you again.

"I am, my dear Sir,
"Yours truly,
"JOSEPH ARMSTRONG.

"F. TREVITHICK, Esq."

"SUNDERLAND, *May* 10*th*, 1870.

"DEAR SIR,

"By this post I have forwarded a tracing of the old locomotive for Mr. Trevithick, and likewise some copies of an extract from the 'Gateshead Observer.' They would have been sent ere now, but I have had some difficulty in procuring an original from which to get a tracing, and have also spent time in hunting up this extract, and having it reprinted.

"Yours truly,
"JOHN ARMSTRONG.
"*Per* JOHN HEMING.

"JOSEPH ARMSTRONG, Esq., *Swindon*."

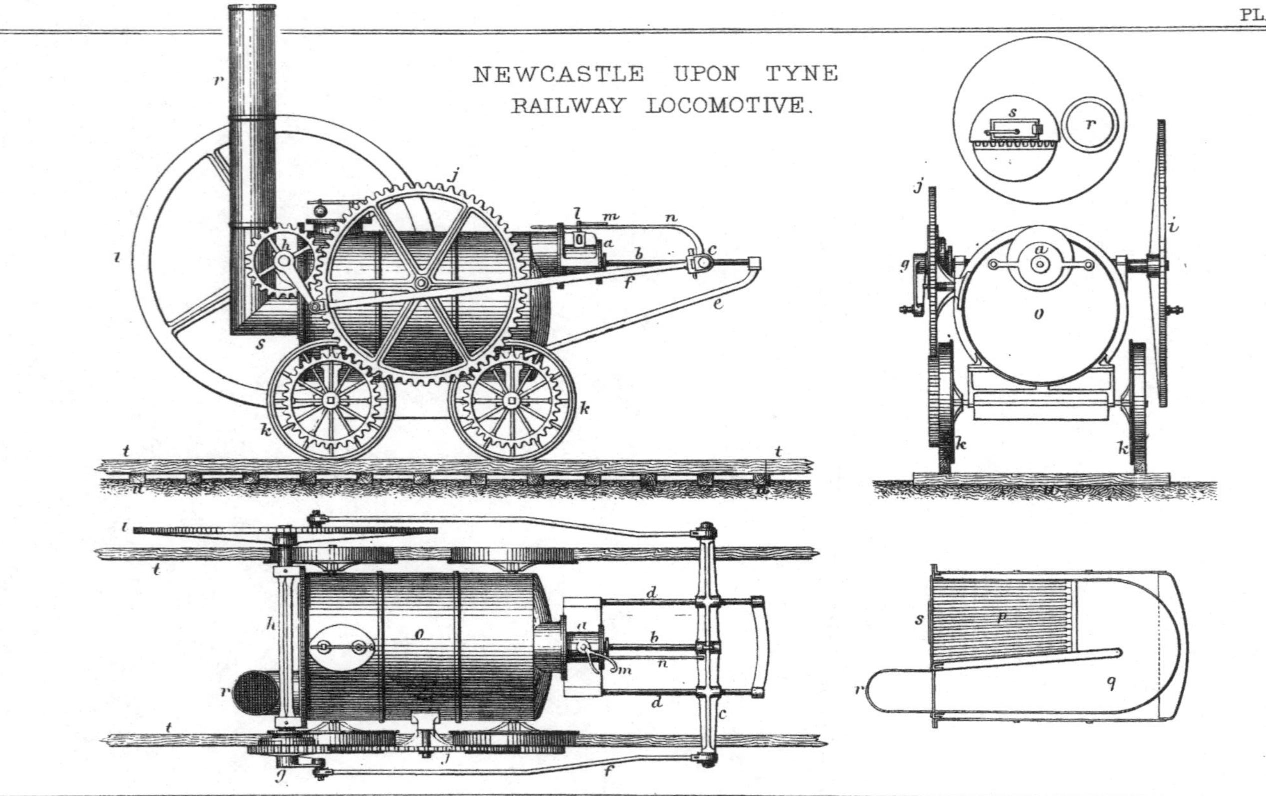

London: E. & F. N. Spon, 48, Charing Cross.

Kell, Bros. Lith. London.

In Plate V., *a* is a steam-cylinder 7 inches diameter, 3-feet stroke, fixed in the boiler; *b*, piston-rod; *c*, cross-head; *d*, guides; *e*, stay; *f*, connecting rod; *g*, crank; *h*, driving axle; *i*, fly-wheel; *j*, gear-wheels; *k*, four driving wheels, 3 feet 1 inch diameter, 4 feet 8½ inches from centre to centre; *l*, four-way cock; *m*, lever for working cock; *n*, plug-rod; *o*, cylindrical wrought-iron boiler, 4 feet diameter, 6 feet 6 inches long; *p*, fire-grate; *q*, return fire-tube of wrought iron, 2 feet 3 inches diameter at the fire-door end, 1 foot diameter at the chimney end; *r*, chimney; *s*, fire-door; *t*, railway of longitudinal timbers, 3½ inches wide, 4½ inches deep; *u*, cross-sleepers, 4½ inches wide, 3½ inches deep; 1 foot 1½ inch apart; gauge between wood rails, 4 feet 10 inches; weight of engine in working order, 4½ tons.

Our First Locomotive.

From the 'Gateshead Observer.'

"The engine erected by Mr. Trevithick had one cylinder only, with a fly-wheel to secure a rotatory motion in the crank at the end of each stroke. An engine of this kind was sent to the North, for Mr. Blackett of Wylam, but was, for some cause or other, never used upon his railroad, but was applied to blow a cupola at an iron foundry in Newcastle.—*Nicholas Wood 'On Railroads'* (third edition, page 281).

"Mr. Blackett was the first colliery owner in the North who took an interest in the locomotive engine. He went so far as to order one direct from Trevithick, to work his waggon-way, about the year 1811. The engine came down to Newcastle; but for some reason or other (perhaps because of the imperfect construction of the waggon-way as compared with the weight of the engine) it was never put upon the road. Mr. Blackett eventually sold it to a Mr. Whinfield, of Gateshead, by whom it was employed for many years in blowing the cupola of his iron foundry.—*Smiles' 'Life of Stephenson'* (first edition, page 74).

"The story told by Mr. Wood and Mr. Smiles, with some little variation in their versions, has been current on the Tyne for a number of years; and it was not till lately that we were led to doubt its accuracy.

"Mr. Robert Wylie, an iron-founder in the Close, Newcastle, having seen a reference, in the 'Gateshead Observer,' to a locomotive engine having been sent to Mr. Blackett by Mr. Trevithick, and afterwards sold to Mr. Whinfield, called upon us with a correction of the statement. He served his apprenticeship with Mr. Whinfield in Pipewellgate, where 'Price's Glass Works' now stand, and remembers the engine. It was made prior to his time, but he can positively say, from what he learned on the spot, that it did not come from Trevithick, or from any other person; it was manufactured on the premises, about the year 1804, by Mr. Whinfield—the engineer of the works being John Steel, a man with a wooden leg (subsequently blown up, by the explosion of an engine in France, and killed). So far Mr. Wylie.

"John Turnbull, of Eighton Banks, aged 80, whom we afterwards made out, informed us that he, too, served his apprenticeship with Mr. Whinfield. He was older than usual when he was bound—'turned of 16'—and it was some time in the present century before he closed his apprenticeship—he could not say when, his memory being now defective; but he perfectly remembered the making of the locomotive engine for Mr. Blackett. It was all made at his master's. The engineer was John Steel, who was regularly employed at the works, 'and a very clever fellow.' When she was finished, a temporary way was laid down in the works, 'to let the quality see her run.' There were several gentlemen present, and she ran backwards and forwards quite well. Mr. Blackett, however, did not take her;—'there was some disagreemency between him and the master, and she never left the works, but was used in the foundry as a fixed engine, to blow the iron down.' He (Turnbull) left Mr. Whinfield when his time was out, and the engine was made long before he left.

"Turnbull's statement, if correct as to his age, &c., would carry back the date of the manufacture of the engine beyond

1804—which, we have reason to believe, was about the year; and we therefore asked him if he could refer us to any other living witness. He at once named John Henderson, a fellow-apprentice, still working as a founder at Messrs. Hawks and Crawshay's, the Gateshead Iron Works. Henderson we found at his work—a hale, intelligent man of 70—with all his faculties in full vigour. He was unusually young (he said) when he went to trade. He could not give the date, but he would be about twelve years old. He remembered the engine. It was made at the works long before he was out of his time, when he and Turnbull were apprentices together; and John Steel was the engineer.

"Thus, then, we have the evidence of more than one living witness, that the engine made for Mr. Blackett, and afterwards used as a fixed engine, was not 'sent to the North,' but was of Gateshead manufacture.

"We have also other evidence. The plans are still extant. They are the property of Mr. Smith, of the Gateshead Park Iron Works (Messrs. Abbot and Co.'s), and have been placed in our hands by Mr. Wylie They comprise:—1. Well-executed perspective views of the engine from various points. 2. 'Drawing of waggon-engine, October 3, 1804.' 3 'Regulating and throttle cocks for engine, No. 1, September 17, 1804.' She had *friction* (not *cogged*) wheels, and was driven by spur-gear, working with a three-way cock, instead of a slide.

"Our case, we think, is now complete. It was in Gateshead, without doubt, that the first locomotive engine was planned and made;—the date being 1804—(the year in which Trevithick was trying his engine in Merthyr Tydfil)—the engineer, John Steel, and the manufacturer, John Whinfield.

"Can any correspondent supply us with facts in corroboration or correction of our statements? If so, we shall be much obliged by the kindness."

Gateshead may claim to have had a hand in putting together a Newcastle locomotive, and also in hiding the usefulness of Trevithick's high-pressure locomotive engine, then a reality of four or five years' standing;

but his foreman, John Steel, who had worked on the Welsh locomotive, superintended its erection. The late Mrs. Trevithick said "about the time her husband was occupied with the engines in Wales, he went several times to Newcastle.[1] John Steel, whom she found as foreman in the workshops in London in 1808, was a Newcastle man, with a wooden leg."

Trevithick's beautifully-designed Newcastle locomotive was perfectly manageable in Whinfield's small, cramped yard, on a temporary railway. The North-countrymen came in crowds to examine and see it work in the presence "of the quality of Newcastle," probably including Mr. Blackett of the Wylam Colliery, for whom the locomotive was made, and his agent, Mr. Hedley, who five or six years afterwards patented a very similar one. Mr. Wood and George Stephenson, with Mr. Wilson, the engineer, and Timothy Hackworth, then a blacksmith, working for Mr. Blackett, afterwards the engineer on the Stockton and Darlington Railway, and maker of the 'Sanspareil' locomotive that, a quarter of a century later, competed on the Liverpool and Manchester Railway.

Robert Hawthorne, then residing near Newcastle, who had employed George Stephenson as engine-boy at Callerton, was also probably one of the lookers-on; and a few years afterwards the two great engineering and locomotive-building establishments in Newcastle, known as Hawthorne's and Stephenson's, began to grow into importance.

The Newcastle drawing shows that Trevithick had made the detail of his locomotive engine fully known to the engineers of Newcastle in 1804, for it illus-

[1] See Trevithick's letter, 23rd September, 1804, chap. xvii.

trates that sent to Blackett's Wylam waggon-way; and its form and date agree with Mr. Wilson's Memorandum of having seen the locomotive at Newcastle, in May, 1805, when Trevithick paid his second or third visit to superintend its use on the railway. Nicholas Wood says, "an engine of this kind *was sent* to the North for Mr. Blackett." Had it been constructed in Newcastle, he would probably have mentioned it.

John Henderson, working for Messrs. Hawks and Crawshay, at the Gateshead Iron Works, recollected an engine. Crawshay, the Welsh ironmaster, and user of Trevithick's engines, and Homfray, his friend, also a large ironmaster, certainly informed Crawshay of Newcastle, and Blackett, of all that was doing at Merthyr Tydfil. Steel, a Newcastle mechanic, was chosen to learn under Trevithick in Wales, that he might become an agent in superintending the use of the high-pressure engine and locomotive in the iron and coal works of the North of England.

The Newcastle locomotive was a smaller but a better engine than the one that had worked so well in Wales; its boiler was wholly of wrought iron. The four driving wheels were constructed to run on rails in place of tram-plates. The cylinder was reduced to a diameter of 7 inches, and its stroke to 3 feet. The gross weight was reduced to 4½ tons.

The men of the North must know why that locomotive, seen by so many, was not allowed to work on the railway and scarcely to be spoken of in public, though it remained in work as a stationary engine almost up to the present time. The not showing the blast-pipe in former drawings nor in this drawing, and the strange perversion of facts bearing on its introduction, are among the mysteries of locomotive history.

In 1854 the writer revisited the Welsh works to inquire more particularly about the blast-pipe. Mr. Ellis's relative, while standing in one of Mr. Homfray's Tredegar workshops, remarked that pieces of Trevithick's early engines had remained for many years in the old scrap-heap, in the corner of the shop; on turning over the surface pieces, and directing Mr. Ellis's attention to one in particular, he replied, "That is Trevithick's first blast-pipe, or a copy of it." In shape and size it was just like the blast-pipe then in use, except that Trevithick's had a casing pipe for heating the feed-water.

Smiles thus describes the invention:—

"The locomotive might have been condemned as useless had not Mr. Stephenson at this juncture applied the steam-blast, and thus at once doubled the power of the engine.

"Although Trevithick, in the engine constructed by him in 1804, allowed the waste steam to escape into the chimney, there was no object in the arrangement beyond getting rid of a nuisance.

"It is remarkable that a man so ingenious as Trevithick should not have discerned its advantages; but it is clear that he could not have done so, for as late as 1815, after George Stephenson had discovered and successfully adopted the steam-blast, Trevithick took out a patent, the principal object of which was to 'produce a current of air, in the manner of a winnowing machine, to blow the fire.'"[1]

Trevithick's patent in 1815 was for an engine so constructed that the waste steam could not possibly be used as blast, and therefore he was obliged in that particular engine to omit the blast-pipe, but this did not annul his former acts.[2]

Goldsworthy Gurney[3] saw Trevithick's locomotive

[1] 'Life of George Stephenson,' by Smiles.

[2] See patent, chap. xvi.

[3] Mr. Goldsworthy Gurney's account of the invention of the steam jet or blast, published 1859.

in Cornwall and in Wales, admits the use of the blast-pipe by Trevithick, and yet claims for himself the credit of the invention.

The last sentences in his pamphlet state, "Mr. Smiles has been sadly misinformed as to the invention of the steam-jet, and also with regard to the locomotive engine. All the facts on record, and the testimony of living witnesses, show that Mr. Richard Trevithick was the inventor of the locomotive engine; and that Mr. Goldsworthy Gurney was the inventor of the steam-jet."

But those sentences were preceded by the following:—

"Mr. Gurney saw Trevithick's first steam-carriage *in* 1804, and all his experiments on locomotion; and remembers, too, the contemptuous way in which he was treated by the engineers of the day. His views were described as 'wild theories,' and his plans ridiculed. Mr. Davies Gilbert, however, thought differently. He aided Trevithick in his calculations, and encouraged him to go on, and Richard Trevithick became the inventor of the locomotive as well as of the high-pressure engine.

"The eduction-pipe at this time entered the chimney half-way up the funnel, and Trevithick, in order that the vapour might more effectually meet and mix with the hot air coming from the furnace, turned it downwards. *This did not succeed, as may be supposed.* The pipe was then turned upwards, as the most ready way for its escape. Thus the waste steam from the engine was thrown upwards into the chimney by Trevithick in 1804, long before any other person was engaged in the subject, exactly in the same way as was afterwards done by all who followed him.

"Many modifications, but very few improvements, if any, were made on Trevithick's engine for many years during his absence in Peru."

From the close of 1801 to May of 1805 Trevithick constructed two Camborne and one London common road locomotives, a Coalbrookdale engine suitable for a railway, a Welsh tramroad-engine, and a Newcastle-on-Tyne railway-engine, all having a blast-pipe.

Locomotive history now makes a jump of three years, when, in 1808, Trevithick constructed, not only a locomotive engine, but also a railway, that the London public might see with their own eyes what the new high-pressure steam-engine could effect, and how greatly superior a railway was to a common road for locomotion.

"THAMES ARCHWAY, ROTHERHITHE,
"*July* 28, 1808.

"MR. GIDDY,

"Sir,—I have yours of the 24th, and intend to put the inscription on the engine which you sent to me.

"About four or five days ago I tried the engine, which worked exceedingly well; but the ground was very soft, and the engine (about 8 tons) sank the timber under the rails, and broke a great number of them. I have now taken up the whole of the timber and iron, and have laid balk of from 12 to 14 inches square down on the ground, and have nearly all the road laid again, which now appears very firm. We prove every part as we lay it down, by running the engine over it by hand. I hope it will all be complete by the end of this week. The tunnel is at a stand.

"Your very humble servant,

"RICHARD TREVITHICK."

The sister of Davies Gilbert named this engine 'Catch-me-who-can,' and her Memorandum "My ride with Trevithick, in the year 1808, in an open carriage, propelled by the steam-engine, of which the enclosed is a print, took place on a waste piece, now Torrington Square," [1] enclosed an engraving of the engine on Trevithick's visiting cards.

[1] Mem. of Mrs. Guilmard, 1808.

The wonderful simplicity of this engine exceeds that of the Newcastle locomotive; Trevithick in those early days knew that the friction of *one pair* of driving wheels was sufficient for his work, enabling him to do away with the gear coupling wheels of his earlier locomotives, and also that the puffs of the steam-blast had so increased the quantity of steam given by the boiler, that a damper in the chimney was necessary.

MR. RICHARD TREVITHICK'S 'CATCH-ME-WHO-CAN,' 1808.

The following interesting account was given by an engineer well known in his day.

Mr. Trevithick's New Road Experiments in 1808.

"SIR,

"Observing that it is stated in your last number (No. 1232, dated the 20th instant, page 269), under the head of 'Twenty-one Years' Retrospect of the Railway System,' that the greatest speed of Trevithick's engine was five miles an hour, I think it due to the memory of that extraordinary man to declare that about the year 1808 he laid down a circular railway in a field adjoining the New Road, near or at the spot now forming the southern half of Euston Square; that he placed a locomotive engine, weighing about 10 tons, on that railway—on which I rode, with my watch in hand—at the rate of twelve miles an hour; that Mr. Trevithick then gave his opinion that it would go twenty miles an hour, or more, on a straight railway; that the engine was exhibited at one shilling admittance, including a ride for the few who were not too timid;

that it ran for some weeks, when a rail broke and occasioned the engine to fly off in a tangent and overturn, the ground being very soft at the time.

TREVITHICK'S LONDON RAILWAY AND LOCOMOTIVE OF 1808. [W. J. Welch.]

"Mr. Trevithick having expended all his means in erecting the works and enclosure, and the shillings not having come in fast enough to pay current expenses, the engine was not again set on the rail.

"I am, Sir, your obedient servant,
"JOHN ISAAC HAWKINS,
"Civil Engineer, London."[1]

---

[1] 'Mechanics' Magazine,' 27th March, 1847.

On this experiment Mr. Albinus Martin gives the following:—

"I am sorry to be unable to give you any useful information about the engine exhibited in London more than half a century ago. I cannot even fix the time; but from comparison with other dates, it could not have been in 1803. William Rastrick was resident engineer on the works of the driftway under the Thames, and he was very kind to me as a boy. I think it must have been to him that I was indebted for a sight of the engine exhibited, for I know I got in without payment, and felt myself thereby recognized as belonging to the craft. The place was at the rear of what were then florists' or nursery gardens, in the New Road, very near if not on the site of the North-Western Railway Station.

"There was a circular railway of about—I can't tell the gauge or the diameter—but perhaps a hundred feet. The engine itself was to me an entire novelty, but not differing in general appearance from that on the Merthyr tramroad, and of which an engraving has been published.

"The space in which this circular railroad was enclosed was surrounded by a fence, made of close-fitting 12 or 15 feet deals. When I saw it the engine was out of steam, and there were no spectators. Trevithick was not there, nor can I recollect who was.

"From my recollection of events it was about 1806."[1]

This London railway of 1808 was near Euston Square, and the site of the present London and North-Western Railway Station, and on it the public were carried at twelve or fifteen miles an hour around curves of 50 or 100 feet radius.

At the end of September, 1808, Mr. Homfray wrote to a friend in London, "I wish you likewise to say if there is anything in the report of *the Racing Engine* being carried into effect, and if so at what time, and the

---

[1] Extract from letter of Mr. Albinus Martin, in 1868.

particulars of the bet, &c., as if it is to take place, I shall be very much inclined to see it."

Trevithick's note of 28th July said that four or five days before he had run very well. This railway experiment, therefore, must have been in operation more or less for two or three months, and during that time his occupations at the Thames Driftway brought numerous engineers around him,[1] and amongst them, Messrs. Stobart and Buddle, from the North, interested in coal works.

"In the year 1805 the late Mr. William Hedley was appointed mining engineer at Wylam Colliery. At that time a railway, five miles in length, communicated with a depôt on the river Tyne. The railway was a wooden one, subject to great undulations. It was worked by the old method, one horse being employed for each waggon. But about the year 1808 the wooden rails were taken up and cast-iron plate-rails substituted.

"Mr. Blackett, in the year 1809, wrote to the celebrated Trevithick on the subject of an engine; his reply stated that he was engaged in other pursuits, and having declined the business, he could render no assistance."[2]

It was at this period that the Thames Driftway failed, and patent law proceedings and bankruptcy, and ill-health drove Trevithick from London, and for a time from active life; but not until after he had achieved the great work of conveying the London public on a railway constructed of longitudinal timbers supporting a rail of iron. A passenger-carriage ran for hire, drawn by a locomotive engine on four wheels, two of them being driving wheels made to revolve by crank-pins in their spokes. The waste-steam pipe from the cylinder to the chimney gave its puffs as in the present day.

---

[1] See chap. xii.

[2] 'Who Invented the Locomotive?' by O. D. Hedley, published 1858.

The cylinder, instead of being horizontal as in the two former railway-engines, was now again vertical, as in the first Camborne common road engine; its weight was 8 or 10 tons.

Mr. Bendy, who had before worked for Trevithick, says, "The engine used in the dredger[1] was of the same size as that fixed in London to run round the circle at the speed of fourteen or fifteen miles an hour—the cylinder was 14½ inches in diameter, with a stroke of 4 feet."[2] The 'Catch-me-who-can,' or 'Racing Steam-Horse,' was to run against a flesh-and-blood race-horse for a fair day's work, the one performing the greatest number of miles to be the winner. Probably the difficulty of settling preliminaries caused the contest to be deferred and forgotten; for there is no record of the race having been run.

Mr. Bendy, who worked as a foreman mechanic on Trevithick's dredger-engine of 1803, and was a spectator, if not an assistant, in the railway experiment in 1808, says the two engines were alike, and this link gives an accurate knowledge of the detail of the engines; remove the fly-wheel and crank-shaft from the drawing of the dredger-engine, attach the two side rods to crank-pins in two driving wheels, place a damper in the chimney, and we have the counterpart of the locomotive on Trevithick's card of 1808.

In 1870 the writer saw in a foundry at Bridgenorth, where those engines were made, one of the original cross-head guide-rods, precisely as shown at K K.

The drawing of the dredger in Rees' 'Cyclopædia' was by Mr. J. Farey. The detail description, omitted by him, is supplied by the writer.

[1] See drawing of dredger-engine, chap. xi.
[2] See Mr. Bendy's letter, chap. xi.

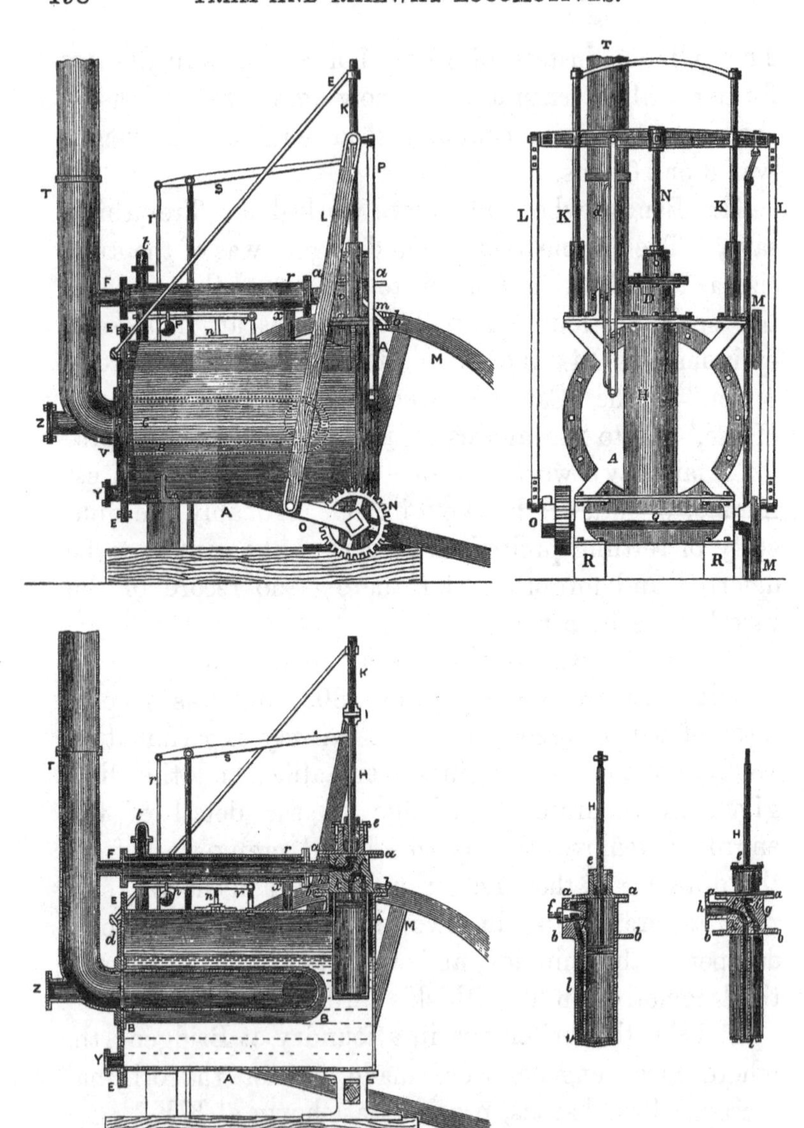

TREVITHICK'S DREDGER-ENGINE OF 1803, AND LOCOMOTIVE OF 1808.

D, steam-cylinder, 14 in. diameter, 4-ft. stroke, fixed in the boiler; N, piston-rod; I, cross-head; K, guides; E, stay; L, connecting rods; O, crank; Q, driving axle and gear-wheel; M, fly-wheel; *f*, four-way cock; *m*, lever for working cock; P, plug-rod; A, cylindrical cast-iron boiler, 4 ft. 10 in. diameter, 8 ft. long; B, fire grate; C, return fire-tube of wrought iron, 2 ft. diameter at the fire-door, and 1 ft. 2 in. at the chimney end; T, chimney; weight of engine in working order, 8 or 10 tons; Z, hole for cleaning fire-tube; Y, hole for cleaning bottom of boiler; *n*, safety-valve; *v*, safety-valve lever; *p*, safety valve weight; *i*, escape-steam pipe, or feed-heating pipe; *t*, feed-pipe; *r*, rod working feed-pump; S, lever working feed-pump.

The following description from Rees was probably written by Robison or Farey, both of them intimately connected with Watt:—

"The high-pressure engines at present in use were introduced by Mr. Trevithick, in conjunction with Mr. Vivian, who obtained a patent for the same in 1802; this was principally for their application of the engine to the purpose of driving of carriages upon railroads. This engine containing no material parts which are not used in other engines, and before described, it may be explained without a drawing. The boiler consists of a large cylinder of cast iron, made very strong, with a flanch at one of its ends to screw on the end or cover, which has the requisite openings for the fire-door, the man-hole, the exit for the smoke, and the gauge-cocks.

"The fire is contained within the boiler, in a cylindrical tube of wrought iron, which is surrounded with water on all sides, in the same manner as the fire in Mr. Smeaton's portable engines; but there is a little difference in the application: one end of this tube is flanched to the end or cover of the boiler, and is divided into two parts by having the fire-grate extended across it; the tube extends nearly to the end of the boiler, where it is reduced in size, then doubles, and returns back in a direction parallel to the first tube or fire-place to form the flue or chimney. On one side of the cylinder, just above the flanch which fixes it into the boiler, and beneath the top flanch, which fastens down its lid, is a protuberance of cast iron, to contain the four passages and the cock (one passage rises directly from the boiler and brings steam to the cock at one side, to be distributed either to the top or bottom of the cylinder, according to the position in which the cock stands). The boiler is supplied with water as fast as it evaporates, by means of a small force-pump worked by the engine; but as it would be a great loss of heat to inject cold water at once into the boiler, it is first rendered nearly or quite boiling by a very simple contrivance. The waste-pipe, which conveys the steam away from the cylinder after having performed its office, is enclosed within an external pipe or jacket leaving a space of about an inch all round; through this space the cold water is forced to enter at one end by the small force-pump, and the boiler is supplied with water by a branch from its other extremity. The velocity of the engine is regulated, or its motion can be entirely stopped if required, by a cock situated in the first passage from the boiler to the four-passaged cock.

"Several very terrible accidents have occurred from the bursting of high-pressure boilers, either from their being made too weak to resist the force they are intended to bear, or from some mismanagement, as loading the safety-valve too much. Some years ago an engine that was employed to drain water from the tide mills while building between Woolwich and Greenwich, was blown up by overloading the safety-valve, when several people were killed."

This description, published in 1819, shows that up to that time Trevithick's patent engine of 1802 was looked on as especially suitable for driving carriages on railways, and though parts are spoken of as not new, no statement is made of who had used them before, except "a similarity to a boiler made by Smeaton with a little difference." The several terrible accidents are not particularized, except that at Greenwich, which was not caused by any defect in the engine or boiler, but by the fastening down of the safety-valve. The bias of the statement is evident, the exploded boiler having been a globular cast-iron boiler enclosed in brickwork, with external fire in contact with the cast iron, while the condemnation is applied to a cylindrical boiler with internal fire in a wrought-iron tube, proving the truth of Trevithick's statement made at the time, that Boulton and Watt had made false reports on the explosion.[1] The waste-steam blast-pipe is shown in connection with the chimney, but the writers in 1819 seem not to have comprehended its value, though it had been in operation for eighteen years.

---

[1] See Trevithick's letter, October 1st, 1803, chap. xx.

The compactness, perfection, and beautiful simplicity of Trevithick's engines of sixty-five years past are by this drawing made manifest to engineers of the present day, while the written description makes it equally evident that it was incorrectly spoken of.

An important evidence of the rapid advance of locomotive mechanism under Trevithick's guidance is in the use of but two driving wheels in the 1808 locomotive, while former ones had four wheels coupled together, making them all drivers. In all probability Mr. Blackett and Mr. Hedley, his agent, rode on this engine, for they were at that time in frequent communication with Trevithick on the subject of railways and locomotives, and yet failed to see the lesson so plainly taught, for shortly afterwards these two gentlemen patented the discovery of sufficient adhesion from smooth wheels.

Mr. Hedley was appointed engineer of the Wylam Colliery about 1805, when Trevithick's locomotive was sent to Mr. Blackett, and was not ignorant of the succeeding trials and improvements by Trevithick, for shortly after the public proof in 1808 of what could be done Mr. Blackett again solicited Trevithick's help, either by engines or drawings; yet in 1813 he patented the invention as his own.

"To operate by mere friction or gravity had not as yet occurred to anyone, until the late William Hedley, Esq., viewer, who had the direction of Wylam Colliery, conceived the idea; and having satisfied himself by a variety of experiments with the waggon-way carriages, he took out a patent for the invention, which bears date March 13th, 1813.

"In 1814 Mr. George Stephenson, having given his attention to the subject, fitted up an engine at Killingworth Colliery." [1]

[1] 'Who Invented the Locomotive?' by O. D. Hedley, 1858.

This is strong evidence of the inability of the intelligent public of that day to comprehend Trevithick or his engines; but it is still more strange that fifty years have failed to entirely remove the mist. Evidences had been given through the length and breadth of England of the sufficiency of grip on common roads, on tram-plates, and on rails. Many years afterwards Dr. Lardner, in a lecture on the locomotive, spoke of George Stephenson as "the father of the locomotive engine." Mr. Hedley wrote the following letter in refutation of Lardner's statement:—

"SIR, "SHIELD ROW, *December* 10, 1836.

"I respectfully beg to call your attention to the following circumstances connected with the establishment of the locomotive engine in this district:—

"In October, 1812, I had the direction of Wylam Colliery. At that period I was requested by the proprietor (the late Mr. Blackett) to undertake the construction of a locomotive engine. The celebrated Trevithick had previously been applied to for one; in reply, he stated that he had declined the business. Amongst the many obstacles to locomotion at that period was the idea entertained by practical men, and which was acted upon, *viz.* that an engine would only draw after it, on a level road, a weight equal to its own.

"Mr. Blenkinsop, in 1811, effected the locomotion by a toothed or rack rail; in December, 1812, W. and E. Chapman, by means of a chain; and in May, 1813, Mr. Brunton, of Butterley, by movable legs. I was, however, forcibly impressed with the idea, and which was strengthened *by some small preliminary experiments*, that the weight of an engine was sufficient for the purpose of enabling it to draw a train of loaded waggons. An engine was then constructed, the boiler was of cast iron, the tube containing the fire went longitudinally through the boiler into the chimney. The engine had one cylinder and a fly-wheel: it went badly, the obvious defect being want of steam. Another engine was then constructed, the boiler was of malleable iron; the tube containing the fire was enlarged, and, in place of passing directly through the boiler into the chimney, it was made to return again through the boiler into the chimney, now at the same end of the boiler as the fire-place. This was a most important improvement. The engine was placed upon four wheels, and went well. A short time after it commenced, it regularly drew eight loaded coal-waggons after it, at the rate of from four to five miles per hour, on Wylam Railroad, which was in a very bad state; in addition to this, there was a great rise in the direction of the load in some parts of it; the road itself was of that kind termed the plate-rail.

"In conclusion, I beg to say that I am the individual who established the principle of locomotion by the friction or adhesion of the wheels upon the rails; and, further, that it was the engines on the Wylam Railroad that established the character of the locomotive engine in this district.

"I trust you will see the propriety in your future lectures of not designating Mr. Stephenson the 'father of the locomotive engine.'

"I beg to subscribe myself, Sir,

"Your obedient servant,

"WILLIAM HEDLEY.[1]

"DR. LARDNER, *Newcastle.*"

About ten years after the date of Mr. Hedley's letter, at the opening of the Trent Valley Railway, Sir Robert Peel, then Prime Minister, proposed the health of George Stephenson, as the "father of the locomotive engine." Mr. Stephenson, in reply, swallowed to himself the whole compliment, causing the writer to leave his seat with the intention of putting in a claim for the memory of Trevithick; the late Mr. Thomas Brassey noticed the intention and pointed out its inconvenience at a convivial meeting.

Mr. Hedley's letter, while it shows the injustice of

[1] 'Who Invented the Locomotive?' by O. D. Hedley, p. 35.

the patent laws, gives the strongest corroborative proof of Trevithick's prior claim as the inventor of the locomotive. It states that in 1812 practical men believed that a locomotive on smooth wheels would only draw after it a weight equal to its own. Mr. Wilson wrote in 1805 that Trevithick's locomotive, weighing 4½ tons, was to pull on the Wylam Railway 10½ tons, at four miles an hour. This was a prudently-careful engagement on Trevithick's part, for he knew that his Welsh locomotive of 5 tons, on smooth wheels, had drawn 24 tons, a load nearly five times its own weight.

Mr. Hedley constructed a second locomotive, which, from his own account, was a closer copy of Trevithick's locomotive, having a wrought-iron boiler and return tube, upon four wheels, and drawing eight loaded coal-waggons, at the rate of from four to five miles an hour, on the Wylam Railroad: the engine had one cylinder and a fly-wheel. This description is very like that by Mr. Wilson, in 1805, of the locomotive sent by Trevithick to the Wylam Railway. At the time that Mr. Hedley was taking a patent for "the principle of locomotion by the friction or adhesion of the wheels upon the rails," Trevithick was writing of his patent engine,—"I am convinced to a certainty that the engine at Hayle will draw above 100 tons of stone from the quarries, and put them into the ship's hold, in one day."[1] And in the preceding year,—"I am now building a portable steam-whim on the same plan, to go itself from shaft to shaft. He may see this at work in a month, which will prove to him the advantage of a portable engine to travel from one plantation to another: the price complete is 105*l*."[2]

---

[1] See Trevithick's letter, 4th February, 1813, chap. xvii.
[2] See Trevithick's letter, 10th March, 1812, chap. xvii.

It may never be known how many of Trevithick's locomotives really worked, or how many of them reached Newcastle.

'Catch-me-who-can' proved the feasibility of steam locomotion on rails of iron resting on longitudinal timbers, and caused Mr. Blackett to write to Trevithick, soliciting another locomotive for the North.

Smiles says:—

"Mr. Blackett had taken up the wooden road in 1808, and laid down a plate-way of cast iron; he went so far as to order a locomotive direct from Trevithick to work his waggon-way, about the year 1811.

"While Mr. Blackett was thus experimenting and building locomotives at Wylam, George Stephenson was anxiously brooding over the same subject at Killingworth.

"Mr. Blackett's engines were working daily at Wylam, past the cottage in which he had been born. After mastering its arrangements and observing the working of the machine, he did not hesitate to declare to Jonathan Foster on the spot his firm conviction that he could make a much better engine than Trevithick. The steam-blast in the chimney was never properly understood, until George Stephenson, adopting it with a preconceived design and purpose, demonstrated its importance and value, as being in fact, the very life-blood of the locomotive engine."[1]

This account of the blast-pipe, and bungling attempt to copy Trevithick's locomotive, may go into the waste-paper basket, with the many other erroneous demonstrations on the same subject; but the lessons taught by comparing the slow progress of the locomotive in the North about 1813, when Stephenson took it in hand, up to the Liverpool and Manchester trials in 1829, with what Trevithick had done, standing by himself, between 1801 and 1813, are melancholy proofs of

[1] See Smiles' 'Life of George Stephenson,' pp. 74–81.

the neglects individual genius has to submit to, and of the insurmountable difficulties placed in the path of progress by ignorance and self-interest.

It has been said that Cugnot, as early as 1769, tried to make a low-pressure locomotive, but failed to accomplish it; and that Watt made a similar futile attempt fifteen years afterwards.

Hebert says:[1]—

"From all the information that we can glean in tracing out the early history of locomotion, this remarkable circumstance constantly presents itself: that when Trevithick's carriages with smooth wheels were employed upon levels, or slightly-inclined planes, invidious comparisons with others having cogs were made against the former, because, as was asserted, they slipped, and could not ascend such acclivities as the latter; and this, notwithstanding Trevithick first suggested, by his 'cross-grooves and fittings to railroads,' the very principle of the cogs, in a less objectionable form, and 'all other appliances to boot,' of the engine and boiler, contained in the said locomotive! Thus Trevithick lost many orders, and they were given to those who adopted all the essentials of his plans, without acknowledgment, and employed them as the basis of their structures; and when, after the lapse of years, it was found out by these *gentlemen* that smooth wheels had sufficient 'bite' of the rail in most circumstances, they made that fact appear to be their own discovery, notwithstanding it is stated in Trevithick's specification of 1802, and was confirmed by his practice; which practice they at first condemned with one general voice; and when at last they were compelled to practise it also, they endeavoured to make it appear as vastly superior to Trevithick's mode of surrounding his wheels 'with heads of nails, bolts, and claws,' which he never used at all! These ungenerous proceedings against the most eminent mechanic of his time, appear to have been going on unchecked from 1802 up to the present time—1836."

---

[1] See Luke Hebert's practical treatise on 'Railroads and Locomotion,' p. 30; published 1837.

The 'Penny Cyclopædia' has the following:—

"The possibility of applying the steam-engine to purposes of locomotion was conceived by several of its earliest improvers; and in 1784 a plan was suggested in one of the patents of Watt; but it does not appear that either he or any other inventor carried their ideas into practice until about 1802, when Messrs. Trevithick and Vivian patented a high-pressure engine, which, by its simplicity and compactness, was admirably adapted for locomotive purposes. Within a few years they

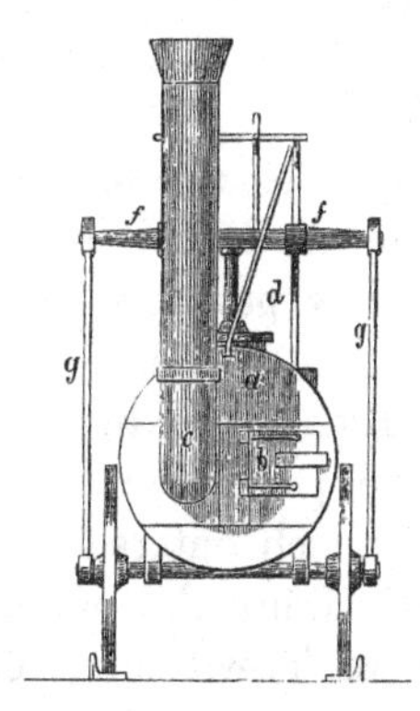

a is the boiler, which is of a cylindrical form, with flat ends. The fire is contained in a large tube within, and on one side of the boiler; one end of this is seen at *b*, and the form is indicated by dotted lines. This tube extends nearly to the opposite end of the boiler, and then being diminished in size, it is turned round and brought out to the chimney at *c*. The fire-tube is completely surrounded by the water, by which arrangement steam is generated with great rapidity, and of a high degree of elasticity. The steam-cylinder is placed vertically at *d*, being immersed nearly to the bottom of the boiler, as shown by the dotted lines. The steam is admitted alternately above and below the piston, by means of a four-way cock in a valve-box at the top of the cylinder, and the waste steam, after propelling the piston, passes by the eduction-pipe *e* into the chimney, where its emission causes a strong draught. The upper end of the piston-rod is attached to a cross-head *f*, which slides up and down on vertical guides, and from the ends of which connecting rods *g g* descend to cramps fixed on axles of the fore wheels, which are thus caused to revolve, like the fly-wheel of a stationary engine: *h* is a safety-valve on the upper part of the boiler. The immersion of the working cylinder in the boiler is happily contrived for compactness and economy of heat, and has been frequently imitated in subsequent engines.

built several carriages, one of which, at least, was for use on a common road. In 1805 they made some interesting experiments with a machine similar to that represented by the annexed cuts, on a tramway near Merthyr Tydfil, and thereby proved the practicability of their plans. It is remarkable that, notwithstanding the extreme simplicity of this machine, it possessed almost all the essential arrangements of the modern engines; and the ideas of its inventors were so complete that subsequent engineers have had little to do beyond improving, and carrying into effect the suggestions of their specification.

"And the admirable arrangement of throwing the waste steam into the chimney, has been almost invariably followed; as it affords a blast always proportionate to the speed of the engine, and the consequent demand for the evolution of steam.

"A supplementary carriage followed the engine, to carry a supply of fuel and water, and a small force-pump, worked by the machine itself, maintained the requisite supply of water in the boiler.

"Being otherwise occupied himself, he did not proceed with his locomotive experiments; but many others entered the field, though they produced few useful contrivances that were not either used or suggested by him.

"Mr. Blenkinsop in 1811 patented a locomotive engine, in which the power was applied to a large cogged wheel, the teeth of which entered a rack laid down beside the ordinary rails. Blenkinsop's engine was in other respects very similar to that of Trevithick; but two cylinders and pistons were employed, working separate cranks at an angle of 90°, so that one was exerting its full force, while the other passed its dead point.

"In 1825 the Stockton and Darlington Railway was opened.

"As the Liverpool line approached completion, the directors took great pains to ascertain the best method of working it. They were soon convinced that horse-power was ineligible, as it was intended to aim at considerable velocity; and the expense of animal power when applied at a speed of eight or ten miles per hour is very great. It was not so easy to decide on the comparative merits of stationary and locomotive engines. Various suggestions were made for the application of fixed engines at intervals of a mile or two along the line, to draw trains by ropes from station to station; but it was eventually determined to use locomotives, and to offer a premium of 500*l.* for the best to be produced, which would fulfil certain conditions, of which some were—that it should not emit smoke; should draw three times its own weight at the rate of ten miles per hour; should be supported on springs; not to exceed 6 tons in weight, or 4½ tons if only four wheels; and should not cost more than 550*l.* The trial was fixed for October, 1829, when four locomotives were produced, one of which was withdrawn at the commencement of the experiment."

This description of work performed by a locomotive at Merthyr Tydfil in 1805, may lead to the conclusion that the Welsh tramroad-engine of 1803 was not the only one worked in Wales about that time; but in the absence of sufficient identity, the writer views this drawing as showing the London railway locomotive of 1808. The mention of several steam-carriages having been tried within a few years of 1802, one at least of which was on a common road, and one on a tramway, throwing the waste steam into the chimney, thereby enabling the boiler to give the required steam supply, is confirmatory of similar statements in Trevithick's letters. The fire-bridge shown in the fire-tube, and the statement that a tender was attached to the engine for carrying water and coke, are both omitted by other commentators; yet we know that all Trevithick's locomotives had the fire-bridge, and if all had not tenders on their experimental trials they would of necessity have had them when in regular work. All the drawings of the early locomotives show the exhaust-steam pipe connected with the chimney, but none give the upturned part; yet it must have been so, because the several locomotives could not have worked without an effective blast. The writer having built and worked numerous locomotives, ventures to make this statement, though in direct opposition to opinions written and expressed by many.

Though the outline of this locomotive is very like the 1803 Dredger-engine, it evidently is not the same one;

for the Dredger has a taper tube in the boiler, while the locomotive has the bottle-neck tube, with the fire-bridge placed at some distance from the contracted part of the fire-tube, and supposing it allowed of the passage of a little air is very like modern patent smoke-burners.

Mr. Rastrick's evidence on the Liverpool and Manchester Railway Bill in 1825 mentions that a locomotive was made for Trevithick in 1807 or 1808 at Hazeldine and Rastrick's works at Bridgenorth, and was in the latter year run on a circular railway in London. "It was stated that this engine was to run against a horse, and that whichever went a certain number of miles was to win." Probably the same engine is described in the 'Penny Cyclopædia' as "generating steam with great rapidity and of a high degree of elasticity; the waste steam, after propelling the piston, passes by the eduction-pipe into the chimney, where its emission causes a strong draught."

A year or two prior to the all-England competition, in 1829, Trevithick had returned, after an eleven years' absence from his native country, to find that what he had performed on bad tramways with sharp curves and inclines twenty-five years before, it was thought unreasonable to expect on a carefully-made railway with easy curves and gradients. The Liverpool and Manchester directors seem to have been guided in their limit of work by Trevithick's 1803 engine, which took more than three times its own weight at a less speed; but when light, moved at a greater speed. The weight of that engine was also their limit; while his 1808 engine was double the weight of the former, and took passengers at twelve or fifteen miles an hour. Sketches of the 'Rocket' and the 'Sanspareil,' the two best engines of 1829,[1] show the latter to have been very like the 'Catch-me-who-can' of 1808; while the former is more like the Newcastle engine of 1804. Each of the 1829 locomotives had two cylinders. The 'Rocket,' whose boiler was much improved by small tubes, averaged a speed of fourteen miles an hour, with 10 or 11 tons of load, its greatest velocity being at the rate of twenty-four miles an hour, or double the speed required by the terms of the race.

THE 'ROCKET,' BY ROBERT STEPHENSON, 1829.

THE 'SANSPAREIL,' BY MR. HACKWORTH, 1829.

It was Trevithick's high pressure that enabled the steam-engine to be used for such purposes, and even the improved tubular boiler enabling the 'Rocket' to win the prize, had in principle been used and patented by him long before.

His patent of 1802 recommends the use of two cylinders, and shows a boiler with three tubes. The difficulty of manufacture confined him in practice to one cylinder

[1] See Luke Hebert 'On Railways.'

and two tubes, called the return tubes. But he patented in 1815 a boiler made of small tubes, and applied it in that or the following year to his screw-propeller engine. There was this difference,—his patent shows the water in the small tubes with the fire around them; the 'Rocket' had the fire through the tubes and the water around them.

The reader will judge of the similarity of the locomotives by Stephenson and Hackworth to Trevithick's earlier locomotives which they had seen. The 'Novelty,' by Messrs. Braithwaite and Ericsson, one of the three best of the competing locomotives, in its outline was not so much like a copy of Trevithick's as the other two; but a closer examination reveals the same family likeness. It has been said that the drawings for the 'Novelty' were made by the late Mr. John Hosking, who had been a pupil of Captain Samuel Grose when he erected Trevithick's engines in Cornwall, and while so engaged was also a fellow-draughtsman with the writer, and was afterwards employed in Stephenson's works at Newcastle.

To Trevithick and his never-ceasing practical exertions, in Cornwall, London, Shropshire, South Wales, and in Newcastle-on-Tyne, are we indebted for the first practical and real evidences of steam locomotion. Yet though he had returned to his native country a year or two before those locomotive competitions on the Liverpool and Manchester Railway, he was not consulted by the competing engineers, all of whom may be said to have taken their first lessons from him. His labours for the locomotive had ceased, but for the general history of its further progress we may trace it for a few years under its modern name of outside-cylinder engine.

In 1840 the directors of the Liverpool and Manchester and Grand Junction Railways, ten years having elapsed since the public locomotive trials, with the assistance of George and Robert Stephenson, and Joseph Locke, came to the conclusion that a new and improved locomotive should be designed, on which, as far as possible, all succeeding ones should be built; this was shortly after the time that Robert Stephenson withdrew from direct interference on the Grand Junction Railway, Joseph Locke having become the engineer-in-chief, whose report to the directors on the engine stock, written on the last day of 1839, states:—

"If a substantial improvement can be made, let it be applied to new engines, or to those already worn out which require to be renewed. This is a field sufficiently wide for the most inventive mind, without permitting it to range over the whole list of engines that are in daily use. In making this suggestion I know that it may be said that '*there is an end to improvements;*' but so convinced am I of the folly and expense of perpetually altering the engines for the sake of some trifling gain, that I would rather submit to this imputation than see those changes so often made. It would be well also to have in view the advantages of making as many parts of the different engines similar to each other as possible. To give you an illustration, I find that notwithstanding the great number of engines and tenders, there is sometimes a want of tenders, arising from the connecting pipes being of different sizes. I lately found an engine standing idle for the want of a valve to the pump, a small piece of brass not more than 3 lbs. in weight, and although there are ten engines of the same class on the line (with two pumps to each engine), there was not one duplicate valve on the establishment. The enginemen should be under the locomotive superintendent, and should take their orders from him. In concluding this report I would take the liberty of pressing on your attention the necessity of preserving with the engines a more uniform rate of speed. All the improvements that experience has suggested, and will suggest, must give way under the effect of overrunning. The only way of avoiding this expense is to make a stand at some given speed, I care not how high it is, so long as the present engines can do it."

The writer was in communication with Mr. Locke prior to sending this report, and during the following seventeen years acted on the advice, though of necessity the strict letter had to be varied with the change of time and circumstance.

Engines were constructed, weighing about 10 tons, with two outside cylinders, each 12½ inches in diameter, supported on six wheels, the two driving wheels being 5 feet in diameter: the average load of a passenger-engine between Liverpool and Birmingham was ten carriages.

In the following year the trains averaged twelve carriages: engines were made to suit the increased work by giving a cylinder of half an inch more in diameter. Again, in 1843 the trains increased to fifteen coaches; another quarter of an inch was added to the diameter of the cylinder, demanding a little more boiler space, and a little more weight on the driving wheels, the more so as the writer was allowed to risk an increase of steam pressure from 50 to 60 lbs. on the inch, to meet the increased size of the driving wheels to 5 feet 6 inches, that the speed might be increased. This improved locomotive had a 13¼-inch cylinder, weighed 12 tons, on six wheels, and would take a train of sixteen passenger-carriages between Birmingham and Liverpool, with the precision of clockwork, at a speed, including stoppages, of thirty miles an hour. In 1844 the driving wheels were again increased to 6 feet; and in 1845, with increasing traffic and speed, and to surmount the Lancaster and Carlisle sharp incline of 1 in 75 for three or four miles, the cylinders were increased to 14⅝ inches diameter, 20-inch stroke, 6-feet driving wheels, and steam of 75 lbs. on the inch; this engine was not much heavier, neither did it work with a higher pressure of steam than Trevithick's London locomotive of 1808, which ran on sharper curves at fifteen or twenty miles an hour, worked by one outside cylinder 14½ inches in diameter, 4-feet stroke, and steam of 100 lbs. on the inch. Such was the slow progress of the locomotive engine.

One of these good little engines of 1845 gave special proof of efficiency. About the year 1846, on a rainy, blowing, autumnal Saturday night, the writer was summoned, from nursing an influenza cold, to the railway station. Her Majesty, Prince Albert, and the rest of the Royal family, had unexpectedly arrived, and desired to be in London by ten the following morning. Continued rain had caused the line to be unsafe in places, except at comparatively slow speeds. Saturday night is proverbially a bad time for finding people wanted in a hurry. However, at six the next morning, in dim light and blinding rain, the Royal train was in readiness, and Her Majesty punctual to the minute, when, after a little animated delay for the lady in waiting, a start was made, and the required speed of forty miles an hour steadily run, until a providential disobedience of orders by the pilot-engine man caused the steam to be instantly shut off, the brakes applied, and the speed reduced to one-half; fog signals exploded in close proximity to the danger; red flags were hurriedly unfurled, and in a moment the engine rolled as a ship in a storm through an alarmed group of a hundred navvies, who, thinking it a quiet day, had raised the rails and sleepers a foot above their bed of soft clay, that a thick layer of ballast might be shovelled under them. For a quarter of a mile did the precious freight pass safely over this bridge of rails supported on brickbats, the only injury being a bent driving axle and broken bearing-brasses, with which the engine kept time to the next relieving station, and then broke down. I believe it was in the same year that the writer, with one of the same engines, took Sir Robert Peel safely over the same railway, before it was completed or open to the public, on his way from Tamworth to Liverpool to deliver his memorable speech on the Corn Laws. That class and size of outside-cylinder engine remained in use for ten years, when in 1856 the growing demand for greater speed caused the driving wheels to be increased to 7 feet and the cylinder to $15\frac{1}{4}$ inches in diameter, retaining the old stroke of 20 inches, but increasing the size of the boiler, using 150 tubes, and giving steam of increased pressure to 100 lbs. on the inch: the gross weight was 20 tons.

Plates VI. and VII. show London and North-Western Railway outside-cylinder passenger locomotive, 1856.

In 1867, a similar locomotive in the Paris Exhibition, from the French engineering works at Creuzot, was labelled Schneider's prize engine. An exceptional engine was built in 1847 to refute a dogma of the broad-gauge advocates that the narrow gauge had reached its limit of speed, because the driving wheels could not be safely increased in diameter. A narrow-gauge engine was therefore constructed by the writer with 8 feet 6 inches driving wheels, being 6 inches more than the largest broad-gauge wheels. This engine still continues to run express trains, the only change being in a more modern boiler. It was sent to the London Exhibition of 1851, and some years afterwards the late Mr. Fairbairn congratulated the writer on the medal awarded for it, and accounted for its non-presentation from the question

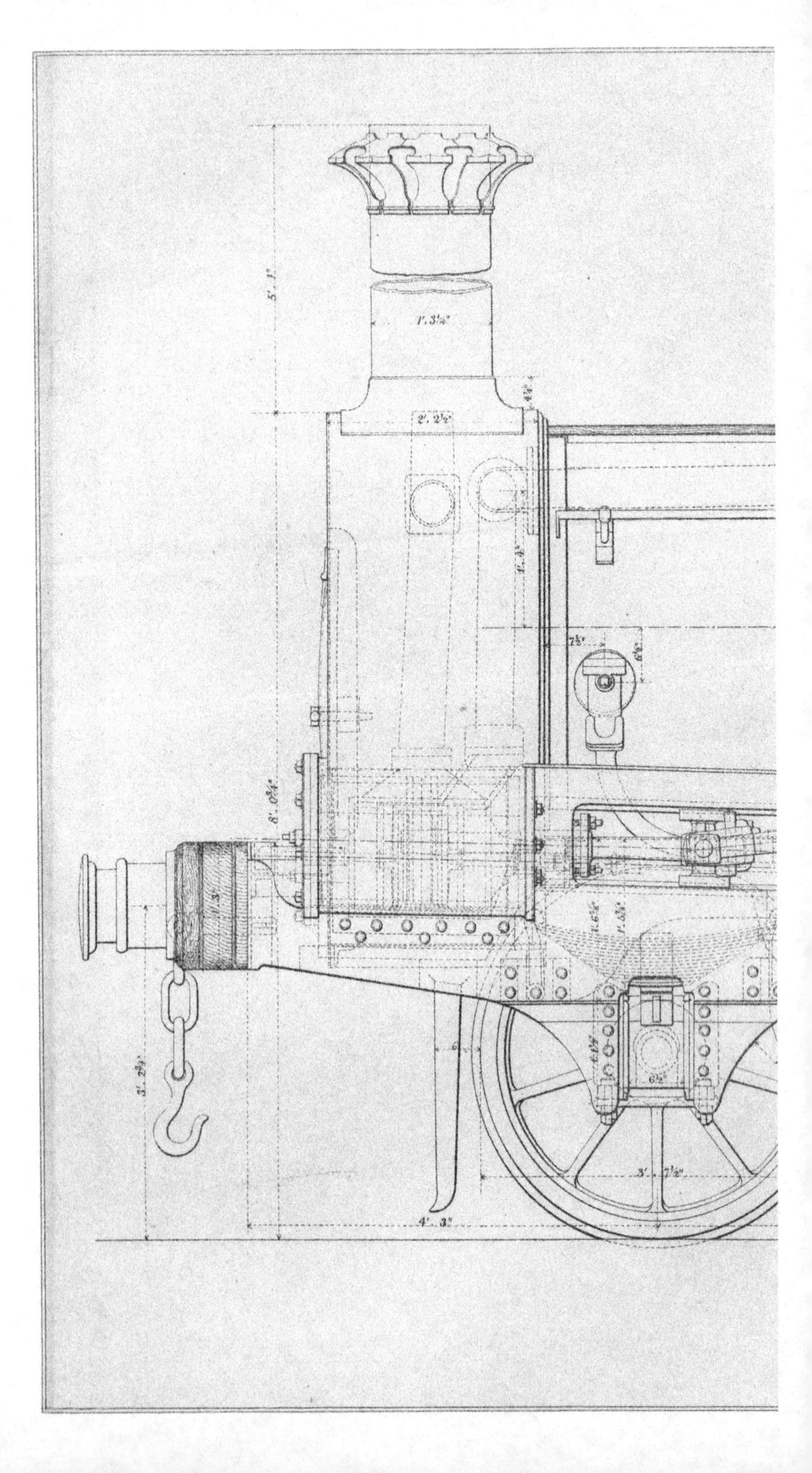
1'. 3¼"
2'. 2½"
6½"
3'. 7½"
4'. 3"

LONDON AND NORTH WESTER

PASSENGER ENGINE, 7 FEET DRIVING WHEE

London: E. & F. N. Sp

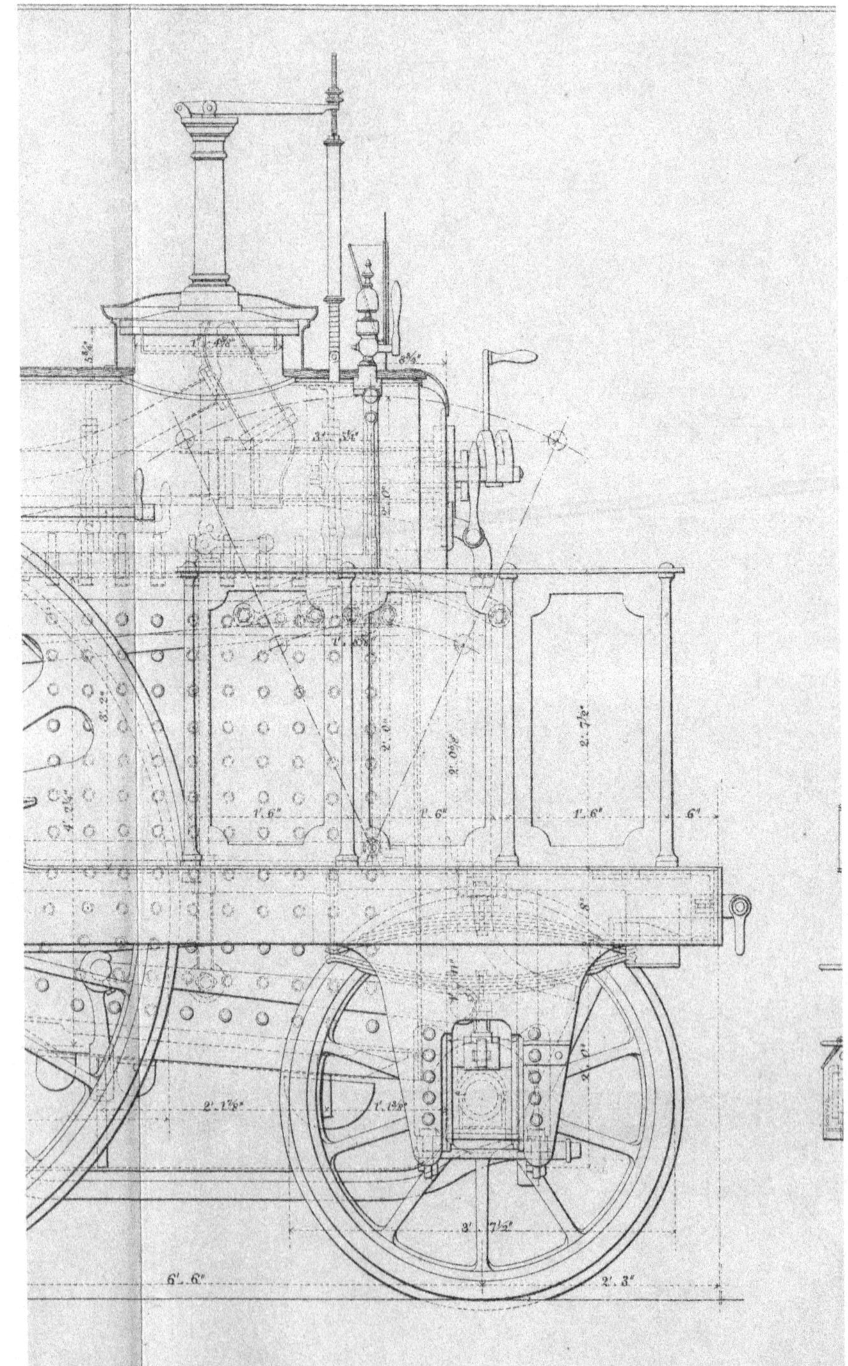
N RAILWAY, CREWE WORKS.
L, 15¼ INS CYLINDER, 20 INS STROKE _ 1856.
n, 48, Charing Cross.

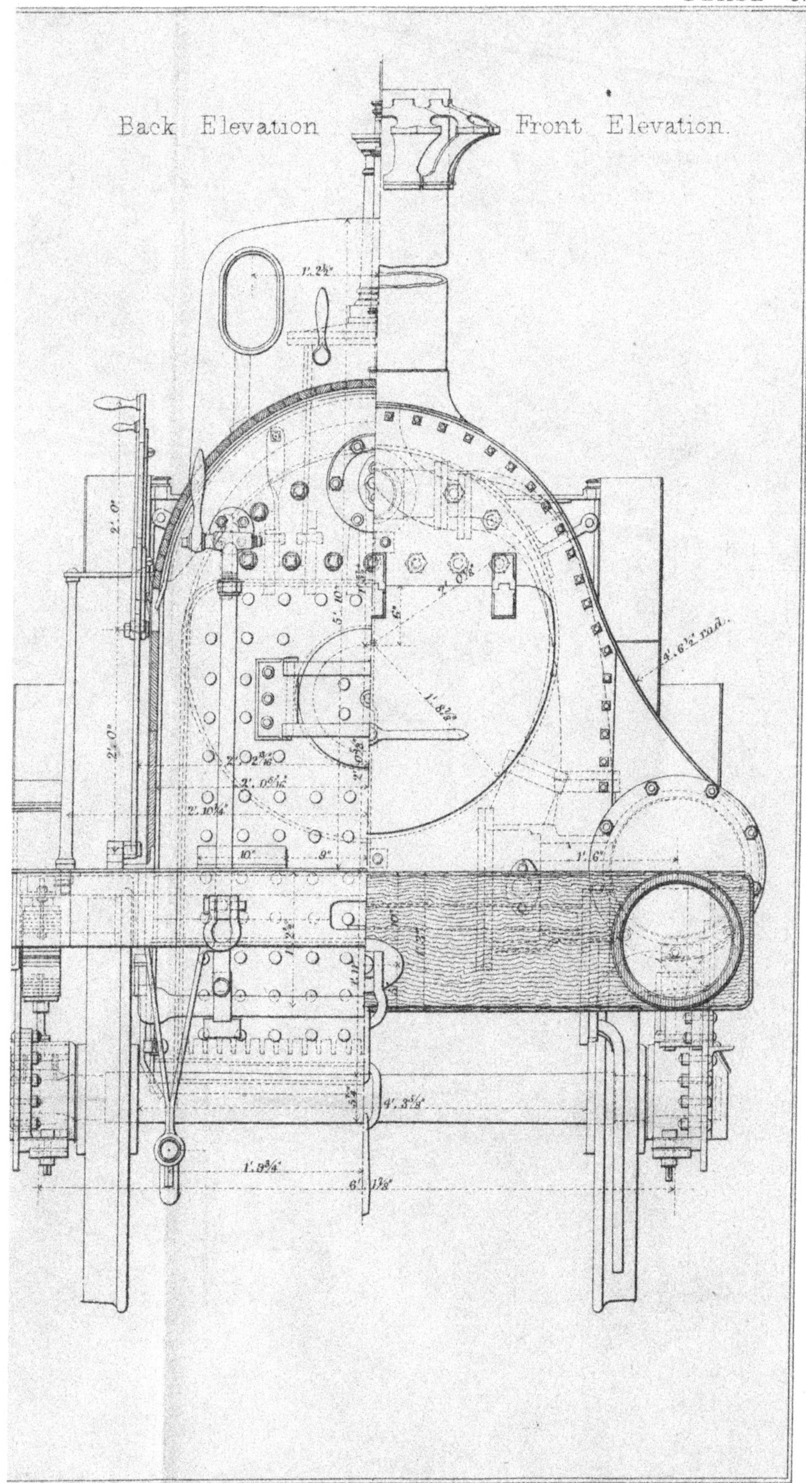

Kell, Broˢ. Lith. London.

having arisen of whether the medal should be awarded to the Railway Secretary, whose name was officially attached, or to the designer, whose name was not officially attached.

Having slightly traced the outside-cylinder locomotive engine from its starting into life in Camborne in 1801 through its chrysalis stages of common road, tram-road, and railway engine up to 1808, then through twenty years of restless sleep to the Liverpool and Manchester period of outburst into general usefulness in 1829, with the prize 'Rocket,' weighing 4¼ tons, on four wheels, having outside cylinders 8 inches diameter, 18-inch stroke, 2 feet 7 inch driving wheels, twenty-five tubes in the boiler, giving steam of 50 lbs. on the inch. Then came during sixteen years a time of steady growth up to 1845, when it scarcely exceeded its weight, size of cylinder, or steam pressure of thirty-seven years before, while the improved form of 1856 is still in use. Shortly after the 'Rocket' period, cranked-axle engines came into use on many railways, but as the Trevithick locomotives were all outside cylinders, that kind alone has been spoken of.

The clear practical understanding of giving motion to a carriage by the friction or grip of its wheels on the road, and of the best construction of a road for such a carriage, was almost as slow in growth as that of the more complicated locomotive engine. In 1801 Trevithick and Davies Gilbert tried experiments on the grip of a wheel on common roads;[1] this was immediately followed by two or three years of comparatively successful experiments with common road locomotives. In 1802 Trevithick wrote, "The Dale Company have begun a carriage

[1] See Davies Gilbert's letter, April, 1839, chap. vii.

at their own cost for the railroads."[1] In 1804 the tram-plate locomotive road was well tested in Wales. "I doubt not we could draw 40 tons at a time very well; 10 tons stand no chance at all with it. It runs up the *tramroad* of 2 inches in a yard, forty strokes per minute with the empty waggons."[2] In 1808 the London locomotive on a railway passed around curves of 50 or 100 feet radius, at fifteen miles per hour: such gradients and curves, manageable with Trevithick's locomotives, were thought impossible thirty or forty years after by engineers of ability.

Tramways and railways for horse-draught were in use before Trevithick placed his locomotive on them, yet he took the improvement of the road in hand, for he says in 1808,[3] "the ground was very soft, and the engine sank the timbers under the rails and broke a great number of them. I have now taken up the whole of the timber and iron, and have laid balk of from 12 to 14 inches square down on the ground." This Trevithick's railway of 1808 is somewhat like the Great Western Railway of the present day. In 1838 the late Mr. Brunel gave the writer a sketch of the permanent way, desiring him to construct a piece of about half a mile in length near Wormwood Scrubs, as a sample of the intended form of construction for the Great Western Railway; the timbers and rails of which remind one of those used thirty years before by Trevithick, except that Brunel attached his longitudinal timbers to piles and cross-timbers, which latter were after a short time removed.

Practical men are too apt to leave facts unrecorded. Some twenty years ago, railway competition caused an

---

[1] See Trevithick's letter, August 22nd, 1802, chap. ix.
[2] See Trevithick's letter, February 20th, 1804, chap. ix.
[3] See Trevithick's letter, July 28th, 1808, chap. ix.

increased speed in the express trains between London and the North. Frequent notes from the manager urged the locomotive superintendent to actively carry out the wishes of the directors. To do this, the superintendent stood by his engine and fire men on a journey south from Carlisle. At the Lancaster station, when the engineman tested the state of the bearings by a slight touch of the fingers, the leading axle-box caused them to emit a smell of burnt skin: buckets of cold water and some grease were hurriedly applied, and, at the guard's whistle, the train proceeded. Before many miles had been run, the sounds of grinding friction gave warning of danger, followed by spurts of blue and white flame, with fizzing sparks from the axle-box. Holding out a little longer would bring the train to the Preston station without loss of time, where another engine was in waiting; the fireman instinctively stood near his brake-handle; the engine-man, with his hand on the steam-regulator, watched anxiously the course of events; time had been kept; the innocent passengers went on rejoicing; and the directors, almost as ignorant as the public, pursued their policy of hard running. The superintendent, on examining the engine, found that the bearing of the leading axle had been raised by friction to a welding heat, causing it to be wrenched from the axle, close to the shoulder or nave of the leading wheel; a small roundish knot, projecting from the shoulder, alone retained it in its place, while the torn-off bearing was imbedded as a solid mass with the fused brass and iron of the axle-box.

About that time, the broad and narrow gauge competition on the extension of the broad gauge to Cheltenham, caused the Great Western Railway directors to travel from Paddington to Cheltenham and back in a special train drawn by their new 8-feet wheel engine. The broad-gauge superintendent invited a narrow-gauge superintendent to ride on the engine with him. On the journey to Cheltenham, on a glorious day, a rate of fifty-five miles an hour was run with comparative ease, sixty miles an hour with difficulty, and sixty-three miles an hour was the extreme limit. The dinner and speeches at Cheltenham were highly approved of, and the specials started on their rapid home journey. On rushing toward the West Drayton station, through blinding darkness, the broad-gauge superintendent hurriedly said, "What's that?" and closed the steam-regulator. A brief reply caused it to be again opened, and a sound, as from compressed space, together with a momentary glimpse of station-lights, indicated that a station had been passed. On inquiry the next day, it appeared that on the approach of the special train, a truck was being removed from the main line to a siding; as it cleared the points, the red signal was turned off; at the instant it had been seen as a confused sensation by those on the flying engine, which in another thirty seconds of time thundered by within an inch of the truck pushed by the station-men, who gazed on the receding tail-lights as scared men reprieved from annihilation.

Many such hairbreadth escapes could be told by those who live on locomotives, and their narration might have checked the headlong race for speed, even when railway accounts were jobbed to cover the growing wear and tear of permanent way and stock from the ever-increasing weight and speed of the engines; not, as the public suppose, for their comfort in saving an hour in a day's travel, but rather that profit may be made by successful competition. There has been a departure from Trevithick's story; we must again seek him with his favourite high-pressures.

CHAPTER X.

PARTNERSHIP, AND EARLY HIGH-PRESSURE ENGINES.

"TREDEGAR WORKS, *June* 22, 1854.

"DEAR SIR,

"I enclose you a tracing of the old puddling engine we have here. I am sorry my young man could not get the motion complete, but as you have better draughtsmen you perhaps could get them to finish one from this. The centres of the parallel motion are all given, and the dimensions of the cog-wheels; there is no beam, only a cross-head, from which there is a connecting rod on each side of the cylinder, down to the crank of the driving wheel. The side beam of the parallel motion is a light casting.

"Yours truly,

"THOMAS ELLIS.

"F. TREVITHICK, Esq."

"About the year 1800 Mr. Trevithick came to Penydarran to erect a forge-engine for the company. I was at this time over-looking the engines at Penydarran. I assisted Mr. Trevithick in the erection of the forge-engine. When this engine was finished Mr. Trevithick commenced the construction of a loco-motive. She was used for bringing down metal from the furnaces to the old forge. After working for some time in this way, she took a journey of iron from Penydarran down the Basin Road. On the journey she broke a great many of the tram-plates. After this she was used as a stationary engine, and worked in this way several years. The boiler was made of wrought iron, having a breeches tube; the cylinder was about 4¾ inches in diameter; the steam pressure about 40 lbs. to the inch. The steam from the cylinder was discharged into the stack."[1]

[1] Recollections of Rees Jones, see 'Mining Journal,' October 2nd, 1858.

"DEAR SIR, "*26th December*, 1868.

"The mill and puddling engine at Tredegar were made by Mr. Aubry and my father at Penydarran. There was another at Llanelthy, near Abergavenny, and several about the collieries. My father was employed in Penydarran from 1800, when Mr. Trevithick put up the puddling engine and a blast-engine: there was also a winding engine worked by old William Richards; he continued to work it for forty years.

"Yours truly,

"THOMAS ELLIS.

"F. TREVITHICK, Esq.

"P.S.—The Penydarran winding engine was taken down in 1855; the Penydarran puddling engine in 1854; the Tredegar puddling engine in 1856."[1]

The variety and practical perfection of mechanical combinations in Trevithick's patent of 1802 are in some measure due to his having before that time proved their suitability.

His Cornish high-pressure steam portable engines of 1798 and 1799 were extended to South Wales, one of which was erected about 1800 or 1801, for giving motion to the large hammer or rolls or other work in the puddling mill of Tredegar Iron Works; remaining at work in 1854, when Mr. Ellis gave an outline description of it.

Rees Jones, who worked as an engine-fitter at Penydarran in 1794, says that Trevithick erected a forge-engine there about the year 1800, and having finished it, commenced his locomotive tramroad-engine. This latter having worked for some time as a locomotive, was removed from the tramway and used as a stationary engine. The boiler was of wrought iron, with an

[1] Letter from Mr. Thomas Ellis, 4th March, 1869.

internal breeches or return fire-tube, working with steam of 40 lbs. to the inch; discharging the waste steam into the stack.

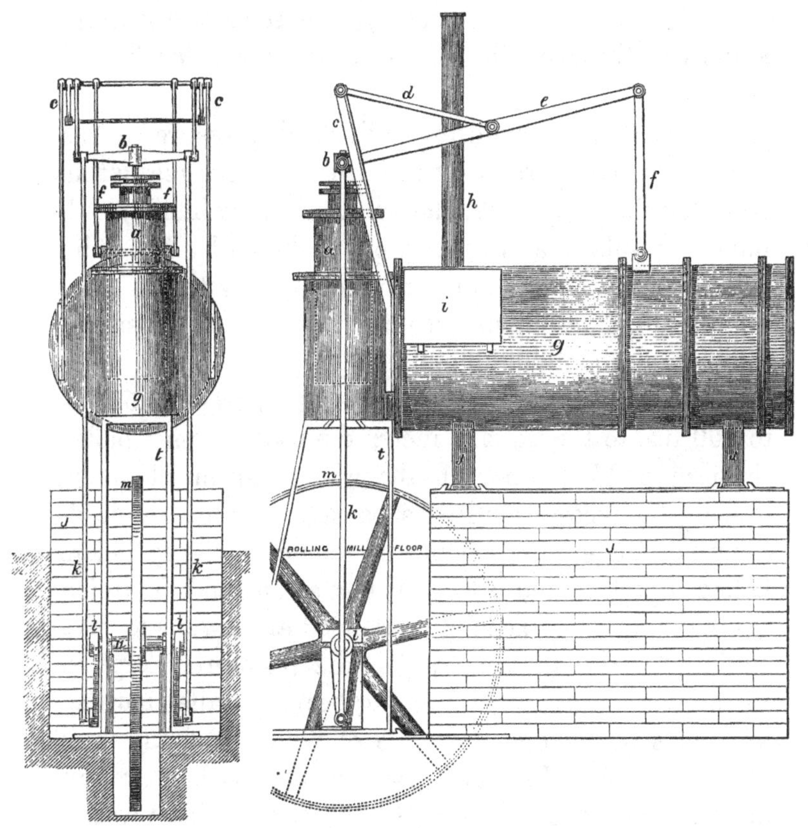

TREVITHICK'S HIGH-PRESSURE STEAM-PUFFER ENGINE, ERECTED AT TREDEGAR PUDDLING MILLS, 1801.

Detailed description of the Tredegar puddling-mill engine:—*a*, the steam-cylinder, 2 ft. 4 in. diameter, 6-ft. stroke, fixed in the end of the boiler; *b*, the piston-rod and cross-head; *c*, supports for the radius-rod bolted to the boiler; *d*, radius rods; *e*, parallel-motion beam; *f*, rocking beam; *g*, cylindrical cast-iron boiler, 6 ft. 9 in. diameter, 20 ft. long, in pieces bolted together (the drawing shows the boiler a little shorter, to suit the paper); *h*, chimney into which the waste steam is puffed as blast; *i*, cistern for heating the feed-water, the blast-pipe going through on its way to the chimney; *j*, foundation and stands for boiler and engine to bring it to the required level for the rolling-mill floor; *k*, connecting or side rods: *l*, cranks; *m*, tooth-wheel, from which motion is given to fly-wheel shaft and machinery; *n*, crank-shaft.

Note from Mr. Ellis:—The tube in the boiler was of wrought iron, 3 ft. in diameter at the fire-door end, becoming less in diameter from the bridge at the inner end of the fire-bars, turning back again at the other end of the boiler, and joining the chimney at the fire-door end, where it was 14 in. in diameter: the steam was worked at any pressure between 50 lbs. and 100 lbs. to the inch, and was puffed through a blast-pipe up the wrought-iron chimney, heating the feed-water in its passage through the feed-cistern. The steam was directed in the right channels by the four-way cock, and another cock served as a steam-regulator. This engine was taken down in 1856.

Mr. Ellis' puddling engine at Tredegar, the Penydarran puddling engine, a blast-engine, a winding engine worked by William Richards for forty years, an engine erected at Llanelthy, and several about the collieries; these continued in constant use for fifty or more years.

The Tredegar puddling-mill high-pressure puffer-engine was comparatively of large size; the regulating cock and four-way cock, blast-pipe in the chimney, feed-pump, fire-place, and flues, are not shown by Mr. Ellis, and therefore I have not ventured to draw them. The practical engineer will readily discern the positions they occupied.

The wrought-iron fire-tube; the steam pressure 50 lbs. to 100 lbs. to the square inch; the used steam "puffed through a blast-pipe, up the wrought-iron chimney, heating the feed-water in its passage through the feed-cistern;" the external boiler of four cast-iron cylinders bolted together, and a kind of box at one end, in which the cylinder was fixed; a cross-head and two side rods giving motion to the cranks, the piston-rod kept in its proper course by a rocking-beam parallel motion, are parts of a whole, not one of which is found in a Watt engine. In 1837 the writer lived near this engine, and frequently saw it at work; the blast was as useful as in the locomotives then in the writer's charge. The old engineman said it had never been altered since its first erection, and had never given any trouble or required any repairs worth mentioning, though working night and day; the boiler was easy to fire and gave plenty of steam; one engineman attended to the varying demands of power and steam supply. The same general principles may be traced in the drawing of the 'Catch-me-who-can' of seven years later, the most marked difference being that the locomotive had piston-rod guides as they still have, while the puddling-mill engine had a rocking-beam parallel motion.

"We have an account of a trial of a small high-pressure engine made in 1804, in Wales, to ascertain its powers to raise water. The cylinder was 8 inches in diameter, and 4½-feet stroke. It worked a pump 18½ inches in diameter, and 4½-feet stroke, which raised water 28 feet high. It worked at the rate of eighteen strokes per minute, and consumed about 80 lbs. of coal per hour. This when reduced is about 17½ million pounds raised 1 foot high for each bushel of coal.

"Many provisions have been made to guard against the bursting of high-pressure steam-boilers by Mr. Trevithick, who first brought the high-pressure engines into use. At first he proposed enclosing the safety-valve in such a manner that no one could get access to it to increase the load beyond what was intended to be employed. Secondly, he drilled a hole in the boiler, which he plugged up with lead, at such a height from the bottom that the boiler could never boil dry without exposing the lead to be melted, and consequently making an opening for the steam to escape."[1]

The engine described by Rees was very similar to the locomotive referred to in Trevithick's letter as applicable to various purposes,[2] for locomotion or for pumping, working with 80 lbs. of steam to the inch, having a locked safety-valve, and soft-metal plug on the boiler. A portion of the boiler of one of those engines may be seen at the Kensington Museum.[3]

Mr. Crawshay believed that Trevithick's Welsh tramroad-engine had a wrought-iron boiler and two cylinders, and certainly such an engine was constructed at Penydarran about that time: whether or not it was

---

[1] See Rees Cyclopædia,' published 1819, under the word Steam-Engine.

[2] See Trevithick's letter, February 20th, 1804, and March 4th, chap. ix.

[3] See chap. ix.

first used as a locomotive, it worked for some twenty years as a hammer-mill engine, and then for about thirty years as a pumping or winding engine in sinking shafts in coal-mines, after which, in 1850, a plate was cut out of the wrought-iron boiler, having two holes which served for fixing the cylinders, clearly showing the use by Trevithick at that early date of double-cylinder engines.

Rees's account of one of Trevithick's high-pressure puffer engines erected in Wales to test its real and economical power as compared with the Watt low-pressure vacuum engine, shows that people did not believe Trevithick,[1] or comprehend his engines. They could trust in an atmospheric or a Watt condensing engine of large size and cost, but not in a small thing put together by inexperienced workmen. Trevithick replied, "I'll show you that it is so." The locomotive engine was turned into a pumping engine, that they might measure the work done. The economical duty of this high-pressure steam-engine as compared with the Watt low-pressure vacuum engine is spoken of in another chapter.[2] A small portable engine with an 8-inch cylinder and 4 feet 6 inch stroke worked a large pump of 18½ inches in diameter and 4 feet 6 inch stroke, raising the water through a column 28 feet high; to perform this required a pressure of steam of from 80 to 100 lbs on the inch in the cylindrical boiler.

In this history of sixty or seventy years ago, applying only to one field of Trevithick's labours, we have puddling-mill engines, forge-engines, blast-engines, winding engines, pumping engines, locomotives, plan-

---

[1] See Trevithick's letter, August 22nd, 1802, chap. ix.
[2] See chap. xx.

ishing-hammer engines, and shaft-sinking engines, with boilers of wrought iron and of cast iron, giving steam up to 100 lbs. on the inch; double-cylinder vertical engines, and single-cylinder horizontal engines, working with beam and parallel motion or guides, with blast-pipe in the chimney—all more or less portable, and in no particular similar to the Watt engine.

While the foregoing was being carried out in Wales, high-pressure engines were also erected in other places.

			£	s.	d.
" 1803.[1]	April	9th. To premium on Wm. Kinman's 6-horse, charged at 4-horse	50	8	0
	May	2nd. Miller and May, 5-horse	63	0	0
	"	" George Russell, 8-horse	75	12	0
	"	23rd. Jas. and Edwd. Sward, 8-horse	75	12	0
	June	4th. Lloyd, 6-horse	75	12	0
	"	" Butt, Jamieson, and Co., 6-horse..	75	12	0
	"	" Anthony Harman, 3-horse	47	5	0
	May	2nd. Josias Spoils, 12-inch cylinder, for Staffordshire	150	0	0
	"	23rd. T. Turton, Esq., 20-inch cylinder, to be erected in Staffordshire	315	0	0
	"	28th. Lord Dudley and Ward, for a whimsey	420	0	0
	June	6th. John Morris, Esq., a whimsey	262	10	0
	Aug.	6th. General Bertham, for Deptford Dockyard, 14-horse, to be erected as per agreement	750	0	0"

In 1803 the Government adopted the new invention, Lord Dudley erected a high-pressure whimsey within sight of Soho, two others were erected in Staffordshire, and many of Trevithick's first high-pressure steam-engines were made at Coalbrookdale, Bridgenorth, and Stourbridge, not far from Soho.

The earliest of Mr. Watt's steam-engines giving rotatory movement were erected in 1784:[2] those used the sun-and-planet wheel; and though the crank in a steam-engine had just then been patented, Watt's

[1] Extract from Andrew Vivian's account-book.
[2] 'Mechanics' Magazine,' August 30th, 1823.

opposition to apply it in engines made at Soho caused its use in a steam-engine to be an unknown thing in Cornwall when Trevithick constructed his first high-pressure models with cranks. The difference between a low-pressure vacuum Watt rotatory engine and a high-pressure steam Trevithick was, that the former had stone foundations, beam and parallel motion, condenser and condensing water, and a large boiler. The latter had none of these things; it was self-contained and portable, and but one-third of the size or cost of the low-pressure of equal power.

Trevithick, Andrew Vivian, and William West were partners in the patent of 1802. Trevithick was the general correspondent, ready for all comers, promising advantages, talking away difficulties, and supplying engines from strange, untried workshops, to be paid for when convenient. Vivian was the commercial man. West was a sensible, steady-going mechanical man, ready to make good his promises to the letter, but somewhat obstinate. Such men under such circumstances could not continue to work in agreement. The large sphere of application of the high-pressure portable engine could not be grasped by three men destitute of workshops, or satisfactory means of construction or supply; and the patent laws were no better than a broken reed for support.

Vivian's account to November, 1804, shows that in the short space of two years since the date of the patent they had received 1250*l.* as patent premium on the new high-pressure engines erected. The expenses had been 1097*l.* Mr. Samuel Homfray had apparently just joined the company.

Extracted from Andrew Vivian's accounts:—

"Dr.				Cr.		
	£			£	s.	d.
Premiums received	1250	1804.	Expenses	887	1	7
		Nov. 25.	By allowance for time to this day	100	0	0
		„	By on account of disbursements since Mr. Homfray is concerned, to be settled next account	100	0	0
		„	By Mr. Harvey, bill for a new cylinder, &c.	9	18	9
				1097	0	4
			Balance	152	19	8
				£1250	0	0

Richard Trevithick two-fifths,
Andrew Vivian two-fifths,
William West one-fifth."

In May, 1805, Vivian was in London negotiating the sale of his share in the patent. Trevithick, then occupied in breaking away the rock in the Thames near Blackwall, was to give Sir William Curtis another offer of Vivian's share for 4000*l.*, with an additional 1000*l.* if the threatened patent lawsuit with Dixon went in favour of the patentees.

"Mr. Trevithick, "Camborne, *May* 22, 1805.

"Dear Sir,—I have yours of the 18th instant, for which I am much obliged. It found me in a most melancholy situation. I returned from London here on Tuesday evening, between five and six o'clock, and found my poor Andrew much weaker than when I left him. He rejoiced to see me, but soon told me that we were soon to part to meet no more on earth; and after taking the most affectionate leave of all, he said he had given up the world, and all that was therein, and resigned himself to will of that God who gave him life. About three or four o'clock the next morning the Almighty was pleased to take him. The dear boy was perfectly in his senses, and appeared to leave the world without pain. This has been a sad stroke to me, and have scarcely been able to write a letter since. My poor wife is still very unwell, and so is the infant child.

"Had I not been so much in want of money at this present time, would not part with my share in the patent for the sum you have offered it at; but my circumstances at present oblige me to do what, in other circumstance, I would not.

"If there is any risk at all in going to law with Dixon, why not avoid it? It is certainly very easy to make a friend of him, as the non-existence of the patent can be but of little consequence to him.

"My finances at present oblige me to empower you to make another offer to Sir William Curtis; that is, my share as it now stands, for 4000*l*., and to receive the other thousand if you succeed against Dixon. But should it not come to a trial, the last-mentioned thousand to be paid at the expiration of one year.

"I have not seen my brother Henry since the receipt of yours, and shall write you on the other parts of your letter in a few posts.

"And am, dear Sir,

"Your most obedient servant,

"ANDw VIVIAN.

"Your dear family are all in good health. I hope my dear friend will do all in his power to serve me, and that if money cannot be obtained for my share, he will do his best endeavour to get about 2000*l*. some other way. Farewell.

"Should you want my assistance on the Rock, I will willingly assist you merely as a friend, without sharing the profits."

"MR. TREVITHICK, "CAMBORNE, *July* 1, 1805.

"Dear Sir,—I have yours of the 27th instant, and am much grieved at the relation of all your distresses, and for my own part, will not agree to give that rascal Davey an inch; and must beg of you not to make it up but on very good terms, for I would rather lose 50*l*. than the fellow should go unpunished. But shall leave it entirely to you, and have no objection to bring it into court again; but hope that will not keep you in town; if it is likely to, would rather you should make it up, but certainly he must pay very handsomely for your false imprisonment. *You are a gentleman of too much consequence in the world to be trifled with, and your time must be valued high.*

"Your dear family are all well and anxious to see you, and mine also.

"Dolcoath is better and better, but cannot say so of Binner Downs, though I believe more than paying cost. I suppose Cardell and Co. have a demand on us, but suppose that will be demanded through Edwards. I am not sorry that Rabey has given up the trial of the patent; and I think it would be right for us to propose to refer the business of his demand and our set-off to some one, two, or three respectable persons. This you may do without advising with Mr. Homfray, as he is not concerned in this part of the dispute, and the other demand for patent premium may be left out of the present question. Pray give this a full thought, and act upon it as your better judgment may dictate. With hopes of hearing from, and seeing you soon,

"I remain, dear Sir,

"Your obedient servant,

"ANDW. VIVIAN.

"Can you with any propriety say anything to Sir William Curtis about purchasing my share now Rabey has given up thoughts of attacking the patent?"

The difficult task of introducing a new thing was increased by legal questions in maintaining the patent right. Rascal Davey had had a trial with Trevithick, and put him in prison, but apparently on false grounds, and therefore ought to be made to pay handsomely for so trifling with a man of such great consequence to the world at large.

Rabey also threatened to contest the patent right for the high-pressure steam-engines. Homfray was then an interested party, but Vivian thought Homfray had nothing to do with the particular engine on which Rabey had raised a question.

This was one of the weaknesses of so all-embracing a patent. The high-pressure steam-engine was new, and so were the particular engines drawn and described in the patent; but Trevithick adapted them to all requirements, many of them of a form and detail not shown in the patent: such a course opened a door for litigation, even though the engines were covered by general terms in the patent, for such and similar purposes, and all on the principle of high steam.[1] Dixon refused to pay patent right, because "the words in Mr. Watt's specification are enough to indemnify him."

The supplying engines to all applicants, through numerous makers, and trusting to their honesty or judgment for payment of the patent premium, continued for a year or two longer, when, in 1807, Mr. Homfray, who had from the first Welsh experiments a watchful eye, if not a pecuniary interest, in Trevithick's labours, came forward as a leading shareholder, having bought Vivian's interest, and negotiated for Trevithick's.

The following papers, illustrating this arrangement, were kindly supplied by my friend Mr. Bennet Woodcroft.

"GENTLEMEN, "PENYDARRAN PLACE, *May* 31, 1807.

"On my return home I find your favour of the 25th inst. It's impossible for me to say if Mr. Trevithick has done any act to encumber the property he has in the engine concern, over which I hold the control by a deed of assignment.

"So far as concerns me I can have no objection to his disposing of his share, provided I am not injured, and placed in a worse situation. I have no wish whatever to prevent him from selling and making the most of his share. There is an

[1] See Trevithick's letter, January 5th, 1804, chap. xvii.

account now not closed between me and him which, by calling upon Mr. Bill, at No. 49, Rathbone Place, he can settle with Mr. Trevithick. There is likewise a note-of-hand of Mr. Trevithick I now hold for 300*l.*, dated 4th January, 1804, which I shall of course set-off against his account of patent-right as it comes due and I receive it. For any information you may wish respecting any legal forms necessary for the completion of Mr. Trevithick's sale, on my part I beg leave to refer you to Messrs. Strong, Still, and Strong, Lincoln's Inn.

"I am, Gentlemen,

"Your most obedient servant,

(Signed) "SAM^L. HOMFRAY.

"MESSRS. BLAGRAVE AND WALTER.

"N.B.—The account I believe is a small one, and easily adjusted; and the note-of-hand I could endorse to your clients, which they could charge to Mr. Trevithick as so much of the purchase-money, and they pay that sum to me.—S. H."

"MR. HOMFRAY, "LONDON, *June* 26, 1807.

"Sir,—The transfer deed is ready for executing, and as it is necessary that you should put your signature to it, have sent it you for that purpose. Hope you will be so good as to execute it and forward it to Mr. West for his signature also. I owe you some patent premium, and wish to settle my account with you as early as possible, therefore would thank you to send me a copy of your account with me from the beginning, and authorize Mr. Bill to settle it with me on your behalf. Please to forward this business, as I am about to leave town.

"I am, Sir, your very obedient servant,

(Signed) "R^D. TREVITHICK.

"*Plough Inn, Kidney Stairs, Limehouse.*"

"SIR, "LONDON, *September* 11, 1807.

"This is to certify that I have sold to you two-fifths of a sugar-mill, now on the premises of Mr. Geo. Bowdeys,

Blackfriars Road, the above share being all my interest in it, for which I have received from you a valuable consideration.

"I am, Sir,

"Your very humble servant,

(Signed) "RICH[D] TREVITHICK.

"MESSRS. HAYNES AND DOUGLAS,
"*Tottenham Court Road.*"

Mr. Samuel Homfray was an influential man of business in the Welsh Iron Works, and had taken an active part in the construction and trial of the Welsh locomotive and other of Trevithick's engines.

The accounts to February, 1807, show that the patent premiums received a little more than covered the expenses. To February, 1808, a profit of 254*l.* 15*s.* 5*d.* was to be divided among four shareholders: Mr. Homfray, five-tenths; Mr. Trevithick, two-tenths; Mr. Bill, two-tenths; Mr. West, one-tenth. Andrew Vivian's share had been purchased by Mr. Bill.

Homfray went to various places to make the high-pressure engines known, lawyers' bills were paid, amounting to a quarter part of the whole premium-money received. Mr. Homfray hesitated to affix his name to the required legal documents; and on his still declining, Trevithick sold to Messrs. Haynes and Douglas his two-fifth share of the patent in a sugar-mill.

"GENTLEMEN, "PENYDARRAN PLACE, *April* 23, 1808.

"In a letter I have received from Mr. Bill, he mentions you have not received the letter I wrote from Messrs. Strong and Still's to you, which fully answered your wishes as to my not calling upon you for more money for supporting the patent than you had expressed to Mr. Bill, and which he mentioned to me. The letter answers every purpose as though I had signed the deed. If you have not received it, some mistake must have happened, and if you will call at Messrs. Strong and Still's, they will show you the copy of it.

"I have an objection to the signing of deeds where a number of parties are concerned if it can be avoided (not more to this than any other), and especially where there appears no kind of necessity for my signing.

"From letters I have received from Mr. Bill, who is now on a journey, I am in hopes many engines will be ordered, and that we shall be put in a train of getting some return of money for the great trouble and expense I have been at in introducing the engines, which are now beginning to be generally understood. Our large engine here goes on wonderfully well; and at any time it suits your convenience to come into this part of the world, I shall be happy to show it you.

"I remain, Gentlemen,

"Your most obedient servant,

"SAMUEL HOMFRAY.

"MESSRS. HAYNES AND DOUGLAS."

"PENYDARRAN PLACE, *24th September*, 1808.

"GENTLEMEN,

"I feel some concern on account of the state of Mr. Bill's health, which, when he wrote last to me from Deal, was so indifferent that, by the advice of a physician, he was down there for sea-air; and he mentioned that if his health did not improve on his return to London in a week's time, that he thought he should decline the conduct of the engine concern. I have written several letters to him, and wished for some information in answer, and as he is silent, and the time of his return to London passed, I am fearful he is very ill. I therefore take the liberty of addressing you (being interested in the concern), and will be obliged by your informing me if Mr. Bill is in London, and if so, what state of health he is in; and if he has mentioned to you that he has thoughts of giving up the agency. I wish, likewise, you to say, if there is anything in the report of the Racing Engine being carried into effect, and if so, at what time, and the particulars of the bet, &c., as if it is to take place, I shall be very much inclined to see it. I have the satisfaction to say all the engines go extremely well; and the packing of our large engine has not been renewed for upwards of four months, and it is unknown how much longer it will go, it being now as good as new. By letting a small quantity of water drop constantly into the stuffing box, it runs down the rod upon the piston and keeps the packing moist. I have written this to Mr. Bill.

"I am, Gentlemen,

"Your obedient servant,

(Signed) "SAMUEL HOMFRAY.

"MESSRS. HAYNES AND DOUGLAS,
"*Tottenham Court Road, London.*"

Such were the pecuniary and legal difficulties attending the introduction of even a good thing. Mr. Homfray was willing to share in the profits of the speculation, but disliked partnership, though he could not get on without Trevithick; while Mr. Bill failed to keep matters square, and wished to give up the agency.

Trevithick had not thought it worth while to inform his late friend of his having constructed the "Racing Engine" and London railway; and judging from the accounts, Trevithick, the inventor, and West, the ingenious workman, reaped the smallest share of profit for their largest share of the work.

In the six years following the date of the patent, from Mr. Bill's accounts, a hundred persons had used the new engine, comprising the Government, business companies, men of rank and influence, and men of science, living in various parts of England. Each particular engine required adaptation to its special work, and had to be constructed under every disadvantage of few-and-far-between manufactories, little better than what may now be seen in every small provincial town, and known as casting foundries.

Watt and Trevithick were rival engineers, working on totally different principles: the one constructed his engines so as to use the pressure of steam to expel the atmosphere, introducing it solely as a means of producing a vacuum as his motive power; whereas the other utilized the steam as a motor, and only used the vacuum when convenient.

Hyde Clarke, who knew Trevithick, and whose father was intimate with him, wrote:—

"The introduction of Trevithick's improvement gave increased power to steam, and it is of that importance that Stuart—no mean authority on historical points, and not likely, from national sympathy, to underrate low pressure, or overrate high pressure—is inclined to date the era of the steam-engine from this invention.[1] In the establishment of the locomotive, in the development of the powers of the Cornish engine, and in increasing the capabilities of the marine engine, there can be no doubt that Trevithick's exertions have given a far wider range to the dominion of the steam-engine than even the great and masterly improvements of James Watt effected in his day."

William West, in his disappointment at not being made rich by his share in the patent, became a clock and watch maker, and produced the best timekeepers in Cornwall, called West's chronometers; but he failed to comprehend Trevithick's account-keeping.

"SIR, "ST. IVES, *7th September*, 1815.

"Your ill-tempered letter I received, and think you conclude on very threatening terms. Now, sir, in the first instance, what right have you to make me debtor to you for 40*l.* received of Wood and Murray? I hold a copy of your

[1] 'Historical and Descriptive Anecdotes of the Steam-Engine,' by R. Stuart; 'Railway Register,' February, 1847; Railway Prejudices and Railway Progress,' by Hyde Clarke, Esq.

answer to them, saying you held no share in the patent at that time when they wrote to you respecting the engine, but recommended them to W. West, whom you sold a share to, saying Wm. West would license to erect engines on the patent. And as to Henry Vivian's charge, I shall not adhere to. I am fully satisfied that Mr. Rabey paid his time and demand while employed about his engine; the 78*l.* is double as much as he ought to have for what he performed. The part of patent money you allude to, received from Mr. Rabey, I suppose is settled in our patent accounts, as I never received a single sixpence from the patent before that from Messrs. Wood and Murray; then I made a present of 1*l.* to your children, because you refused making a charge for the drawing sent to Leeds, so I beg to remind you, in a comparative sense, you have heretofore been, like the caterpillars on the island of St. Helena, eating up the product of industrious labour; but as time has set free your demands of blunders from the command of law, I will guarantee to come forward, under bonds of word, to leave our account to arbitration, you appointing one and I appointing the other, whenever you like; and if I owe you any balance, I'll pay you, let it distress me as it may. Certainly it must distress my mind to think of paying for your blunders. Though I have no 20,000*l.*, I am not afraid to meet you, and if I have no thousands, I have no hundreds demanded on equal accounts, so I conclude in peace, and hope you'll be reconciled the same as an honest man.

"Your humble,

"WM. WEST."

The result to the active agents of this most comprehensive patent of 1802, making practical the high-pressure steam-engine and the locomotive, was years of labour without reward, annoyance and dread from ill-defined patent laws, and ill-will and loss of friendship between all the members, whose early acquaintance had led to mutual respect and esteem.

CHAPTER XI.

HIGH-PRESSURE STEAM-DREDGER.

THE late Mrs. Trevithick said that her husband was good-tempered, and never gave trouble in home affairs, satisfied with the most simple bed and board, and always busy with practical designs and experiments from early morning until bed-time. He sometimes gossiped with his family on the immense advantages to spring from his high-pressure steam-engines, and the riches and honours that would be heaped on him and his children, but thought little or nothing of his wife's intimations that she barely had the means of providing the daily necessaries of life.[1]

Captain John Vivian, at the period of steering the London common road locomotive in 1803, "saw Trevithick breaking the rock at the East India Dock entrance to the Thames at Blackwall; using a water-wheel worked by the tide, and also a small high-pressure engine for driving or turning large chisels and borers, and other contrivances for breaking and clearing away the rock to increase the depth of water."[2]

A little more than thirty years ago, the writer lodged in the house of a Mr. and Mrs. Bendy, at Slough, who took an especial interest in his comfort, because he was the son of the inventor of the steam-dredger and the locomotive.

Mr. Bendy had worked on the first steam-dredger,

[1] See Trevithick's letter, May 2nd, 1803, chap. ix.
[2] Statement by Captain John Vivian in 1869 residing at Hayle.

and thought that Mr. Trevithick was unkindly and unfairly treated, for the machinery was complicated, and when anything had gone wrong he perhaps spoke sharply to those about him. The ballast-men on the Thames did everything they could to prevent the use of the steam-dredger; and there were other more secret enemies, who offered Mr. Bendy a bribe to so fix the particular work in his charge as to cause a breakage.

Things went very well for several days after the first putting to work of the steam-engine dredger, and then there was a breakage, caused by a nut jammed in the cog-wheels: many believed it was purposely done.

The following letter was written twelve years after those conversations:—

"DEAR MR. TREVITHICK, "*Nov. 8th*, 1840.

"I will furnish you with all the particulars I recollect respecting the dredger-engine and machinery made by your father for Mr. Bough. It was fixed in the year 1803, and was altered by Mr. Deverill (the patentee of the double engine) in the year 1805. The cylinder was 14½ inches in diameter, the stroke 4 feet, the chain-ladder 28 or 30 feet. The largest quantity of stone and gravel lifted in one tide was 180 tons. The reason for using the word stone is from its being part of Blackwall rock. The engine was cast at Hazeldine's at Bridgenorth, but finished at Mr. Rowley's factory in London, by some men from Cornwall, and a part of the machinery by Jackson, a Scotch millwright. The working time between tides was from six to eight hours, in from 14 to 18 feet of water, at the entrance of the East India Dock. The expense of the engine and machinery, a little more than 2000*l.* I should think the time she worked was about ten or eleven years.

"The other engine you mentioned was the property of the Trinity Company and Government, but by whom made I cannot ascertain. It was worked in the river, near Woolwich entrance dock-gates, but not to much account, only lifting mud to clear the entrance. No new materials were taken there by Mr.

Deverill; all that was done was to refix the old, and repair the engine. This engine was the same size as that fixed in London to run round the circle, at the speed of fourteen or fifteen miles per hour, commonly called locomotive engine.

"Yours truly,

"T. BENDY."

Trevithick having seen his engine boring and dredging on the Thames in 1803, went to Wales to superintend the tramroad-engine of 1804, then to Newcastle-on-Tyne with the railroad-engine, and in the early part of 1805 he was again with the steam-dredger and rock breaking at Blackwall.[1]

Mr. Bendy says 180 tons of stone and gravel from the Blackwall rock were raised in a tide of six or eight hours, from a depth of 14 or 18 feet.

This rock breaking and dredging in deep water was a severe test for a new invention; and the Trinity Board, seeing its usefulness, engaged with Trevithick to erect and work steam-dredgers in other parts of the river.

In February, 1806,[2] he was about entering into an engagement for twenty-one years with the Trinity Board to lift ballast from the bottom of the Thames, at the rate of 500,000 tons a year, for a payment of sixpence a ton. Dredging had been done by hand for about eightpence a ton; but the required quantity could not be raised by men working small dredge-bags attached to long poles.[3]

In May, 1806, his partner in the patent wrote :—" I am very happy to find that you have so far continued your agreement with the Trinity gents. I think the bargain is a good one; must still beg leave to remind

[1] See Andrew Vivian's letter, 22nd May, 1805, chap. x.
[2] See Trevithick's letter, February 18th, 1806, chap. xviii.
[3] See Trevithick's letter, March 4th, 1806, chap. xviii.

you not to proceed to show what your engine will do, till the agreement is fully drawn up and regularly signed."[1]

In July, 1806, Trevithick wrote:—"This day I set the engine to work on board the 'Blazer,' gun-brig. It does its work exceedingly well. We are yet in dock, and lift up mud only. I hope to be down at Barking shaft in a few days, at our proposed station, when I will write to you again. I think there is no doubt of success."[2] An account-book in Trevithick's writing, headed "Expenses on the ballast machine from 1805 to 1807," shows that those ballast-dredging engines, like the earlier dredging engine at the Blackwall rock, came from the Bridgenorth Foundry, and many of the mechanics erecting it came from Cornwall. One of the engines was fixed in the 'Blazer,' the other in the 'Plymouth Barge.'

Watt honoured Trevithick with a visit to inspect the steam-dredgers, having been introduced by Mr. John Rennie, of whom Smiles wrote:—"What is called the Humber Dock was begun in 1803 and finished in 1809 It was in the course of executing the Hull Harbour Docks that Mr. Rennie invented the dredging machine as it is now used. Mr. Rennie carefully investigated all that had previously been attempted in this direction, and then proceeded to plan and construct a complete dredging machine, with improved cast-iron machinery, to which he yoked the power of the steam engine."[3]

Rennie's investigation prior to invention probably refers to what he saw Trevithick doing, whose dredger is unmistakably described and delineated by Rees, though the inventor's name is omitted.

---

[1] See Vivian's letter, May 30th, 1806, chap. xviii.
[2] See Trevithick's letter, July 23rd, 1806, chap xv.
[3] 'Lives of the Engineers,' by Smiles, vol. ii.

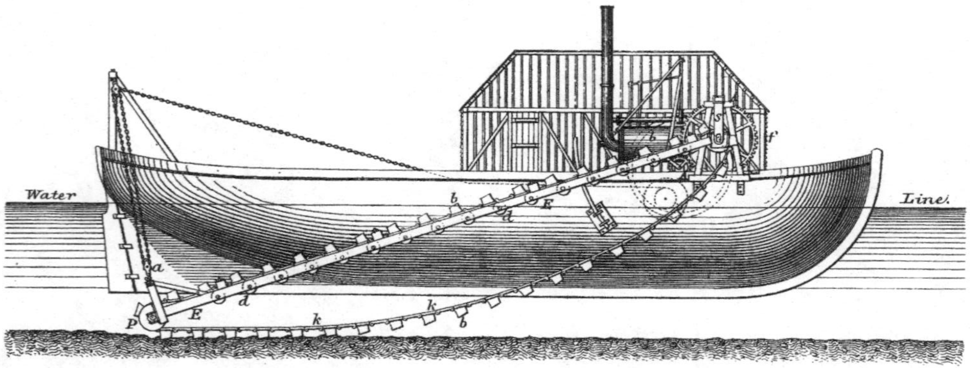

Water Line.

Fig: 2.

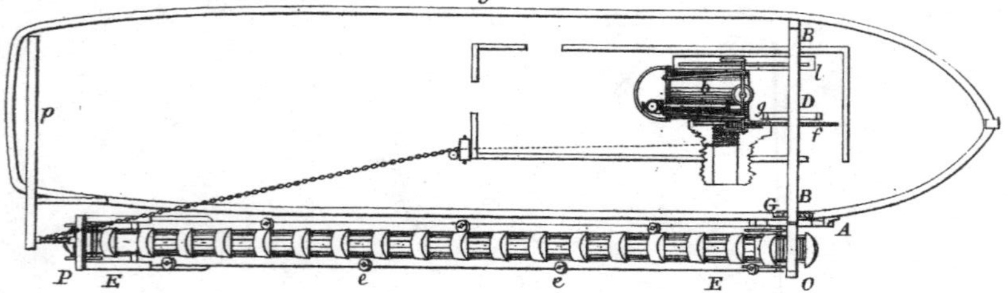

DREDGING MACHINE, 1803.

London: E. & F. N. Spon, 48, Charing Cross.

Kell. Bros Lith. London.

"The convicts at Woolwich upon the Thames perform the ballast heaving, or dredging, which they are condemned to labour at as a punishment: the above method of manual labour became so expensive, that a large machine, worked by horses or a steam-engine, is usually employed. Two such machines have been some time in use in the river Thames. It is erected in the hulk of a dismasted ship. Plate VIII.:—A A, Fig. 1, is a frame of timber bolted to the starboard gunwale, to support a large horizontal beam, B B, Fig. 2; another similar frame is fixed up in the middle of the ship at D, Fig. 2, and the end of the beam is sustained by an upright post bolted to the opposite gunwale; the starboard end of the beam projects over the vessel's side, and has an iron bracket, S, fastened to it, to support one of the bearings for the long frame, E E, composed of four timbers bolted together; the other end of the frame is suspended by pulleys, *a a*; from a beam, F, fixed across the stern, the upper ends of the outside beams of the frame, E E, have each a stout iron bolted to them, which are perforated with two large holes to receive two short cast-iron tubes; one fastened to the iron bracket, S, at the end of the beam, B, and the other to a cross beam of the frame, A. These tubes act on the pivots of the frame, E, upon which it can be raised or lowered by the pulleys, *a a*; they also contain bearings, for an iron axis, on which a wheel or trundle, O, is fixed, containing four rounds. Another similar trundle, P, is placed at the bottom of the frame E E; and two endless chains, *k k*, pass round both, as is seen in the plan. Between every other link of the two chains, a bucket of plate-iron, *b b b*, is fastened, and as the chain runs round, the buckets bring up the soil; a number of cast-iron rollers, *d d*, are placed between the beams of the frame to support the chain and buckets as they roll up. Four rollers, *e e*, are also placed on each of the outside beams, to keep the chains in their places on the frame, that they may not get off to one side. The motion is conveyed to the chains by means of a cast-iron wheel at G in the plan, wedged on the end of the axis of the upper trundle, O. The wheel is cast hollow, like a very short cylinder, and has several screws tapped through its rim, pointing to the centre, and pressing upon the circumference of another wheel enclosed within the hollow of the first, that it may slip round in the other when any power greater than the friction of the screw is applied; the internal wheel is wedged on the same shaft with a large cog-wheel, *f*, turned by the small cog-wheel, *g*, on the axis of the steam-engine.

"The steam-engine is one of that kind called high pressure, working by the expansive force of the steam only, without condensation. *b* is the boiler containing the fire-place and cylinder within it; *i* is one of the connecting rods; and *l* the fly-wheel on the other end of the same shaft as the wheel *g*.

"The pulleys, *a*, which suspend the chain-frame are reeved with an iron chain, the tackle-fall of which passes down through the ship's deck, and is coiled on a roller, as in the plan, and represented by a circle in the elevation; on the end of the roller is a cog-wheel, *p*, turned by the engine-wheel, *g*; the bearing of this wheel is fixed upon a lever, one end of which comes near that part of the steam-engine where the cock which regulates the velocity of the engine is placed; so that one man can command both lever and cock, and by depressing that end of the lever, cause the wheel *p* to gear with *g*, and, consequently, be turned thereby, and wind up the chain of the pulleys; *g* is a strong curved iron bar bolted to the vessel's side and gunwale, passing through an eye bolted to the frame, E, to keep the frame to the vessel's side, that the tide or other accident may not carry it away.

"A large hopper or trough is suspended beneath the wheel, *o*, by ropes from the beam, B, into which the buckets, *b b b*, empty the ballast they bring from the bottom; the hopper conveys it into a barge brought beneath it. This hopper is not shown, as it would tend to confuse parts already not very distinct. The motion of the whole machine is regulated by one man.

"The vessel being moored fast, the engine is started, and turns the chain of buckets; the engine tender now puts his foot upon a lever, disengages the wheel *p* from *g*; and by another takes off a gripe which embraced the roller *m*. This allows the end, E, of the frame to descend, until the buckets on the lower half of the chain drag on the ground, as shown in Fig. 1, when he stops the further descent by the gripe. The buckets are filled in succession at the lower end of the frame, and brought up to the top, where they deliver their contents into the hopper before mentioned; as they take away the ballast from the bottom, the engine tender lets the frame E down lower by means of the gripe-lever, and keeps it at such a height that the buckets come up nearly full. If at any time the buckets get such deep hold as to endanger the breaking of the chain or stopping the engine, the coupling box at C, before described, suffers the steam-engine to turn without moving the chain of buckets, and the engine tender pressing his foot upon the lever which brings the wheel *p* to gear with *g*, causes the roller *n* to be turned by the engine, and raise up the frame E, until the buckets take into the ground the proper depth, that the friction of the coupling box at C will turn the chain without slipping in any considerable degree.

"The steam-engine is of six horses' power, and is so expeditious that it loads a small barge with ballast in an hour and a half."[1]

Trevithick's expeditious steam-dredging engine was "of that kind called high pressure, working by the expansive force of the steam only, without condensation," very like Mr. Wilson's description of the Newcastle locomotive engine, for in principle if not in outline the two engines were alike.

If Rennie commenced his dock operations in 1803 Trevithick was then breaking and clearing the rock from the dock entrance to the Thames at Blackwall, when Captain John Vivian gave his spare days to ride on the London common road locomotive, and to the works at the dredger. In 1805, while the dredger was still going on,[2] Rennie's report to the directors of

[1] Rees' 'Cyclopædia,' published 1819.
[2] See Vivian's letter, May 22nd, chap. x.

the Thames Driftway was put aside, and Trevithick was appointed as the engineer;[1] in 1806 the dredgers were in full operation, or the Trinity Board would not have proposed a contract for a term of years; and in 1807 he was still in daily attendance on the dredging schemes:[2] thus for several years, during the most inventive and active period of his life, did he give his time to the perfecting the steam-dredger

This machine is memorable as the first steam-dredger, and also as leading to the locomotive of 1808; and, like many other of Trevithick's practical inventions, was near perfection at birth. Hundreds of steam-dredgers are now at work throughout the civilized world, and although their construction has occupied the time and knowledge of numerous practical engineers during a period of more than sixty years, yet this first production is the type of them all.

The small and compact high-pressure steam-puffer engine was placed in a wooden house; a main shaft extending over the side of the boat gave motion to an endless chain of dredging buckets kept in the required position by a long wooden frame, having a roller or guide-wheel at its lower end, round which the revolving chain of buckets passed, and this end of the frame was raised or lowered at will, causing the lower sweep of the chain of buckets to press with greater or less weight into the bed of the river, and thus cause the buckets to raise a greater or less quantity of mud or gravel. The chain and loaded buckets in going up the incline formed by the wood frame passed over rollers in order to lessen the friction; on passing the top of the incline the buckets turned short round over a guide-

[1] Report of directors of the Thames Driftway, June 1805, chap. xii.
[2] See Trevithick's letter, August 11th, 1807, chap. xii.

wheel, similar to that at the bottom of the frame, with their bottoms upwards, causing their contents to fall into a barge or guide launder as was most convenient for the disposal of the ballast. The wheel giving motion to the bucket-chain was not wedged fast to the driving axle, but was bored to fit loosely on the turned part of the axle, or rather on a turned cylinder fixed firmly on the driving axle. Tightening screws in the bucket-wheel with their inner ends pressing on the cylindrical driving axle formed a friction band, and should the buckets have come in contact with an unusual obstruction, allowed the driving axle to revolve while the obstructed bucket-wheel remained stationary.

Another safeguard was provided by the rope from the blocks supporting the lower end of the bucket-frame being connected with the engine, enabling the engineman without moving from his station at the engine to raise or lower the bucket-frame by putting his foot on a lever. "The steam-engine was one of that kind called high pressure, working by the expansive force of the steam only." "It is of six horses' power, and is so expeditious that it loads a small barge with ballast in an hour and a half." The steam-engine and dredging machine were so admirably arranged that "one man can command both lever and cock for regulating either the machine or the steam-engine." The feed-pump was worked with a lever, giving a shorter stroke to the feed-pole. The position of the engine is seen in the drawing of the dredger boat; the detail of the engine is in chap. ix., that it may illustrate the progress of locomotion, for similar engines served both purposes.

A letter by Trevithick, years after the event, corroborates the work done by those dredgers.

"SIR, "CAMBORNE, *4th February*, 1813.

"I have your favour of the 2nd inst. respecting an engine for lifting mud from the bottom of Falmouth Harbour.

"I made three engines with machinery for lifting mud at the entrance of the East and West India Docks, and also for deepening the water at the men-of-war's mooring ground at Woolwich. One of these engines was a 20-horse power, erected on an old bomb-ship of about 300 tons burthen, which machine cost (exclusive of the ship) about 1600*l.*

"This engine would lift and put into barges near 100 tons of mud per hour. Another engine of 10-horse power I erected on board an old gun-brig of about 120 tons burthen, which cost (exclusive of the vessel) about 1000*l.*, which lifted about half the quantity of the large one; and another engine of 10-horse power I erected on board a barge of about 80 tons burthen.

"Unless the mud will pay for bringing on shore for manure, I should think that a better plan than this might be adopted to clean the harbour.

"I remain, Sir,

"Your very humble servant,

"RICHD. TREVITHICK.

"JOHN GOULD, Jun., Esq., *Penryn*."

The first steam-dredger at Blackwall succeeded, or two others would not have been constructed. One dredger with the 20-horse-power engine lifted and placed in barges near 100 tons of mud per hour, making good his promise as to quantity. Others were used to deepen the entrance to the East and West India Docks, and also at the man-of-war's mooring ground at Woolwich, in such deep water as fully proved the correctness of their design.

The three engines and dredgers, exclusive of the vessels, cost 3600*l.*, a less sum than such machines would now be made for.

CHAPTER XII.

THAMES DRIFTWAY.

WHEN Trevithick was dredging in the Thames, the Thames Archway Company were anxiously seeking new plans and a new engineer; not being satisfied with the proposals of Mr. Rennie and Mr. Chapman they sought out Trevithick, who soon found himself in a position most trying, and unsuitable to his energetic temperament.

" At an annual meeting of the proprietors on the 4th of May, 1808, the directors refer to their second meeting in June, 1805: 'That it appears to this meeting that the well constructing the driftway is of the highest importance in the future progress of the works, involving in it the success or failure of the undertaking.

"'That, therefore, the works relating to the driftway be suspended until the opinion of a professional man of eminence be taken on the various matters respecting it.'

" Mr. Rennie and Mr. Chapman were accordingly consulted; but as their opinions did not coincide, nor indeed were stated upon all the points on which the directors chiefly wished for information, they felt themselves bound to resort to some other source; and Mr. Trevithick was introduced to them by their resident engineer, Mr. Robert Vazie, as a person skilled in mining. After a due examination into his character, as appears by the minutes of the directors, and having received the strongest testimonies in his favour from several quarters as to his skill, ingenuity, and experience, the directors were induced to contract with him for superintending and directing the execution of the driftway, such as he proposed it to be; for which they agreed to pay him 1000*l.*, provided he succeeded in carrying it through to the north shore; or 500*l.* if the directors ordered it to be discontinued in the middle, which they reserved to themselves the power of doing; but to receive nothing in case he did not succeed.

"The driftway was accordingly commenced on the 17th of August.

"On the 5th of September following Messrs. Vazie and Trevithick, in a joint report to the directors, strongly recommended the immediate purchase of a 30-horse-power steam-engine. The directors did accordingly purchase the same, and it is now ready to work. The driftway proceeded till about the beginning of October, when it appeared that the works had been very considerably interrupted and delayed in their progress. The directors therefore, on the 8th October, resolved to institute an inquiry into the cause; and the consequence of this investigation and disclosure of facts was the removal of Mr. Robert Vazie from his office as resident engineer, on the 19th October, by which time the drift had been extended 394 feet, that is, at the rate of 6 feet 2 inches per day, through a dry sand.

"The works now proceeded without embarrassment, and with considerable less cost; as from this time (the 19th October) to the 29th November, the ground continuing as nearly as possible of the same quality, it was extended 421 feet, or 11 feet 2 inches per day, which is nearly a double rate (deducting three days and a quarter that the works were suspended while the directors determined on the turn the drift should take).

"From the 29th November to the 19th December the drift was extended only 138 feet, or 6 feet 10 inches per day, in consequence of the drift now running in a stratum of rock, great part of which was so hard that it could not be broken up without the use of chisels and wedges.

"By the 21st December the drift had proceeded 947 feet from the shaft; and it was observed that the strata through which it passed dipped to the northward about 1 foot in 50, in consequence of which the rock that at one time formed the whole face of the breast. now only reached within 2 feet of the top, and which was occupied by a sandy clay, mixed with oyster and other shells, and containing some water.

"On the 23rd, notwithstanding the workmen were proceeding with the utmost precaution, the roof broke down and discharged a great quantity of water from a quicksand, which was afterwards ascertained to be about 5 feet 6 inches above the roof.

"By the 26th January the drift was extended 1028 feet, having been worked through a considerable part of the quicksand; and at this period the river made its way into the drift by a fall of earth, which made a considerable orifice in the bed of the river, which has been filled up at several times with earth, carried there for that purpose, and the drift has since then been extended to 1040 feet, which is the present length of it.

"Resolved, that the money paid or claimed by the engineers, Mr. Rennie, Mr. Wilcox, Messrs. Stobart and Buddle, Mr. Thomas Cartwright, Mr. James Barnes, and Mr. William Chapman, whom the directors have consulted with as to the prosecution of the works, be allowed.

"Resolved, that Mr. Richard Trevithick be, and is hereby appointed engineer to the said company, and be directed to proceed forthwith with the works."

At the second meeting of the proprietors in 1805 differences had arisen between the directors and their engineer, Mr. Robert Vazie, on the method of constructing the driftway. Mr. Rennie and Mr. Chapman were called in to advise a plan for the guidance of the directors. Their advice was not approved of, and after much delay Mr. Vazie introduced to their notice Mr. Trevithick as a skilful miner and engineer. This was about the middle of 1807. On the 10th August, 1807, the directors engaged with him to make the proposed driftway from the south shore of the Thames, at Rotherhithe, to half-way across the river, from which central point it was contemplated to commence forming the permanent tunnel, of the size to be then determined, either for foot-passengers alone, or if the nature of the work gave promise of success, to enlarge it sufficiently for the passage of carriages.

"Mr. Giddy, "Limehouse, *August* 11*th*, 1807.

"Sir,—Last Monday I closed with the tunnel gents. I have agreed with them to give them advice, and conduct the driving the level through to the opposite side (as was proposed when you attended the committee); to receive 500*l.* when the drift is half-way through, and 500*l.* more when it is holed on the opposite side. I have written to Cornwall for more men for them. It is intended to put three men on each core of six hours' course. I think this will be making 1000*l.* very easily, and without any risk of loss on my side. As I must be always near the spot, to attend to the engines on the river, an hour's attendance every day on the tunnel will be of little or no inconvenience to me. I hope nine months will complete it. From the recommendation you gave me, they are in great hopes that the job will now be accomplished; and as far as Captain Hodge and myself could judge from the ground in the bottom of the pit, there is no doubt of completing it speedily. I am very much obliged to you for throwing this job in my way, and shall strictly attend to it, both for our credit as well as for my own profit.

"I am, Sir,

"Your very humble servant,

"Rd. Trevithick."

Within a week of his engagement he had commenced the driftway. The sinking of the shaft on the south shore, from the bottom of which the driftway was to commence, was the extent of progress during the two or three years before Trevithick's appointment. It was 68 feet 4 inches in depth from the surface; a wooden platform a slight distance above the bottom of the shaft served to form a reservoir for the drainage water from the drift, from which it was pumped to the surface by a steam-puffer engine. The bottom of the drift was even with this platform. In size it was sufficient only for the passage of a workman with his barrow, being 5 feet high, 3 feet broad at the bottom, and 2 feet 6 inches at the top, inside measurements; the four sides were kept in form by woodwork and strong 3-inch planking. The working end required great care, and frequently the application of close planking across the end or face of the work to prevent a sudden fall of water and sand. As the drift proceeded the increasing quantity of water and earth or sand to be raised to the surface led the two engineers, Vazie and Trevithick, to make a joint report recommending a 30-horse-power steam-engine. Before two months had passed, this joint engineering broke down, and Trevithick was left to carry out his own plans on the work, but subject to annoying remarks from without.

The drift progressed at the rate of 6 or 10 feet daily, and in a little over four months, or up to the 21st December, it had advanced 947 feet, being more than three-quarters of the whole distance, and within about 200 feet of high-water mark on the north side.

"MR. GIDDY, "LONDON, *August 28th*, 1807.

"Sir,—Tuesday last was a week since we began to drive our level at the bottom of the engine-shaft at the archway.

"The level is 5 feet high, 3 feet wide at the bottom, and 2 feet 6 inches wide at the top, within the timbering.

"The first week we drove 22 feet. This week I hope we shall drive and timber 10 fathoms. As soon as the railway is laid I hope to make good 12 fathoms a week. The distance we have to drive is about 188 fathoms. The ground is sand and gravel; it stands exceedingly well, except when we hole into le-areys, and holing into such houses of water makes the sand very quick. We have discovered three of these holes which contained about 20 square yards.

"It is very strange that such spaces should be in the sand at this depth. When we cut into such places we are obliged to timber it up closely until the sand is drained of the water, otherwise it would run back and fill the drift and the shaft.

"I cannot see any obstacle likely to prevent us from carrying this level across the river in six months. The engine throws down a sufficient quantity of air; and the railway underground will enable us to bring back the stuff, so as to keep the level quite clear, and the last fathom will be as speedily driven as the first. There is scarcely any water in the level—not above twenty gallons per minute—and not a drop falling from the back of the level (the end). Therefore, I think we may expect that the land-springs will not trouble us.

"The spring that came down around the outside of the walling of the shaft is rather increased. The directors are in wonderful spirits, and everything goes on very easily and pleasantly. The engines on the river go on as usual.

"I remain, Sir,

"Your very humble servant,

"RICHD. TREVITHICK.

"*Plough Inn, Kidney Stairs, Limehouse.*

"P.S.—The 2-inch iron air-pipes that were provided before I took the work in hand are too small. The smith's bellows has nearly two hundredweight on the top plank, yet at the bottom of the shaft it will scarcely blow out a candle. I shall put down larger pipes next week."

It is remarkable that in 1807 Trevithick familiarly spoke of a railway in connection with underground work, and of supplying miners with pure air by the use of the steam-engine, while in 1870 we talk of enforcing the latter by Act of Parliament.

"MR. GIDDY, "LONDON, *September 12th*, 1807.

"Sir,—This day I received yours, and am very much obliged to you for your attention to my welfare. Last week we drove and secured in the tunnel 25 yards, and I see no doubt of getting on in future with the same speed. We are now about 180 feet from the shaft, and as we approach the river the ground is better, and the water does not increase; but to be prepared for the worst, it is agreed to have a 30-horse-power engine in readiness, to assist in case of cutting more water than the present engine can cope with.

"The distance from the shaft to the spot on the opposite shore, where we intend to come up to the day, is about 1220 feet. This distance I hope to accomplish in a short time, unless some unfavourable circumstance turns up, which at this time there is no sign of. Should the ground prove softer we can drive horizontal piles; should the water increase we shall have three times the power in reserve as we now occupy.

"The directors are highly pleased with our present proceedings. In consequence of this job I have been called on to take the direction of a very extensive work, the nature of which I am not yet fully informed; but am to meet the party on Thursday next for further information, when I shall communicate the plan to you for your investigation; at the same time I must beg your pardon for so often troubling you on matters that cannot advantage you, and hope you will excuse my freedom, being driven to you as a source of information that I cannot be furnished with from any other quarter.

"The engines continue to get on as usual on the river. The great engine is not yet at work; but hope it will soon be completed.

"I remain, Sir,

"Your very humble servant,

"RICHD. TREVITHICK.

"*Plough Inn, Kidney Stairs, Limehouse.*"

In the midst of hoped-for success, a special meeting of proprietors was held on the 24th December, 1807, and it was put to the meeting that Mr. Trevithick does not possess the confidence of the company, and that the directors be requested to put an end to all engagements with him. The proposal was negatived by 137 voices against 61; the directors supporting their engineer. The continuance of this opposition (this was the second time this question was discussed during the four months) not only diverted the attention of the directors from the daily consideration of the progress of the work, but roused Trevithick's naturally impetuous action to drive ahead for the other shore.

The drift had been carried on a level line from the shaft until past the middle of the river, to facilitate drainage, when it ascended at about the same slope as the river-bed towards the northern side, the roof of the drift being about 30 feet under the bed of the river. The strata passed through from the shore to the middle was firm sand and gravel, while for the last 200 feet it had been rocky This strata of rock had been seen in the shaft on the south side 7 feet 6 inches thick, and was known to dip going north about 1 in 50. While the driftway was under it, it helped to keep the water out and gave firmness to the ground; but as going north it was deeper below the surface than on the south side, and as the driftway had to rise to the surface within a certain distance from high-water mark on the north side, it had of necessity to pass through this strata of rock. The rise of the drift was then about 1 in 9. The strata of rock had been cut through on the 21st December. The face or breastwork at the end showed that the lower half or bottom of the drift was still in rock, while the upper half or roof was in treacherous quicksand.

Then was the time for calm and thoughtful consideration of evidences in the drift by the directors, when a wise word would have averted failure and secured success. Unfortunately the directors were occupied in preparing to meet the dissatisfied proprietors at the special meeting to be held within three days; the drift followed its doomed course for two more critical days, when on the 23rd the water and quicksand rushed in, in spite of every precaution; and the next day, the 24th December, the directors, who should have been in quiet consultation at the driftway, were in noisy dispute with their opponents at the 'George and Vulture' tavern, Cornhill.

There is no trace of wise retreat under the shelter of the rock—in which course another 100 feet would have brought them beyond low-water mark on the south shore—except in the drawing made by Trevithick, showing a test-hole bored through it and strata above, just before reaching the point where the fatal rock was cut through. He calls it the *test-hole*, and says, "The stratum of sand with water showed that the break was the consequence of not boring in time. The boyer hole near the shaft on the south shore, made by Robert Vazie, shows the evil consequence of communicating the two springs."

From Trevithick's words in his own writing there is no doubt that he feared to break through the rock, and foresaw the rush of sand and water as in the earlier experiment in the shaft. His opponents used this breakdown as an additional reason for his dismissal.

"DAVIES GIDDY, ESQ., "LONDON, *December 30th*, 1807.

"Sir,—Your politeness in answering my inquiries respecting Mr. Trevithick's engines some time ago, emboldens me to request a further favour of you, relating to that gentleman. Since the directors of the Thames Archway Company contracted with Mr. T. to execute the driftway, he has had the misfortune to incur the displeasure of a considerable proprietor (not a director), and who has had influence enough, by partial representations indefatigably urged, to form a strong party against him, so much so that a special assembly of the company was convened by this proprietor and his friends, for the special purpose (as it turned up in argument) to throw a degrading stigma on Trevithick's character, not only as an incompetent miner and engineer, but as a man of integrity; and indirectly to censure the directors for employing him. A resolution was proposed by this gentleman that Mr. T. does not possess the confidence of the company, and that he be discharged from all his engagements with it. Though the author of this proposition was induced by the sense of the company to withdraw it, yet his assertions, which were very bold and numerous, and unsupported by his facts, seemed to retain their weight on the minds of his own party, notwithstanding the folly and weakness of the proposition; folly, because Trevithick could not *legally* be discharged, and wicked, because if he could have been, it must have been by a breach of honour and good faith in the directors, to the great injury of Mr. Trevithick.

"Among the assertions of the proprietor alluded to, he represented that T. had imposed on the directors by magnifying beyond the truth the importance of his engagements in other objects, so as to induce the directors to believe that they could only secure his services by a bargain extravagantly favourable to him, and that all his representations to this effect were false in fact. Although the directors declared that these circumstances had no influence on their minds in making their contract with T., which they considered fair and reasonable, and do still, yet it seems highly desirable now, to the vindication of Mr. T.'s moral character, so grossly attacked, that some evidence be adduced at a future opportunity of the existence of some of those engagements which Mr. T. did certainly mention; and as I think you stated to the directors your knowledge or belief in some of them, it would be highly pleasing to me, and I believe to every other director, if you would take the trouble of informing me what you know or recollect about them.

"It was said that he had represented himself to have large concerns or engagements in Cornwall, or in the neighbourhood of Bristol, an engagement with the Trinity House for raising ballast, and an offer of some engagement of importance with a Mr. Trotter. It was roundly asserted by the said proprietor that Trevithick had no foundation whatever for these representations. Speaking of him as an engineer, he treated him in the most degrading terms, and as a miner, a very quack.

"Notwithstanding the directors have had no reason in any instance whatever to change the very favourable opinion they formed of him, at your recommendation, yet, sir, it would be eminently satisfactory to them to be able to refute, by such respectable evidence as yours, at some future period, those base calumnies attempted to be cast upon him—certainly in no respect justified by his conduct; and as I recollect you spoke of him in very flattering terms as a miner and engineer of great skill and experience, and as a man of great integrity, I venture to hope that you will indulge me with your candid opinion of him on all these points, as well for the sake of my own and co-directors satisfaction, as for doing justice to a man against whom a very illiberal attack has been made, and which if not refuted may very sensibly affect his future comfort.

"I have the honour to be, Sir,

"Your very obedient servant,

"JOHN WYATT.

"P.S.—The particulars of the observations made at the meeting on Mr. T. are not known to him, nor is he acquainted with this letter to you; but he has informed me that Mr. Robert Vazie once addressed a letter to you respecting him, and of your favourable answer. If you will have the goodness to favour me with copies of these, they may be used very advantageously in favour of Trevithick."

Mr. Wyatt was one of the few appreciative men that now and then crossed Trevithick's troubled path.

"MR. GIDDY, "THAMES ARCHWAY, *January 5th*, 1808.

"Sir,—About ten days since I received your letter, which was wrongly directed, and did not come to me in due course. I have sold the ballast business to a company, which is carrying it on. I should have written to you on the receipt of your letter; but at the time it arrived we were in a quicksand, and I wished to get through it before writing. The drift is driven 952 feet from the shaft, and is about 140 feet from high-water mark on the north side of the river, and hope to be through in a fortnight. Some weeks we drove 20 fathoms; but for fifteen or twenty days before we met the quicksand we had a very hard lime rock, which much impeded us. During the last fortnight scarcely anything has been done in the end on account of the quicksand; but now we are again in a strong clay ground, and getting on quickly. The drift is 72 feet below high-water mark. When we commenced to drive from the shaft it was a firm green sand; soon a bed of gravel of about 8 inches thick came down from the back. As we drove north we found all the strata dip about 1 foot in 50. On the top of the gravel there is about 3 feet of clay, then a limestone rock (made from water) about 5 feet thick. It is evident that this is a made rock, because the green sand that we began to drive in is several feet under the rock, and there is a large quantity of branches of trees in it turned into stone. On the top of the rock there is a proper bed of oysters, mud and oysters about 2½ feet thick. Above this is about 5 feet of clay mixed with sand; above that is the quicksand of 2½ feet thick; and above that is a strong dry clay, which we suppose holds up to the bottom of the river, which is about 20 feet. We bored on the north side, and found this quicksand was above the back of the drift, and a great quantity of water in it.

"I proposed to bore up in the back of the drift, to tap it and let down the water gently through an iron pipe, but I could not be permitted.

"In course of working in the drift the water and quicksand broke down in the back of the drift, drowned the engine, and threw a great quantity of sand down into the drift. This sand is nearly as fine as flour, and when in water is exceedingly quick. We had the good fortune to stop up the drift tight with timber before the water got up to prevent us, and we have drained the sand, so that the engine is again complete master of it. I had a machine made to drive very long flat iron bars close up to the back of the drift, through the stopping boards in the end; and then cleared out the drift under them, and timbered very closely under the bars. By driving a great number of bars of iron I again reached the end of the drift, which is now again in good course for driving. After we had stopped the end of the drift to prevent the quicksand from coming back, I bored 14 feet up in the back of the level, and forced up a 2-inch iron pipe above the quicksand, which prevented the quicksand from coming away with the water, and also took off the weight or pressure of water from the breaking ground in the back of the drift.

"It was strongly proposed by one of the proprietors (when the drift was half-way in) to open out from that place the tunnel to 16 feet high and 16 feet wide. I refused to do it, knowing that this water and quicksand was over our head, and that as soon as we began to incline the bottom of the drift up toward the surface on the north side we should be into it. It was with the greatest difficulty that we could stop it in the drift, only 2½ feet wide and 5 feet high.

"Had we opened the tunnel to the full size, every man that might have been underground at the time must have been lost, and the river through into the tunnel in ten minutes, for the water would have brought the 2½-feet stratum of quicksand into the tunnel, and then the clay roof would have sunk under the weight of the river; for it would have been impossible to stop it over a space 16 feet high and 16 feet wide; the engine would have been drowned in one minute, and the sand would have constantly come away until the roof fell through into the drift.

"This proprietor has been very much exasperated against me ever since, because I would not open out the tunnel from the middle of the drift up to the full size. This gent was never in a mine in his life, neither does he know anything about it. He called a general meeting to discharge me, but he was taken no notice of, and the thanks of the meeting given to me for my good conduct, and his friend Mr. Vazie discharged. They have offered me 1250*l.* more for my attendance, to open the drift up to the full size of the tunnel, and wish me to engage with them immediately, before the first contract expires.

"The quicksand when drained is very hard, and after the drift is through the ground it will be so completely drained that I cannot see any risk in opening to the full size.

"When you come to town I shall have several things to lay before you, and shall be very much obliged to you to say when you expect to be here.

"I am, Sir,

"Your humble servant,

"RICHD. TREVITHICK."

In Trevithick's time geology had not become a familiar study, and probably he had never read a page on the subject: yet his reasoning is simple and logical; for the infiltrated limestone rock, "made from water," resting on a strata having in it branches of trees, was a proof that it was not an original limestone rock.

The propriety of forcing an iron pipe from the roof of the drift up into the top of the bed of quicksand for the purpose of drawing off the water was evidently a much-disputed point, and was carried against Trevithick, though after the water had forced its way into the drift his drain-pipe was approved of, and was used by him and others; but the strata had then become broken, and the pipe less effective than it might have been at an earlier period. Trevithick has been called an obstinate man because of his improper use of this drain-pipe, but this letter proves quite the contrary, for it shows that he obeyed orders, and without a tinge of recrimination.

By a continuous working of the extra pumping engine which he had provided at the shaft in case of such an emergency, the water and sand were so far cleared out of the drift that men could go in. How were the numerous inpours of water and sand through every crevice to be stopped? The timbering at the top of the end was the most leaky; but in the cramped width of 2½ feet it was impossible to remove and replace timbers with celerity. Iron plates were therefore driven through the stopping boards, close to the roof, in the end of the drift, into the disturbed strata, forming a roof of iron. This enabled the wood end to be moved, and to be advanced without bringing down the roof. Unfortunately the machine invented and made for driving those bars is not described.

He was also right in declining to undertake to enlarge the drift when half-way through, as proved by the great difficulty afterwards experienced by Mr. Brunel in driving the larger tunnel. The object was to perfect the drift that it should serve as a drain to free the larger tunnel from water during its construction, as well as a means of knowing the strata.

Sketch of Miner in Drift.

By the 26th January, 1808, in a month's work since leaving the rock strata, they had only advanced 81 feet when the water and sand again rushed in in such increased quantities through the openings for advancing the end or breastwork, that the workmen could no longer resist it, and with difficulty reached the shaft, with but just sufficient space for breathing room between the surface of the rising water and the roof of the drift. Trevithick was one of the lot; and as the water filled the drift and rose in the shaft, the miners floated with it until rescued. Mrs. Trevithick spoke of her husband coming home through the streets without hat or shoes, half drowned, and covered with clay and dirt, but not discouraged by this first really serious break-down, the drift being 1028 feet in length. The drift-way and shaft were filled with water and sand, and the bed of the river, 30 feet above it, sank into a dish form over the hole, from the quantity of sand carried by the rushing water into the drift; causing the thick strata of clay forming the bed of the river to sink more and more. A crack or hole through this clay would be fatal, therefore more clay was thrown into the river, to fill the hollow and prevent if possible any communication between the water and the beds of quicksand surrounding the drift. A large canvas sheet loaded with clay, and bags filled with clay, were also thrown on to the river-bed.

The following shows that even then he was sanguine of success:—

" MR. GIDDY, "THAMES ARCHWAY, *February 2nd*, 1808.

" Sir,—I have been for some time expecting a note from you, of your arrival, which is the reason for not having written to you long since. I am glad to find that both you and the old gentleman enjoy good health. I enjoy the same; but have a great deal of exercise both of body and mind about this job here.

" Last week the water broke down on us from the river, through a quicksand, and filled the whole of the level and shaft in ten minutes. I have stopped it completely tight, and the miners are at work again. We are beyond low-water mark on the north side with the drift; if we have no further delays we shall hole up to the surface in ten or twelve days. I cannot give you as full a description, by the pen, as I wish; but will see you on Friday morning.

"On Thursday, at twelve o'clock, there will be a meeting of the directors, on the spot. If they knew you are in town, you would be pressed very hard to attend. If you can attend I shall be very happy to see you on the spot, as I have a great deal to communicate to you.

"I have no doubt of accomplishing my job.

"I am, Sir,

"Your humble servant,

"RD. TREVITHICK."

The discontented proprietors had now a real cause for talking, and made such a stir as reached the ears of the Lord Mayor, who accused the headstrong engineer of making a great hole in the bed of the river; and then the tide turned, and he was charged with forming a great bank in the river. The Lord Mayor laid an embargo of pains and penalties on all and every who should interfere with the river-bed without his permission; and being dutifully asked, allowed the engineer to proceed. Trevithick struggled on slowly, and, at imminent risk of health and life, forced the drift forward 1000 feet from the shaft, that narrow passage being the only means of escape, should any one of the numerous bits of planking break away and allow the inrush of sand and water. This was enough to prevent ordinary men from going on; and in addition was the impure, dull, unchanged air in that wooden drain, in which the dim light of a candle that could scarcely be kept burning, and the constant drip of water, where the workmen could never stand erect, or squeeze past one another without difficulty, and could only work one at a time in the narrow end, were more than sufficient to daunt the most determined.

Iron pipes were driven through the roof into the sand-beds, in the hope of drawing off the water; but they became choked with clay, and the slightest opening in the planking of the end was followed by a rush of sand and water. Only 70 feet remained between the end and low-water mark on the opposite or north shore: a small barrier which was fated to be impassable. The directors stopped all proceedings, and considered what should next be done. Trevithick supplied the following plan and opinion: the Plate is copied *from his original drawing*, except that, for want of space, only the part bearing on the stoppage is shown. His plan, some 5 feet long, gives the whole length of the work.

This proposed plan (Plate IX.) for enabling the drift to be continued illustrates his genius; for he had never been taught to deal with river-beds, yet principles are here introduced which have since been largely applied, and spoken of as modern discoveries.

The day following the date of Trevithick's last note to Davies Gilbert (then Giddy), another inrush of water and sand prevented progress. The drift, 1028 feet in length, was through a considerable portion of the bed of quicksand, when a fall of earth in the roof filled the drift with sand, clay, and water, causing a sensible orifice in the bed of the river. Trevithick persevered, and again draining the drift, forced his way forward another 14 feet, making a total length of 1040 feet.

The proprietors called a meeting, and desired two engineers to examine and report, and also to act.

Trevithick recommended a wood caisson filled with clay, with a wooden shaft through it, to be placed over the sunken hollow in the bed of the river, the shaft to be carried down to the roof of the drift; another wooden shaft, hooped with iron, was also to be sunk through the bed of the river, a little

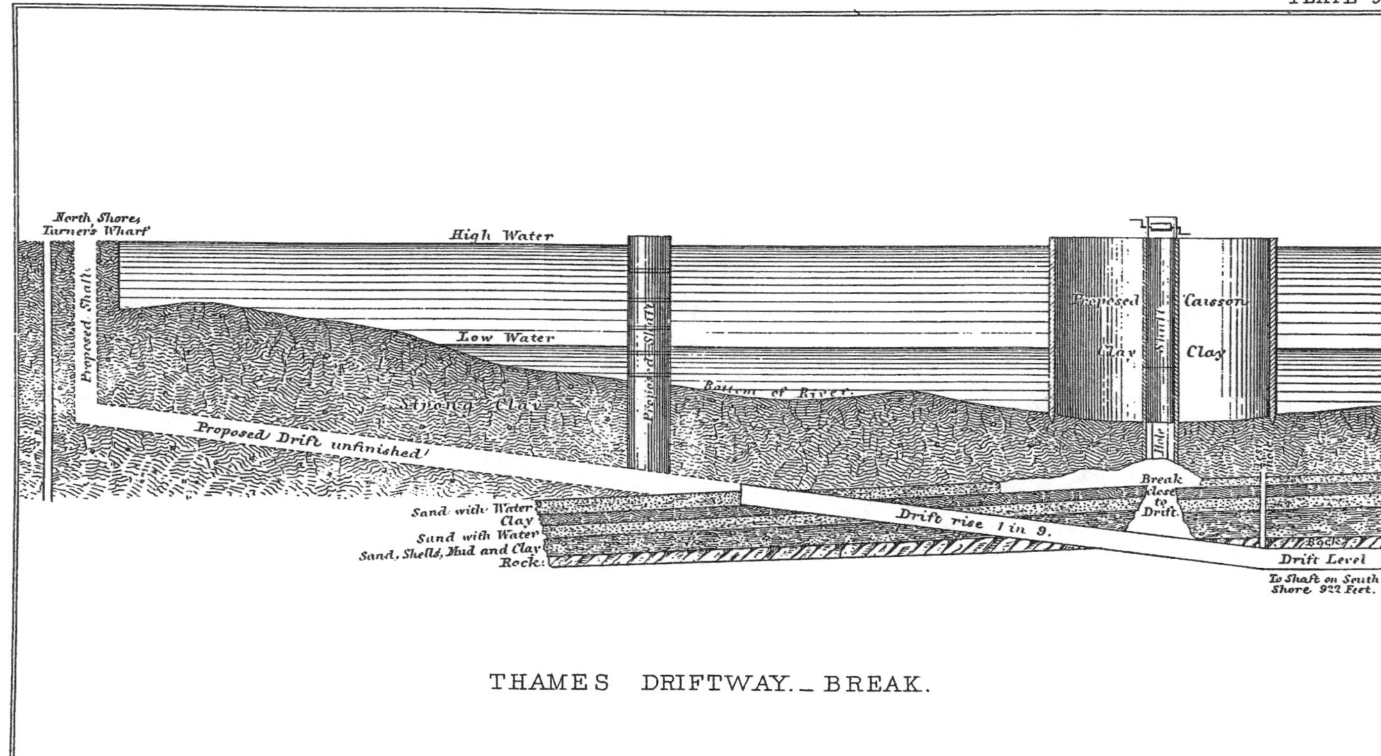

THAMES DRIFTWAY._ BREAK.

London: E.& F.N. Spon, 48, Charing Cross.

Kell, Bro.s Lith. London.

in advance of the end of the drift, to facilitate the carrying the drift through the remainder of the quicksand.

Directors, proprietors, and engineers were now at variance, and from a meeting of directors on the 12th April, 1808, the following particulars were sent to each proprietor :—

"Since the removal of Robert Vazie from the office of resident engineer to the Thames Archway Company, several proprietors in that concern have expressed their dissatisfaction at the proceedings of the works under Mr. Trevithick, and have taken many steps to impede their progress and darken Mr. Trevithick's reputation; yet, as no act of Mr. Trevithick's incompetency was ever shown, though many were falsely alleged, and as the directors never observed any instance either of neglect or want of skill in him, but that on every occasion where his knowledge, his intelligence, and experience in his profession were questioned and examined by competent persons, his talents appeared very superior to the common level, the confidence which the directors reposed in him was not shaken. But the directors, upon the suggestion of the dissatisfied proprietors, and with a view to gratify their wishes, consented on the 29th February, 1808, to apply to two professional miners of high reputation in the North of England, approved by themselves, to come and examine the state of the works. These gentlemen, namely, Mr. William Stobart, of Lumley Park, Durham, and Mr. John Buddle, of Walls' End, Newcastle-upon-Tyne, arrived in London on the 5th instant, and inspected the works on the 7th; and on that day attended a meeting of the directors, when the written questions hereafter stated were delivered to them for their consideration and answers; and the meeting of the directors being adjourned to the 9th instant to receive the answers, those gentlemen attended on that day, and delivered to the Board their answers in writing.

I extract only those portions bearing directly on Trevithick and his acts.

"2nd.—Do you approve of the methods he has pursued for stopping the water from the river?

"We do approve of the methods Mr. Trevithick has pursued of stopping the water from the river.

"4th.—Do you approve of sinking a shaft, as proposed by Mr. Trevithick, for securing the ground injured by the fracture in the bed of the river; or would you advise any other method?

"Taking all the circumstances into consideration, we do not think that any better plan can be adopted for the security of the ground injured by the fracture in the bed of the river, than that suggested by Mr. Trevithick.

"11th.—Is it your opinion that this fracture has been occasioned by any unskilfulness on the director of the works, Mr. Trevithick?

"This fracture, in our opinion, has not in any degree been occasioned by the unskilfulness of the undertaker, Mr. Trevithick, although we might have recommended a continuance of the drift in a horizontal direction for about 140 feet farther, previous to the commencement of its ascent. We might have recommended this 140 feet additional, only to have given equal width between the drift and the river; but, if this had been done, the tunnel itself must have been so much longer, or else the rise quicker in proportion.

"12th.—Have you in any part of the works discovered a want of knowledge and intelligence in Mr. Trevithick in the business of this undertaking?

"No part of the works, in our opinion, exhibits any want of knowledge or intelligence in Mr. Trevithick in the business of this undertaking. As a practical miner, this work does credit to any man who has performed it; and he has great merit in the performance, and need not be afraid of anyone viewing the work.

"17th.—From what you have observed of Mr. Trevithick, on this occasion, are you of opinion that he is a proper person to conduct the undertaking?

"We have not the least hesitation in saying that he is. He has shown most extraordinary skill and ingenuity in passing the quicksand; and we do not know any practical miner that we think more competent to the task than he is. We judge from the work itself, and until this occasion of viewing the work, we did not know Mr. Trevithick.

"18th.—Do you approve of the borings which Mr. Trevithick made in the roof of the drift; and putting up the pipes to draw off the water?

"It was the most judicious method he could pursue."

"After perusing the opinion of men of such eminence in their profession as Messrs. Stobart and Buddle, the directors, whose confidence in Mr. Trevithick has been unvaried, made no doubt that the proprietors at large, whose minds are not infected by prejudice or invidious motives, will concur with them, that that confidence was not only well founded, but that Mr. Trevithick's character stands even higher in his profession than their discernment of his merits had led them to judge of him.

"Not only did a Member of Parliament, highly respected for the purity of his conduct and scientific requirements (Mr. Davies Giddy), attend the Board of Directors, and give so high a character of Mr. Trevithick, both as to his talents as a miner and integrity as a man, that not a single director hesitated on that head, but also, a proprietor, Mr. Butt, who is a distinguished individual of the discontented party, did himself bring to the Board, from another quarter of great respectability, a report as favourable to Mr. Trevithick as that given by Mr. Giddy."

The directors received these opinions of Messrs. Stobart and Buddle on the 7th April, 1808, and directed Mr. Ryan to make borings on the north shore; but being still in doubt, on the 19th April sought advice from Mr. Charlton, from whose report a few questions and answers are selected to show the confusion of management.

"3rd.—Do you conceive that the drift corresponds with the plans?

"Yes; I see nothing to the contrary.

"4th.—Is the drift made in such a way as you would have advised?

"For a temporary drift it is.

"5th.—Is it carried on in such a way as you would have advised as to its level and distance from the river?

"Any man would naturally take it as near the river as he could with safety, and I would have advised it.

"8th.—Whether you would have advised making a drift in the way this has been done, for the purpose of enlarging that drift to the extent of 11 or 12 feet diameter?

"I would not; it is my opinion that you should have finished as you went on.

"9th.—Whether you would advise in the present state of the drift to continue it to the north side, or abandon it and make another drift?

"I think that as it is gone so far, if it was driven through it would give great advantage for air, and convenience for bringing out and taking in materials, and safer for the workmen, and therefore I would advise it to be carried through.

"15th.—Would you advise the borings that are now making to be completed?

"No."

"Resolved, from the evidence adduced at this meeting, that Mr. Trevithick was acquainted with the borings already made in the north shore of the river, notwithstanding the representation to the contrary appears in Mr. Braithwaite's report."

"At a meeting of directors on the 21st April, 1808:—

"Resolved, that the borings on the north side of the river be discontinued till further orders, and that notice thereof be given to Mr. Rastrick."

"At a meeting of directors, 9th May, 1808, Mr. Trevithick attended this meeting and stated that the new engine was completed and at that work; and he now thought it prudent to attempt proceeding with the drift, previously to sinking a caisson, as he expected to get through the quicksand, after proceeding with the drift 5 or 6 feet farther."

"Resolved, that Mr. Trevithick be directed to open the breast, and proceed with the drift without delay. Mr. Trevithick delivered to this meeting the following claim upon the company, for the time he has lost since the works were last suspended, namely, thirty-four days at two guineas per day."

"Resolved, that the consideration of the above claim be suspended for the present."

"At a meeting of directors on the 16th May, 1808, Mr. Trevithick attended, and delivered the following report:—

"On opening the breast this week we found the wood piles rather loose, which were driven up and fastened; after which the miners attempted to proceed, but were driven back several times by water and sand. Opened several old pipes that were put in the roof, out of which water flowed strongly leaving the breast; this enabled the miners to proceed 22 inches."

"At a meeting on the 6th June, 1808, Mr. John Rastrick's report is received by the directors:—

"Your workmen were employed last week in taking up 40 tons of clay and gravel, which formed a ridge on the bed of the river, and in putting a sail over the broken ground, getting, loading, and unloading 160 tons of clay, which clay covers the sail about 3 feet thick. The drift is quite clear of sand up to the door; which door is 18 feet from the face, and is full of sand. The water at the tin and door flows strong and clear.

"Resolved, that Mr. Trevithick be directed to put the old engine in repair."

In June, 1808, Mr. Rastrick was the engineer of the driftway, and on the 11th made a report to the directors:—

"Monday, men were employed in driving iron piles, and loading a barge with clay.

"Tuesday, boring a hole 98 feet south of the incline: this pipe, like all the former ones, discharged water free of sand for half an hour, during which time the end was dry, when the sand flowed and plugged up the pipe. The rock at this hole is 5 feet thick.

"Wednesday, boring a hole 109 feet south of the incline, out of which we could get no water, though it was bored 16 feet high. Opened pipe 98 feet south of the incline; this eased the breast, and miners proceeded in driving piles. They had not been above half an hour at this work before the pipe stopped, and sand and water flowed strong through pipe 109 feet.

"Thursday, miners opened the above pipes, which, as before, discharged sand and water; plugged up, and opened the breast, which appeared favourable, but as soon as the first pile was driven, the whole of the piles came back about an inch and half. Secured the breast and left the drift.

"Friday, opened the hole 109 feet, and put up pipes 3 feet long, making pipes in this hole 15 feet. Water flowed strong and clear for two hours, during which time miners opened breast and drove piles, after which sand flowed very strong through the pipe; plugged up, and left the drift.

"Saturday, bored a hole 173½ feet south of the incline. This hole was bored 19 feet high, through which the sand flowed very strong; plugged up, and drove four iron plates in the breast 18 inches long and 10 inches broad.

"Sunday, opened breast, cut out piles, and prepared for a set of timbers. At this moment the piles appeared loose; sand and water flowing strong out of the bottom of the drift, about 8 feet from the end. Secured breast, caulked the joints, and left the drift."

Without attempting to define Trevithick's position during this knocking together of heads of eminent engineers, directors, and shareholders, it is evident that for two or three months Mr. Rastrick had power to act, closing his short authority by the significant words, "secured breast, caulked the joints, and left the drift."

Mr. Rastrick and other engineers acted on Trevithick's plans without even suggesting any particular change. They would scarcely have been employed, except on the understanding that they could do better than Trevithick; yet they made no progress, and admitted that he was the only man for the work.

During this short respite from daily attendance

on the driftway, Trevithick had constructed the first London railway, and had conveyed or was about conveying the first railway passenger train by the locomotive 'Catch-me-who-can.'

The drift having been idle for nearly two months, Trevithick recommended a new plan, and a special meeting for its consideration was called.

"THAMES ARCHWAY, ROTHERHITHE,
"MR. GIDDY, "*July 28th*, 1808.

"Sir,—I have yours of the 24th, and intend to put the inscription on the engine, which you sent to me.

"About four or five days ago I tried the engine, which worked exceedingly well, but the ground was very soft, and the engine (about eight tons) sunk the timbers under the rails, and broke a great number of them. I have now taken up the whole of the timber and iron, and have laid balk, of from 12 to 14 inches square, down on the ground, and have nearly all the road laid again, which now appears very firm. We prove every part as we lay it down, by running the engine over it by hand. I hope it will all be complete by the end of this week.

"The tunnel is at a stand, and a special meeting is called for Saturday, in consequence of my proposing a new method of carrying it on; and if put into effect (of which I have no doubt) the driftway is long enough, and would not be of more service if wholly across.

"The plan I have laid before them is to make a caisson, 50 feet long, 30 feet wide, and as high as from the bottom of the river to high water. It is made of whole balk, the joints caulked tight as a ship. On the caisson is placed an engine for driving the piles and drawing them again, and also for lifting bricks out of the barges, and the stuff out of the caisson into the barges, by a crane worked by the engine. The piles must be driven within this caisson, in every square, and as deep as the bottom of the tunnel. Then remove the earth and construct 50 feet in length of brick tunnel; then remove the caisson, draw the piles, and refix them 50 feet farther on in the river, and add another piece to the tunnel, and so on, until the whole is finished.

"By this plan only 50 feet of the river would be occupied at one time, or less than a 400-ton ship at anchor in the river. We must bore up from the roof of the drift and put up a pipe to tap down the water from the caisson every time we move it.

"The caisson and engine, with all its materials, will float together, and almost all the work will be done by the steam-engine.

"The first plan was for a tunnel of 11 feet in diameter, for foot passengers only, and was to be 14 feet lower than this present plan.

"This plan is two tunnels side by side, 12 feet diameter, each to have a waggon-road of 8 feet wide, and a foot-path of 4 feet wide; one tunnel to admit persons going forward, the other backward, so as to prevent mischief in passing.

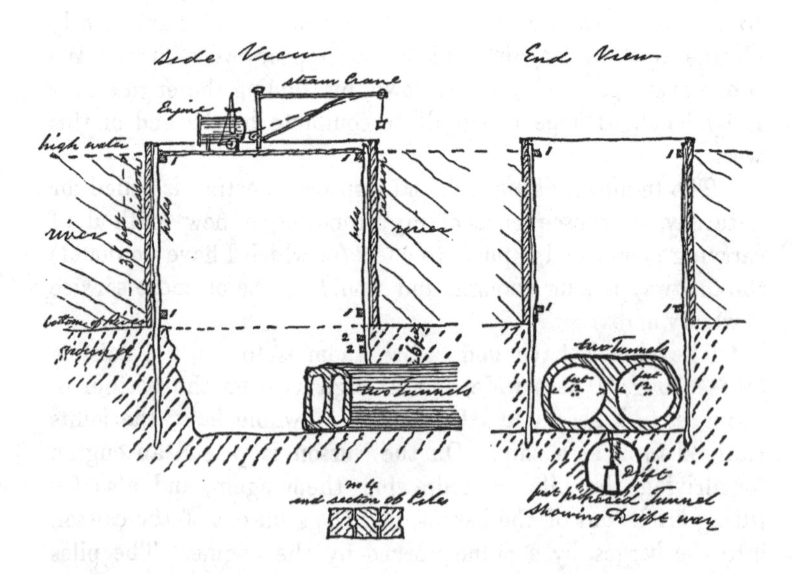

"This plan can be completed for 10,000*l.* less than the original foot-path of 11 feet wide. I think this a safe and sure plan, while the other would be very expensive, dangerous, and uncertain; and if ever executed would be but a foot passage.

"No. 1 is the frame that guides the piles straight.

"No. 2, pieces of timber across over the back of the arches, to keep the feet of the piles firm when we move the caisson.

"No. 4, showing the piles, with a half-circle cut out of each, about 3 inches in diameter, down which oakum is driven to make the piles water-tight in the joints.

"This plan will shorten the tunnel; it is not so deep as the other, therefore will come to clay much sooner on each side of the river, and the timber to support the roof will be saved. Your opinion on this plan will very much oblige

"Your very humble servant,

"RICHARD TREVITHICK."

The proposed caisson for repairing the drift seems not to have been adopted, and was followed by a recommendation to build two brick tunnels, each 12 feet in diameter, to accommodate the to and from streams of passengers and wheel-carriages, in lieu of the original intention of one tunnel of 11 feet in diameter: it was to be 6 feet below the bed of the river, or 14 feet nearer the surface than the original design; thus reducing the stiff and very objectionable incline to and from the tunnel from its low level.

The double tunnel was to be constructed on a new plan. Instead of mining through the quicksand, a water-tight box of planks and piling was to be placed in the river, the enclosed earth excavated to the required depth, a length of brick tunnel built, the river-bed replaced, and the box removed ahead, for an additional length of brick tunnel, while the materials were to be moved, and the piles drawn, by a steam-crane, introduced by Trevithick a few years before,[1] and the surplus water drained by a pipe into the old driftway.

---

[1] See Trevithick's letter, 10th January, 1805, chap. xv.

This plan for an improved double way, estimated to cost 10,000*l.* less than the tunnel or mining process, came before the directors on the 4th August, 1808, when "they resolved that the works be not proceeded with until further orders;" and on the 16th November,

"Resolved" (Mr. Trevithick and Mr. Rastrick attended the meeting) "it having been stated to the directors by Mr. Trevithick that from unforeseen difficulties arising in continuing the present drift through the quicksand and broken ground on the north shore, the same will be attended with much greater expense than was originally estimated; and he having at the same time suggested another mode of making the tunnel at a much less expense, the directors think it expedient that such plan should be taken into consideration."

"Resolved, that the following gentlemen be requested to take the said plan into their consideration: General Twiss, Lieut.-Colonel Mudge, Lieut.-Colonel Shrapnell, and Sir Thomas Hyde Page, with a request that they will report to the Board their opinion thereon."

"On the 18th March, 1809, Mr. Healing, the attorney of Mr. Trevithick, attended this meeting, and stated he was authorized by Mr. Trevithick to accept 800*l.*, in addition to what he had received, in full of all demands on the company, or to consent to leave his demand to arbitration; but the directors not consenting to so large a sum, Mr. Healing withdrew, having no authority to accept a less sum."

The following plan, together with a description of the work done, was issued by the directors in March, 1809, soliciting opinions from fresh engineers; but the writer believes this attempt to tunnel the Thames never advanced another inch:—

"Fig. 1 is the section of the shaft sunk on the south shore, lined with 9-inch brickwork laid in cement impervious to water.

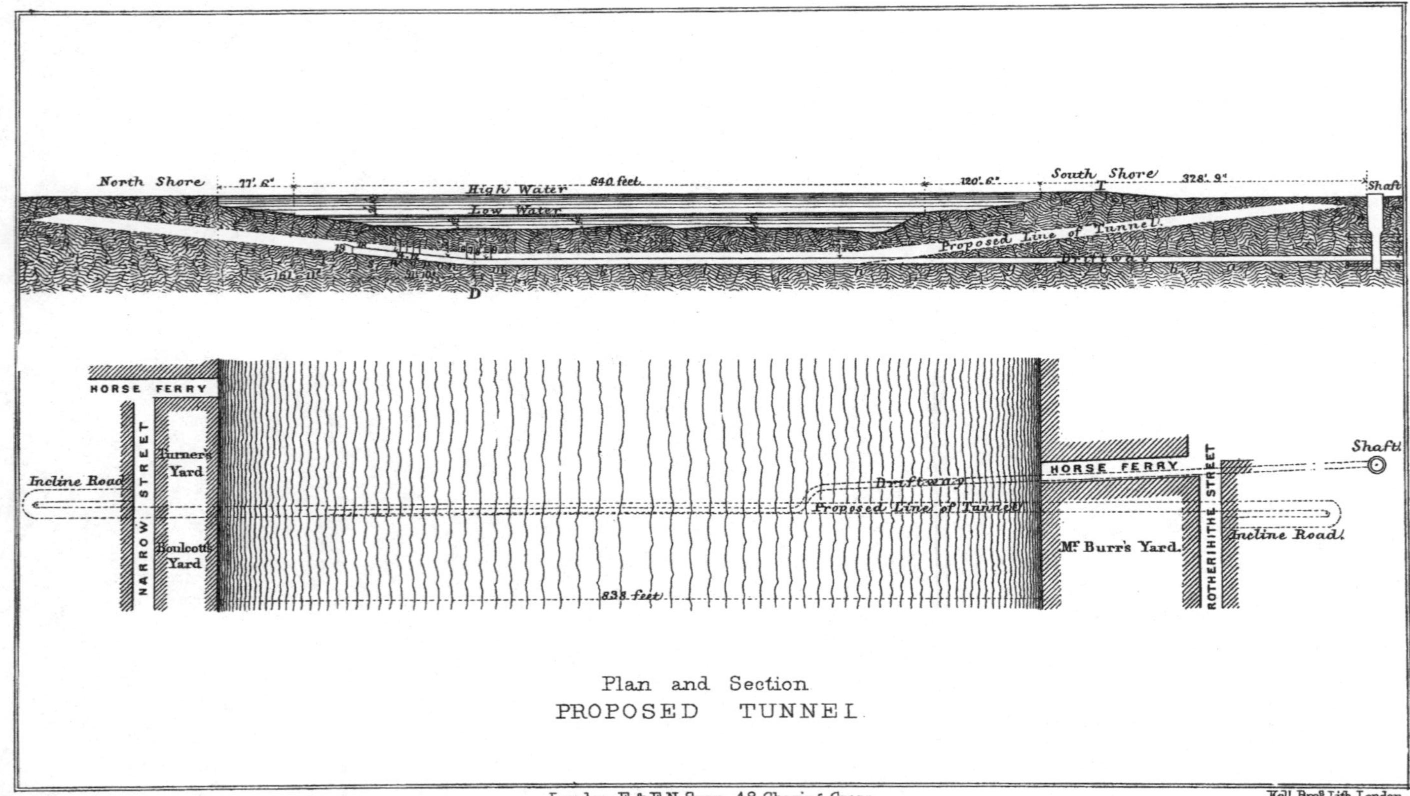

Plan and Section
PROPOSED TUNNEL

London: E. & F. N. Spon, 48, Charing Cross.

Kell, Bros Lith. London.

Description of the Strata through which it passed.

		ft.	in.		
No. 1.	Stratum consisting of brown clay	9	0		
„ 2.	Loose gravel with a large quantity of water	26	8		
„ 3.	Blue alluvial earth inclining to clay	3	0		
„ 4.	Loam	5	1		
„ 5.	Blue alluvial earth inclining to clay, mixed with shells	3	9		
„ 6.	Calcareous rock, in which are imbedded gravel stones, and so hard as to resist the pickaxe, and to be broken only by wedges	7	6		
„ 7.	Light-coloured muddy shale, in which were imbedded pyrites and calcareous stones ..	4	6		
„ 8.	Green sand with gravel and a little water ..	0	6		
„ 9.	Green sand	8	4		
				68	4
From the surface of the ground to high-water mark				8	0
	Depth of the shaft from high-water mark ..			76	4

"The gravelly stratum No. 2 in the shaft extends about 400 feet into the river from high-water mark at T, to V: at this latter place it is about 2 feet thick, and underneath is alluvial earth approaching the nature of clay.

a The entrance of the driftway (further described in Fig. 2), 5 feet high, 3 feet wide at the bottom, 2 feet 6 inches at the top in the clear.
b b The framing of the drift, consisting of 3-inch plank.
c The platform.

"Fig. 2 (shown in Plate X.) is a section, on a smaller scale, of the river, with the shaft and driftway.

"Fig. 3 is a plan of the same.

"In proceeding with the driftway towards the north shore, the strata were constantly varying. The following is a description of the strata as they appeared at the face of the drift, at the several places specified. The variations in the intermediate spaces were not noted, but the surface of each stratum was nearly even for the greater part of those spaces.

	ft.	in.	ft.	in.
Face of the drift at the entrance from the shaft, measuring from the bottom upwards,—				
Green sand	4	6		
Gravel	0	6		
			5	0
At *a*, being 177 feet from the shaft,—				
Green sand	4	0		
Gravel	0	6		
Blue muddy shale	0	6		
			5	0

		ft.	in.	ft.	in.
At *b*, 234 feet,	Green sand	3	9		
	Gravel	0	3		
	Blue muddy shale	1	0		
				5	0
At *c*, 295 feet,	Green sand	3	7		
	Gravel	0	3		
	Blue muddy shale	1	2		
				5	0
At *d*, 317 feet,	Green sand	3	5		
	Gravel	0	4		
	Blue muddy shale	1	3		
				5	0
At *e*, 321 feet,	Green sand	3	3		
	Gravel	0	4		
	Blue muddy shale	1	5		
				5	0
At *f*, 333 feet,	Green sand	3	3		
	Gravel	0	4		
	Blue muddy shale	1	5		
				5	0
At *g*, 350 feet,	Green sand	2	8		
	Gravel	0	4		
	Blue muddy shale	2	0		
				5	0
At *h*, 493 feet,	the green sand ends.				
At *i*, 730 feet,	Hard calcareous rock, mixed with loamy land			5	0
At *k*, 799 feet,	Hard rock			5	0
At *l*, 858 feet,	Ditto			5	0
At *m*, 901 feet,	Ditto			5	0
At *n*, 931 feet,	Rock, with a little sand and shells, and water in the roof			5	0
At *o*, 945 feet,	Hard rock	2	6		
	Clay and shells	2	6		
				5	0
At *p*, 996 feet,	Rock	0	3		
	Clay	0	4		
	Shells	2	0		
	Clay	1	0		
	Cockle shells	0	4		
	Clays and shells	1	0		
	Sand	0	2		
	Clay	0	6		
	Sand	0	5		
				6	0
At *q*, 972 feet,	Clay and shells	4	0		
	Sand	1	0		
				5	0
At *r*, 992 feet,	Clay and shells	0	8		
	Sand	4	4		
				5	0
At *s*, 1011 feet,	Sand	3	6		
	Clay	1	6		
				5	0

"The quantity of water in the gravel, No. 2, was so considerable, that a 14-horse engine could only keep the water a few feet below its natural level, and the shaft was sunk through by far the greatest part of this stratum into the blue stratum No. 3, with the water standing in it to the depth of several feet. It is well ascertained that this stratum of gravel extends through a considerable part of the adjoining country; but borings being made in the shaft from the bottom of this stratum, no water was met with in the substrata to the depth of 86 feet from high water, where a spring was discovered, which rose in a few hours, through pipes inserted for that purpose, to a higher level than that in the gravelly stratum No. 2. The shaft was therefore sunk only to the depth of 76 feet 4 inches.

"The drift was then carried forward in a horizontal direction to the north, 559 feet. And, in order to explore the ground in the northern part of the line of the then proposed tunnel, the drift was turned to the west 23 feet 6 inches from the centre of the former line to the centre of the new direction, and then to the north (intended to be enlarged afterwards to the size of the tunnel), and carried forward 341 feet, making the distance from the shaft to the beginning of the rise at D 922 feet. Through the whole of this line no material interruption occurred; the strata consisted of firm sand, calcareous rock, and concreted gravel, with no more water than was easily kept under by a 14-horse engine.

"At this point D the drift was made to incline upwards at the rate of 1 foot in 9. In prosecuting this part of the drift, at the distance of 23 feet from the beginning of the incline, the earth in the roof broke down, and discharged a great quantity of sand and water into the drift. At the time this circumstance happened, a space of only 6 inches by 30 of earth in the roof and none in the face was left untimbered; and through this space the earth kept falling by degrees, until a hole was formed capable of letting a man stand up in it, who perceived a quicksand, about 3 feet thick, and about 4 or 5 feet above the roof of the drift. The stratum between the drift and sand was clay; and water flowed from the sand. The hole was after some difficulties filled up, and the works proceeded.

"From the observations which had been made in the progress of the drift, the engineer found that the strata dipped slightly from the south to the north, and concluded that the gravelly stratum No. 2 in the shaft would end in quicksand. This inference was confirmed by borings in the north shore at E, and by the fact that the wells there are much deeper than on the south. In expectation therefore of drawing off the water from the face of the work, borings were made at Nos. 6, 7, 8, 9, and pipes were forced up to the top of the quicksand at those places, which had the desired effect. The water came free from sand for a considerable time; but when the sand began to come through any of the pipes they were plugged up, and others occasionally inserted in different places, to the south of these, with the same object in view, and which kept the face of the work dry. By this means, and by using the utmost precaution in all other respects, the drift was afterwards extended 70 feet beyond this fracture; where the roof broke down a second time, and sand and water entered the driftway with great violence, and to an alarming degree; so that in about a quarter of an hour the water rose in the shaft nearly to the top of it. On examining the river an opening or hole at *w* was discovered in the bed of about 4 feet diameter and 9 feet deep, and its sides nearly perpendicular. Into this hole clay, partly in bags, and other materials were thrown sufficient to fill it up; and which succeeded in stopping the communication between the river and the drift. The face of the drift was again opened, but the men could make but little progress, as the water and sand frequently burst in upon them, and drove them away. The pipes Nos. 1, 12, 13, and 14, were put up, and the drift was extended 20 feet 6 inches farther, in nearly a horizontal direction, through the quicksand. At pipe No. 14 the first 14 inches above the roof was clay, and 3 feet of sand above it. The face was then timbered up, to prevent any further fall of earth or sand; and the pipe No. 18, 9 feet long, forced upwards diagonally, and the pipe No. 19 horizontally. The first 8 feet through which the pipe 18 passed was blue clay, and the last foot quicksand, of which a considerable quantity immediately flowed into the drift. This pipe soon became clogged up, it is presumed with clay, as some lumps came through nearly as large as the diameter of the pipe. The pipe No. 19 was 8 feet 6 inches long, and discovered nothing but blue clay; no sand nor water came through it.

"At this period the engineer reported that he had examined the bed of the river, and found the hole at *w* considerably increased both in width and depth, and the earth at *x* very much sunk; and that he had no doubt these two fractures communicated underneath. He then gave it as *his opinion* that an *underground* tunnel could not be made in that line, unless the fractures were covered by caissons, without which the further progress of the drift would be useless; but that he had no doubt of being able to make a tunnel over the *same* line through the river, sufficiently deep into its bed, by means of movable caissons, or coffer-dams, and at a less expense considerably than the original estimate for the underground plan, and *without any impediment to the navigation of the river.* Under these circumstances the further progress of the works was suspended.

"It is an important consideration with the company, that the size of the tunnel be large enough to admit two carriages to pass each other; or two of smaller dimensions, each to admit a carriage.

"The company contemplate a foot tunnel, only in the event that a larger one should appear to be impracticable.

"The plans must be formed with regard to the tunnel being lighted.

"N.B.—That plan whose line is the shortest, and ascent the easiest, will have great claims to preference, if equal in merit in other respects.

"By order of the directors,

"S. W. WADESON, Clerk.

"AUSTIN FRIARS, LONDON, *March* 30*th*, 1809."

Trevithick had not counted on this abandonment of the work; for at the time when the directors put aside the plans of their engineer, "and thought proper to invite ingenious men of every description to a consideration of the best means of completing so useful and so novel an undertaking," he was in communication with his brother-in-law, Mr. Henry Harvey, to arrange the preparing and sending to London "300 tons of scantled Cornish granite fortnightly," and his old friend William West was to superintend the cutting and shipping of the stone.

As the free movement of ships in the river prevented the use of a bridge, the chief object in this underground passage was to avoid the great inconvenience of the boat-ferries across the river, between Rotherhithe and the large commercial docks on the south side of the Thames, and the narrow street on the north side, near the river entrance of the Regent's Canal at Limehouse, not far from the West India Docks, and about 2½ miles below London Bridge.

Trevithick was engaged on this driftway a year and a half; during less than half of which active operations were in progress, while the remaining portion was taken up with vacillations and weakness in the directory, and interference and discord from the discontented shareholders.

After a lapse of nearly twenty years, Brunel, in 1825, commenced the Thames Tunnel, which, after great difficulties and unlooked-for expense, was opened to passengers in 1843, from Rotherhithe across the river to Wapping, near the London Docks, about a mile and a half below London Bridge; it consisted of a double tunnel made of brick, something like Trevithick had recommended; but instead of using the proposed caisson, Brunel followed the principle applied in the drift, of underground working through the river-bed, securing the sides with brick or other material, and the working end by wood or iron, or other means, allowing small openings through which the earth was removed, which could be closed when quicksand or water overpowered the workmen. Brunel called this end apparatus a movable shield; it was a large complicated structure, allowing about thirty-six men, each in a small compartment, to work in the end. In 1837, Sir Isambard Brunel informed the writer that the plans and papers on the driftway left by Trevithick had given him the highest opinion of his character and genius.

In 1869 the Thames subway was constructed by Mr. Barlow between Tooley Street and Tower Hill, about half a mile below London Bridge; consisting of a cast-iron tube, 7 feet in diameter, put together in pieces.

The general outline of procedure was similar to that in the driftway, except that the work was secured by cast iron in lieu of wood. Trevithick's drift was of wood, Brunel's tunnel of brick, Barlow's of iron.

The many scientific and practical men brought into contact with Trevithick during the progress of the driftway helped very much to diffuse the knowledge of his high-pressure engines and railway locomotion.

Among the seventeen gentlemen who reported on his work were Rennie, Braithwaite, and Rastrick, all afterwards eminent in the engineering world, and others equally well known. Lieutenant-General Twiss, Lieutenant-Colonel Mudge, Lieutenant-Colonel Shrapnell, helped with their particular experience; and Mr. William Stobart, of Durham, and Mr. John Buddle, of Newcastle-upon-Tyne, experienced miners, were also present. All these persons undoubtedly saw Trevithick's high-pressure engines at the Tunnel and elsewhere in London; the dredging machines then at work in the river; and the locomotive and passenger railway at that time carrying passengers at a shilling a ride near Euston Square. The Newcastle and Durham coal proprietors in particular must have seen the engines, as they knew that Trevithick had sent a colliery locomotive to Newcastle and to Wales, and were especially interested in such matters.

CHAPTER XIII.

IRON TANKS.

Trevithick and Dickinson's Specification, 31*st October*, 1808.

"INSTEAD of the packages, cases, chests, trunks, casks, vessels, or other receptacles heretofore and commonly used for the purpose of containing, enveloping, preserving and securing from damage the several articles of merchandise and other goods, whether in the solid and consistent or in the liquid form, which are taken and stowed on board ships and other vessels to be transported or consumed, or otherwise used and applied, we do construct, make, use, and apply certain other packages, vessels, or receptacles of iron, made by casting the said metal, or by forging, laminating, and riveting together plates or portions of the said metal, with covers adapted to and capable of being secured by screws, bolts, and other known means, to the said vessels or receptacles, so as to render the same (when shut) close, secure, and impenetrable to the external air or moisture, or other hurtful matters and things. And further, we do make our said packages, vessels, or receptacles of such figures or forms that they fit exteriorly to each other without that waste of space which takes place in the stowage of wooden casks. For this purpose different forms may be used, but we prefer rectangular or hexagonal prism forms to all others. Where the same economy of room is not requisite, we employ the cylindric form, but whichever form be employed a much larger quantity of goods can be stowed by means of our Invention than can be stowed in an equal space when the goods are put into packages made of wood, the sides and ends of which are necessarily of a great thickness compared with those made of iron. Nor is saving of room the only advantage which results from our aforesaid Invention. Water, oil, and various other fluids, as well as provisions of different kinds, will be better preserved from waste, putrescency, leakage, depredations from insects and other living creatures, in iron vessels than in vessels made of wood. For some purposes we have our said iron packages, vessels, and other receptacles tinned on the inside, or coated with a varnish suited to the commodities they are destined to contain.

"By our said Invention great advantages as to space and comfort will be afforded to those on board ships, without any detriment, but rather with advantage, to the proper trim of carrying sail, and the safe and expeditious performance of their respective voyages.

"In witness whereof, we, the said Richard Trevithick and Robert Dickinson, have hereunto set our hands and seals, the Twenty-eighth day of April, in the year of our Lord One thousand eight hundred and nine.

"ROBT. DICKINSON. (L.S.)

"RICHARD TREVITHICK. (L.S.)

The dredgers, the Thames Driftway, the London locomotive railway and engine, were all in hand about this time, but were not sufficient to give full scope and occupation to the never-tiring energy and fruitful invention of Trevithick. The few lines of his patent specification, though bearing mainly on tanks of iron for holding liquids on board ship, have also reference to the general question of conveying small packages liable to damage, stowed in cases of iron, tinned or varnished, sometimes in the form of a cylinder, to be rolled to or from the ship to the storehouse,—even the trim and sailing qualities of the ship were to be improved by their use; and the agreeability of the drinking-water on board ship was to be greatly increased by being stored in iron instead of in wood vessels. This latter idea originated when observing the purity of water in an old boiler that had been used at the Thames Driftway.

"When a little boy, shortly after reaching London from Cornwall, about 1808, my father, on coming into the house on a Sunday morning, desired me to fetch a wineglass, and taking me by the hand, walked to the old yard near the Tunnel works. There was an old steam-engine boiler in the yard; my father filled the glass with water from the boiler gauge-cock, and asked me to tell him if it was good water. We used to speak of this as the origin of the iron tanks."[1]

This happened in 1808, when removing certain machinery at the Thames Driftway, to make room for one of his 30-horse-power high-pressure engines. The water which had found a resting-place on iron was very clear and fresh-looking,—an old boiler was full of clear water, without unpleasant taste or smell. In a moment it was plain to him that an iron water-tank was preferable to one made of wood, and that by its use a serious evil of the day might be removed, for at that time sailors were obliged to drink stinking water, loaded with living and dead animalculæ, and other impurities, and to have even of this a smaller allowance than was required for washing and cleanliness. Iron tanks could be fitted into the spare corners of vessels in a way that wooden casks could not, and both the quality and quantity of water would be increased. "The greater advantage as to space and comfort afforded to those on board ships, without any detriment, but rather with advantage, to the proper trim for carrying sail; and the better preservation of water, oil, and various other fluids from putrescency, leakage, and depredations from insects, in iron vessels, than in vessels made of wood."

In February, 1809, the patentee published a pamphlet describing the advantages derivable under the patent.

[1] Recollections of Mr. Richard Trevithick, residing at Pencliffe, Hayle, 1869.

"The wood alone in a provision cask of Mr. Jellish's, containing 41 gallons, occupies a space equal to 19 gallons. A water-cask of one tun, or 252 gallons, has a loss from wood and space equal to 76 gallons. An iron cask being only $\frac{3}{16}$ths of an inch thick, containing the same quantity, occupies the space of only 7 gallons, giving an advantage of 69 gallons of room saved in every tun of water, or 26 per cent., exclusive of the space occupied by the fathom wood, and the difference between squares and circles in stowage.

"The improvements now submitted consist in *the adaption to the holds of vessels of a system of wrought or cast iron cases of a square, polygonal, or any other form, by which they may adapt themselves to each other without leaving any interstices, or made cylindrical where room can be spared.*

"A ship engaged in the whale fishery, belonging to Messrs. Bennet and Co., lately returned from a successful voyage, with what is called a full ship. This vessel is of the burthen of 208 tons admeasurement, and yet when her hold was completely full of wooden casks, the whole quantity of oil they contained was only 140 tons, being full one-third less than what she might have stowed in differently-formed packages.

"The patentees now propose to furnish this ship with a ground-tier of wrought-iron tanks, 4 feet square by 8 feet long; each of these will contain 4 tons. The lower part of the hold will then stow 9 of these casks lengthways and 5 breadthways, being 45 casks, which together will contain 180 tons. This quantity of iron stowage will, of course, occupy a depth of only 4 feet, at the bottom of the hold, and her full tonnage remains to be made up with an additional quantity of oil, leaving a space in the hold unoccupied, from the top of the tanks to the deck, of 6 feet high, which space would stow considerably more than her admeasurement of 208 tons.

"With respect to the Royal Navy, the patentees presume to think that the adoption of iron stowage would be advantageous in a more than common degree. If ships of war were provided with metal tanks for containing their beer, water, provisions, and stores, would not the necessity for ballast be in part done away?

"This alteration would be attended with but little expense,

as the iron ballast at present employed might be speedily converted at the foundries into iron tanks, adapted for the ground-tier of the hold. In a ground-tier iron cases might be made of a greater thickness for the sake of securing a proper sailing-trim when at sea; they will contain a much larger supply of fresh water in the same bulk than wooden casks, and when it is necessary to lighten the vessel in naval manœuvring, these iron tanks may be speedily emptied by means of pumps and hose adapted to their orifice, and afterwards filled with sea-water, when circumstances require an addition of ballast.

"As an illustration of the saving in point of expense, it may be mentioned that a first-rate man-of-war generally takes on board wooden casks for 600 tons of water; if supplied with iron stowage, there will be a saving of 2400*l.* To this saving may be added the annual expenses of hoops, cooperage, &c., which in vessels of these dimensions are never less than 500*l.* a year; in short, from their present material, a total saving would accrue to Government of 500,000*l.* per annum, by the adoption of iron casks alone, exclusively of the preservation of the stores, health of the crew, and advantage in stowage, which is incalculable."

A ship of 208 tons, returning from the whale fishery with a full-stowed cargo in wood casks, gave 140 tons of oil. The same vessel, fitted with a ground-tier of wrought-iron tanks, each of them 4 feet square and 8 feet long, carried 180 tons of oil, and still had 6 feet in height unoccupied between the top of the tanks and the deck of the vessel.

Ships of war should carry liquids in iron tanks fitted to the form of the bottom of the hold, either to serve as drinking-water or as ballast; so that in "naval manœuvring" the vessel might be deepened, thereby exposing a less amount of vulnerable surface to an enemy, and again lightened by the nautical steam labourer pumping the water out of the tanks.

Should the Government object to the cost of wrought-iron tanks, their cast-iron ballast, "Seely's pigs," should be recast into tanks suitable to the form of the hold.

Trevithick was requested to meet some scientific and medical men at a Navy Board that those great advantages might be explained. Rising from the consultation in a huff, he exclaimed, "I had expected to find gentlemen who would understand the matter, but you seem to me to be a lot of old women." He then put on his hat and walked out.

On another occasion he recommended his iron tanks at the Admiralty Office, and was told that water from them would poison all the sailors in His Majesty's service. The question was referred to Sir Joseph Banks, who reported that "water kept in iron tanks looked very well, smelt very well, tasted very well, and produced no injurious effects on those who had drunk it." Trevithick's two eldest children[1] spoke of having fastened together pieces of cardboard to represent the forms for iron tanks and buoys, and were promised a pretty model of a ship, about 2 feet long, having little blocks of wood, representing iron tanks, fitted to the hold of the vessel, and kept in their places by wooden pins. All the deck part could be lifted off, that the tanks in the hold might be more easily seen.

This model was sent to the Admiralty, and neither Trevithick nor his family ever saw it again.

"Capt. Aldridge,[2] commanding a Government vessel, solicited permission from the Navy Board to have some iron tanks placed in his vessel: it was refused. He procured a small one at his own expense."

"Capt. Alder knew the admiral had some iron tanks sent on board, in opposition to the orders of the Victualling Department at Deptford."

---

[1] Residing at Hayle in 1869. [2] Lately residing at Ilfracombe.

"Mr. Cockburn sailed on board the 'Amphion,' 32-gun ship, in 1813. Iron water-tanks were fitted in the hold; a pump at the main hatchway filled the deck water-tank and the cook's coppers; by this means the ship's company were saved a great deal of labour. Before the iron tanks were used, the watch was employed the greater part of the afternoon in hoisting up and lowering down water-casks for the supply of the ship's company."

Colliers having iron tanks in the bottom of the hold would by their use avoid the expense and delay of ballasting, and consequently perform a greater number of voyages. This wise plan, so fully made public sixty years ago, both by practice and precept, has now come into general use.

Ballasting of Ships.

"This is effected in a cheap, efficacious, and convenient manner, by employing tanks filled with water. The tanks used for this purpose are made of iron, and of shapes and forms adapted to the vessels or trade in which they are to be employed, actually occupying no more space when not employed for the purpose of ballasting than the bulk of the iron itself—of which they are made—namely, in a collier, about the space of half a chaldron of coals, for they are filled with coals when not used for ballasting. By this contrivance the expense and delay occasioned by taking ballast on board, and discharging it from time to time, are entirely done away, and a collier will be enabled to perform an extra voyage, or probably two extra voyages every year.

"Greenlanders go out in ballast, and are exposed to considerable risks from its shifting in stormy weather. By adopting iron ballast tanks, the expense and danger of common ballast will not only be avoided, but the expense of wooden casks for the home cargo will be lessened by filling the tanks with oil, while the ship will also be enabled to bring home a much larger cargo, and with less waste, the iron occupying much less space than wood, and absorbing none of the contents."

The Admiralty were moved to use the water-tanks,

and Trevithick asking for an acknowledgment of their usefulness was answered by the following letter, written on the 9th April, 1811:—" I am commanded by my Lords Commissioners of the Admiralty to acquaint you that no report has been yet received relative to the iron tanks invented by you for stowing water."

During 1808, while Trevithick hoped for continued occupation on the Thames Tunnel, and profit from his patent inventions, he strongly advised his wife to leave Cornwall, and with her four little children, to join him in London. This troublesome journey was undertaken in the autumn of that year. A house had been selected at Rotherhithe, in a dingy situation, near the mouth of the driftway. There had been much correspondence about the wisdom of this move. Mrs. Trevithick's brother, Mr. Henry Harvey, advised her not to leave her home and friends, until things were more settled and more certain in London. Trevithick's notes to his wife, however, made everything easy and agreeable. More than 300 miles had to be travelled in a post-chaise, occupied by herself and her four little ones, the youngest of them a baby. The contrast between her clean and fresh Cornish home, and the habitation at Rotherhithe, did not help to remove the fatigue of the journey, and a further disappointment awaited her. In her husband's pocket were two of her last letters unopened What reasons could possibly be offered for such hard-hearted ingratitude? Trevithick's answer to the charge was simply, " You know, Jane, that your notes were full of reasons for not coming to London, and I had not the heart to read any more of them."

On the breaking down of the Thames Driftway, they removed to No. 72, Fore Street Limehouse In the yard Trevithick set up workshops for constructing iron tanks and buoys, and model iron ships; to his wife this noisy residence could not have been much more agreeable than gloomy Rotherhithe.

During these trials Mrs. Trevithick had the consolation of making the acquaintance of a friend in adversity. Mr. Vigurs, a Cornish acquaintance, had married a lady, driven by the French revolution from luxury in her native land to comparative poverty in London. The two ladies consoled one another over a cup of tea and a Cornish pasty at Limehouse; and on the return visit to Bond Street by a sample of French cookery.

The tank manufactory was under the foremanship of John Steel, with the wooden leg, who had been with Trevithick in South Wales and Newcastle-on-Tyne on the locomotive experiments. Mr. Savage and Thomas and Rudge were also making iron buoys under the patent, twelve of which had been supplied to Government. Samuel Hambly, who had worked on the Camborne locomotive and succeeding schemes, kept Steel company, both often having to wait for their wages until it was convenient to pay; this fraternity between Trevithick and his men extended through the Hambly family to a brother, old James Hambly, who taught the writer to use mechanics' tools, because Captain Dick was such a wonderful man.

"72, Fore Street, Limehouse, London,
"Mr. Trevithick, "*April 20th*, 1810.

"Sir,—Above you have a sketch of an end section of a vessel I measured this morning, for a ground-tier of iron tanks to go into the South Sea Whale Fishery, belonging to Mr. Blyth, sail-maker, Limehouse.

"The whole length of the tanks will be about 50 feet; across the upper surface, 14 feet 6 inches; the other two sides forming a triangle (to suit the bottom of the vessel), the perpendicular line of which is 5 feet; the whole containing about 55 tons.

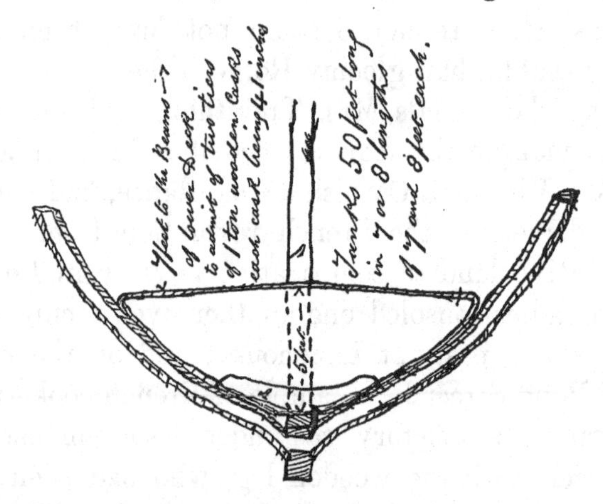

The tanks will have to be made in lengths of 7 and 8 feet each, so as to stand in between the flooring pieces A and uprights A, which are ranged all the way along the keelson of the ship, at the distances of 7 and 8 feet.

"The vessel being so extremely sharp, the tanks must be made of a peculiar form to fit it; of course will not easily fit any other; on that account it will be necessary to form a contract with Mr. Blyth for the time when to be completed, and the price per ton, together with the mode of payment. A contract must also be made with the manufacturer as to price and time of completion; otherwise we are running a great risk in having them thrown on our hands, the ship being in so great a hurry for them.

"On Thursday last Mr. Savage received a letter from Mr. Dickinson recommending an improvement in the ends of the buoys, for the better purpose of (as he calls it) *keel-hauling*, by having them made round at the ends, and a staple for making the rope fast to, one of which they have made, the staple of which is made fast with four small rivets, scarcely sufficient to bear the weight of the buoy itself, exclusive of its buoyancy.

"After descanting a good deal on the superiority of this

improvement, he tells him not to finish or send any more down here, but to have the twelve for Government ready for his delivery on his arrival home, and then concludes with an insinuation *that these are not his only reasons* why he does not wish any more to be sent here; and it was with the greatest difficulty and through the greatest persuasion I was enabled last week to procure the two promised to Mr. Pickering before you left home.

"Mr. Savage has made one with rounding ends brazed on, and the rings riveted in, the same as in the old method.

"This I conceive would answer a very good purpose, provided they had made the end pieces of stronger iron than the buoy; but says Mr. Savage if he pursues this method the prices will have to be advanced, and in my opinion they are already as much as the buoys will bear.

"Seven of the largest buoys for Government are finished, the other five want a few hours' work. Five more are about three-parts finished. And in this situation have things stood all this week, in consequence of Mr. D.'s leaving orders on Monday morning at Mr. Savage's not to proceed any farther till he saw them again; and the Lord knows when that will be! as he has not been seen there since.

"Mr. Savage was at Mr. D.'s house on Tuesday, and was told he would be in to dinner.

"I called yesterday, and was told he was gone into the country the day before; but where he was gone to and when he would return again I could get no intelligence.

"I shall go again to-morrow to Somerset House, and shall call on Mr. D., and if he does not order to the contrary, I will call at Thomas and Rudge's and hear what they have to say about the making of tanks already mentioned.

"I have to see Mr. Blyth again when he returns from Gravesend and Deal, which will probably be on Saturday or Sunday; when I shall very likely write to you again.

"In the meantime I should like to hear what you have to say on the subject.

"I remain, Sir,

"Your most obedient servant,

J. STEEL."

This was followed by a letter apparently from Dickinson. The name and date are torn off.

"MR. TREVITHICK,

"I have only just received the letter sent you by Cockshott's attorney so long ago as the 6th, and which you had detained till the 11th, although it states that the expenses *are going on, &c.* This is most unaccountable.

"Steel has never yet furnished me with Thomas and Rudge's account, made out as I desired a week ago. This is a conduct not to be allowed.

"I have a letter before me just received from John Steel, Horseferry, for 90*l.* for barge hire. These with the acceptances I am under, and the sums I have paid already, make me begin to look about me, and for once to express myself dissatisfied generally.

"It is but of little use I know, and for me to attempt to take any part in the management were ridiculous: and even Mrs. T. has flatly objected even to my sons being in the house; so that on the whole, to know anything about it, I must give myself wholly up to it.

"I am much dissatisfied with the neglect of Steel as to the keeping accounts, as well as other things; seeing that half his time, I shall say *nine-tenths,* he does little or nothing. It were easy for him to have *made* a temporary set of books for the present, and to have entered every transaction, had it been only for my satisfaction.

"The way we are going on I shall tire of, finding sums of money, and *nothing done.*

"It were better that Penn's engine had never been undertaken, than that it should keep back, as I suspect it does, *all other things.*

"A coal engine, with the caboose, might and should have been *ready.*

"The *first* towing machine should have been *got on with* immediately on the others failing; not an hour ought to have been lost in applying to your friend Davies Giddy, as a stepping-stone to Lord Melville.

"Since the tanks came home from the unfortunate Margate expedition, Steel might have been delivering the printed letters to the captains, and trying, at least, to get them on board; but really this man appears to me to be of very little use in getting things forward. The buoy scheme might, through him, have been forwarder; in short, he might if he had been active, have been doing many things which have stood still.

"On the whole things begin——"

Steel had to obey both Dickinson and Trevithick, and did not approve of the former's interference in engineering matters. Trevithick put the lawyer's letter in his pocket, and allowed it to remain there unnoticed. Steel, like Trevithick, was a poor hand at furnishing accounts. Things were going wrong; money went out faster than it came in. Dickinson wished his sons to look after what was going on at the house and workyard, but they were refused admission to the house by Mrs. Trevithick. An account-book in Trevithick's writing shows that John Steel, Samuel Hambly, and Samuel Rowe were in constant pay in the workshop. Steel was a Newcastle-upon-Tyne man: he had been with Trevithick for seven or eight years; the first mention of his name being in connection with the Welsh tramroad locomotive as engine foreman in 1804.

Samuel Hambly, a cousin of Mrs. Trevithick's, and Samuel Rowe were still older hands, having worked upon the first high-pressure experiments and the Camborne, London, and Welsh locomotives: both Hambly and Rowe were Cornishmen; these men performed an important share in bringing those inventions into use, and often continued to work when there was not the means of paying them their wages.

The unfortunate Margate expedition is very characteristic of Trevithick's readily applying himself to new requirements; two vessels and sixteen men left London for Margate to raise a sunk ship. After nearly a month's labour, an attempt was made to fasten iron tanks to the vessel: one man refused to wet his feet, another was willing to fasten the tanks, but refused to put the bolts into their places for the purpose of fastening; others would not go unless they could be brought back to their lodgings on shore, without the risk of being obliged to sleep on board a gun-boat. It was a case of mutiny. The account goes on to show that during forty-two days the men only worked on the sunken wreck six hours each, at an average cost per man of 23*l.*

The late Mrs. Trevithick informed the writer that the vessel was really raised, and was being floated into shallower water, when a dispute arose about the payment. Her husband wished an immediate "yes" or "no" to his bargain. The shipowners wished to defer the answer until the ship was safe. In a moment the tanks were cast loose, and the vessel was again a sunken wreck.

"DESCRIPTION OF THE PROCEEDINGS AT MARGATE.

In Trevithick's handwriting:

"*January* 30*th*, 1810.—Trevithick and Dickinson engaged sixteen men with two vessels and crews to go from London to Margate to recover a sunken ship.

"These men were engaged at 10*s.* 6*d.* per tide when at work, and 7*s.* 6*d.* when idle or not able to work. They were provided with victuals and drink and lodgings.

"On the 4th February, it being low water at five o'clock in the evening, Mr. Trevithick called on George Nicholls and four of his partners to go off to the sunken ship, which lay about four miles from the land; a gun-brig being stationed by order of the Admiralty, near the ship, for the purpose of protecting those men and the wreck, and assisting us.

"About three o'clock in the afternoon of the 4th February, the boats belonging to the gun-brig were sent on shore to take these men off to the wreck: George Nicholls and four of his partners refused to go; one of them said he should not wet his feet, another said he did not come down to bore holes, and put screws into the ship's sides; that when the work was done, ready to put on the tanks, that then he should have no objection to go off to the ship and lash on the tanks, but he would be d— if he did any more.

"The other three said they would not go, unless the gun-brig boats would bring them back again to their lodging that same evening.

"My answer was that I would request the officer in the boat to take them back that evening, and I did not doubt but that he would; but if he would not take them back, the tide would not permit us to work on the wreck later than seven o'clock that evening, and they would then go on board the gun-boat and wait until ten o'clock, at which time there would be water to float one of our own ships out of Margate Harbour, and then they could go to their own hammocks on board our own ships. Their answer was, we will be d— before we will go on board the gun-brig."

This application of iron tanks must have been widely known, for shortly afterwards he was consulted by Lloyd's on the means of raising, under similar circumstances, a sunken ship.

"CORNWALL, CAMBORNE, 26*th March*, 1812.

"MR. BRACKENBURY,

"*Lloyd's Agent, Liverpool.*

"Sir,—I have yours of the 14th, enclosed to me by Mr. Giddy, in answer to which, respecting the lifting of sunken vessels by wrought-iron tanks, the practice has been tried in London, and fully answered the purpose of raising vessel and cargo.

"I can make and send to Liverpool any number or size of tanks you wish. The tanks made for London would lift about ten tons each, but I should prefer a larger size, about 7 feet diameter and 20 feet long. Eight tanks would lift a 500-ton ship, unless she was loaded with iron or some very heavy cargo. The expense would be about 200*l.* each tank.

"Four tanks will lift a ship of 300 tons, unless loaded with an unusually ponderous cargo.

"If you wish to have a drawing or a model, to show the effect in a small way, please to write to me, and I will furnish you with every necessary for the purpose.

"I remain, Sir,

"Your very humble servant,

"RICHARD TREVITHICK."

The tanks used at Margate lifted each of them 10 tons, but larger ones would be preferable, 7 feet in diameter and 20 feet long. His mind was running on his contemplated Cornish boilers, the first of which on a large scale was made in the latter part of that year, and did not differ much from the dimensions of the proposed tanks. The number and cost of the tanks required to raise the sunken vessel, and a drawing of their application, or a model of the ship with the tanks attached, was offered, making the whole business plain and distinct as though the designing it was mere child's play, and the putting it into execution the work of an intelligent boy.

Trevithick made no profit by the valuable invention and practical introduction of iron buoys and tanks. Mrs. Trevithick said that when at that period her husband was imprisoned for debt for the money he had expended in making known the usefulness of the tanks, Mr. Maudslay visited him, and expressing his concern that so clever a man should be shut up in prison, asked what the debt was, and what he intended doing with the patent. Trevithick replied that the patent had done him more harm than good, though he was sure it was an excellent thing; and on a sum of money being tendered in exchange for the patent, he was once more a free man.

Harry Maudslay was then spoken of as an excellent smith, and was the founder of the engineering firm of Maudslay, Son, and Field; his handywork helped some of Trevithick's schemes, and he is said to have profited by the manufacture of the patent iron tanks, the right to which he had bought.

Penn at the same time, now more than sixty years ago, was also employed by Trevithick to construct a coal-discharging engine, with steam-cooking apparatus, and a steam tug-boat, under the patents of 1808 and 1809, together with some other marine steam-engine, then in hand, probably the adapting to steamboat propulsion the locomotive 'Catch-me-who-can' of 1808, whose first application had been in the dredging barge of 1803, and was now in 1810 to propel a Lord Mayor's barge. Trevithick's eldest son still recollects seeing in 1809 or 1810 in Cockshott's yard an old Lord Mayor's barge in which his father was placing his high-pressure steam-puffer engine.

Lord Melville's influence was solicited for the encouragement of the Government in making useful such novel schemes, but the want of support and debts incurred ended in an attack of brain fever and postponement of the useful marine steam-engine.

CHAPTER XIV.

SHIPS OF WOOD AND IRON, AND IRON TANKS.

Trevithick and Dickinson's Specification of 1809.

"THE first of our said inventions or improvements is, a movable caisson or floating dock, made of wrought-iron plates riveted together, or fastened together by screws, or in any other secure manner for docking a ship or other vessel of considerable size while riding at her moorings in any suitable depth of water, by means of which the said ship may be supported as in a common dock, and the water pumped out so as to leave her keel dry in a few hours, without removing any of her stores, masts, or furniture. The said caisson or floating dock may be made of any convenient size or dimensions, but for the sake of precision in our description thereof, we shall mention certain dimensions fit to receive the largest first-rate, at the same time that we shall describe its form and structure. The internal figure of the said caisson or floating dock resembles that of a boat, and it may be made of wrought iron half an inch thick, and 220 feet long and 54 feet wide and 30 feet deep, with a flanch 6 feet wide extending horizontally outwards from the upper edge for the workmen to stand upon, and also to strengthen the caisson; the weight of such a caisson or floating dock is nearly 400 tons, and it is surrounded by an air-chamber or by air-chambers riveted water-tight to its external surface, which renders it so buoyant that it will draw only 9 feet water when the said air-chamber or air-chambers is or are empty; and the said air-chamber or air-chambers doth or do consist of wrought-iron plates riveted together, so as to form a semicylindrical hollow protuberance extending along the sides of the caisson horizontally, and of a figure which in any of the vertical sections thereof would present a semicircle, or outline nearly approaching to a semicircle, which is the best outline, although other outlines may be used; and the said air chamber or chambers contribute much to strengthen the caisson or dock, and support the principal shores from the ship, and the strength of the caisson may be still further augmented, if necessary, by edge bars within. The method of using the said caisson consists in taking it to the ship intended to be docked, and then the water is to be let into the caissons by an opening with a valve at the bottom, and the caisson is to be suffered to sink until the upper part is even with the surface of the water, the air chamber or chambers still keeping it buoyant; a small quantity of air is then to be discharged by opening a plug-hole, or a valve in the air chamber or chambers, until a quantity of water has been let in just sufficient to sink the caisson, which is then to be drawn under the ship's bottom. This being effected, the caisson (of which the residual gravity or preponderancy for sinking is very little) is to be raised to the surface of the water by ropes made fast to the caisson from each quarter of the ship. A pump placed within the caisson is then to be worked by manual labour, or any suitable power; but we prefer for this purpose a steam-engine; one of 12-horse power, placed in a barge alongside the caisson, will empty it in three hours, and reduce the draught of a first-rate ship of war 8 feet, namely, from 26 feet to 18 feet, and then she may be carried up into shoal water if required, because the caisson will float with the draught of only 18 feet, while the ship she carries would have required 26 feet. As the ship is to be supported by props or shores as if in a dock, it is manifest that the external pressure of the water against the caisson will support the ship, and she will ride with all the stores on board, and masts standing, nearly as easy as when floating in the water. The advantages of this invention for inspecting the bottoms of ships in an extremely short time, and of docking them in places where that process would else have been impracticable, and of carrying them over bars and shoals, are too obvious to require any detail; and if at any time the caisson should, from blowing weather or any other cause, be thought inconvenient, it may be cast off and suffered to sink to the bottom, and may be afterwards weighed with as little inconvenience as raising an anchor.

"The second of our said inventions is, to build in the same manner as the floating dock, not small craft fit for canals and inland navigation, such having been long in use, but ships of war, East Indiamen, and other large decked vessels fit to navigate the ocean. In such ships we mean to make the decks as well as the sides of plates of wrought iron, riveted or joined by screws, not only to each other, but to the sides of the ship, and to support the decks with wrought-iron beams. Ships so constructed will possess the following advantages: they can be built at much less expense than wooden ships of equal dimensions, the material is not only a home production but inexhaustible; they will be much lighter and of greater capacity than a vessel of equal oustide dimensions built of wood, and will consequently carry a larger cargo, while they will be also much stronger, safe against fire, tight in a gale of wind, require neither caulking nor copper sheathing, will defy all mischief from vermin, will never become unhealthy, and will be extremely durable; they will also have this advantage, in a battle there can be no splinters; a shot will carry away no more of the ship's side than the diameter of the ball, the hole may be instantly plugged, and the damage can be repaired at sea.

"The third of our said inventions or improvements doth consist in making masts, bowsprits, yards, and booms, of wrought iron, out of plates riveted or screwed together in hollow or tubular forms. These masts being hollow tubes, the upper mast may be made to slide into the lower mast, as the inner tubes slide into the outer tube of a pocket telescope, or they may be struck exactly in the same manner as the wooden masts are at present, but we prefer the former method, because by making the masts in a sufficient number of sliding pieces, the whole may be lowered by means of ropes and pulleys successively, till all be as low as the deck; an arrangement which affords to a ship in the case of a storm that advantage which is endeavoured to be obtained by cutting away masts made of wood. When the storm is over, these masts may be raised again to their first situation and original utility with great ease. The bowsprits, when so wanted, may also be made in sliding pieces like the masts.

"The fourth of our said inventions or improvements doth

consist in the preparation of timber, by seasoning the same most effectually and in a very short time, and also by giving any required curvature during the seasoning process which shall afterwards continue perfect and invariable. For these purposes we place the said timber in iron chambers, disposed over a long horizontal flue which communicates between a fire-place or furnace and an upright or ascending chimney. The heated air and smoke which has passed through the fire, and is incapable of producing or maintaining any combustion in the wood, is admitted into the said chambers by regulated openings with dampers, and so thoroughly heats the said wood that the crude sap and pyrolygnic acid become evaporated and driven out, and may, if required, be collected for use in manufacturing processes; and by this treatment the wood becomes firm, hard, thoroughly dry, and well seasoned throughout its whole substance; and we do construct some of our said chambers with a curved side of such a figure as it may be required to give to the wood by bending; and we do, by means of screws, chains, or bearing pieces suitable to the figure intended to be produced, the form and application of which will be readily apprehended without any further instruction by any competent workmen, gradually compress and bend the said wood against the said curved side, during the time of the said seasoning heat, until the desired flexure is produced therein; and we do suffer the said wood, so bended and seasoned, to remain, until cold, before we proceed to slacken or relax the actions of the said chains, screws, or bearing pieces; or otherwise, instead of making a curved side to the chamber, we do place or fix within the said chamber such curved pieces or pins or reaction pieces as may be suitable to the effect and figure required to be produced.

"The fifth of our said inventions or improvements doth consist in a method of framing or putting together the hulls of ships or vessels. The said method consists in substituting, in the place of those pieces which are disposed between the outer and inner planking, and are called ribs or timbers, a double range of straight-grained pieces of half the substance, more or less, and as long as can be conveniently obtained, the same being bended to the proper figure by the means before described, and placed contiguous to each other between the outer and inner planking, with the joints, as far as possible, from being opposite each other, and the whole is secured by bolts or treenails as usual or otherwise. The said pieces may be disposed, not upright or at right angles to the keelson, but oblique, the inner range of these pieces being inclined forward on the one side and aft on the other side of the ship, as they rise from the keel, and the outer range of these pieces being inclined in the contrary direction to that of the inner range which they cover, by which arrangement the said ranges do act with regard to each other as diagonal spurs.

"The sixth of our said inventions or improvements is, buoys made of cast-iron or of wrought-iron plates, screwed, brazed, soldered or riveted together, so as to form a hollow water-tight vessel of any shape or size that may be requisite. The advantage of these buoys is, that they cannot be injured by worms or imbibe water to make them lose any of their buoyancy, and consequently that they will at all times float higher and be better seen than buoys made of wood.

"The seventh of our said inventions or improvements consists in a method of working an arm or lever attached to a steam-engine for the purpose of hoisting, pumping, rowing, or other similar works in naval affairs. To effect this the piston-rod of a steam-engine is racked on one side, and drives a wheel, or a portion of a wheel, which has a horizontal axis, with an arm of considerable length attached thereto. The whole of the said apparatus, and the steam-engine for working the same, which we usually construct of about the weight of one ton, may be placed on wheels and conveyed to different parts of the ship; and the said arm, which may be varied in its length as well as in the inclination of its positions at the ends of its stroke, may be employed in hoisting a basket of coals successively out of the hold of a ship, and conveying the same by means of a rope of suspension, to the gangway; or goods may in like manner be hoisted and carried over the side, and, in cases where the hoist may prove too long for the circular motion of an arm to carry the goods sufficiently high or with due perpendicularity, we avail ourselves of a runner or rope, one end of which is made fast to the deck, and the other serves to seize the goods, while the bight or bend of the rope passes over a sheave or pulley at the extremity of the arm; and by this contrivance every seaman and mechanic will be aware that the hoist is rendered more direct, and the rise and fall nearly double; and further, we do apply the said arm to pumping ship and other like uses; and when the said arm is passed over the side it may be employed to work a paddle or rowing board of the usual construction, or a rowing trunk of the construction hereinafter described.

"The eighth of our said inventions or improvements is, a rowing trunk, tube, or prismatic cavity. When the stroke of an oar or any similar implement is made against water, for the purpose of obtaining a reaction as nearly as may be similar to that of the resistance of an immovable body, part of the force is lost in producing a lateral motion in the water which escapes sideways, and the blade of the oar is far from being stationary. If the oar were to pass in a channel to which it is fitted, and the channel were of such dimensions as to present a great length of water before the oar, which could not otherwise move than by passing out of the channel, it is demonstrable that the said mass of water may be assumed of such a magnitude as to render the actual motion of the oar, by a given force and during a given time, less than any assignable quantity. We have availed ourselves of the foregoing considerations in our rowing trunk or tube. It consists of a tube of considerable length disposed horizontally in the water, and the stroke of rowing is made by means of a piece of the nature of a piston with valves, which is acted upon by the rowing arm and shuts the valves; but in the return of the stroke the valves open, and the said rowing trunk or tube being attached to the vessel, and its contents having too much inertia or resistance to receive much velocity from the piston, the vessel itself and the tube are carried along with much more power and effect that by common rowing. If the tube itself be made movable and attached to the rowing arm, and there be a stop within of the nature of a valve to shut against the stroke, and open with the return, the effect will be the same.

"And lastly, the ninth of our said inventions or improvements consists in the adaptation of various particulars or additional parts made of iron, and conducing to the comfort and better subsistence of mariners and others, unto the boiler of the steam-engine that may be used for the forementioned purposes, that is to say, in the upper part of the said boiler we insert or fix a vessel, either single or divided into compartments, for boiling provisions in water, or cooking the same in steam; and in the part of the said boiler opposite the fire-place we insert or fix an oven surrounded by a space communicating with the chimney, so that the walls of the said oven shall be surrounded by heated air, and not in contact with the water; and further, we do adapt and apply to the boiler, or to the place of escape at the steam-cock, any of the apparatus heretofore in common use for condensing steam, in order to procure fresh water; and we do apply and use the said last-mentioned addition to the boiler either separately or together as may be required.

"In witness whereof, we, the said Richard Trevithick and Robert Dickinson, have hereunto set our hands and seals, twenty-eighth day of October, in the year of our Lord this one thousand eight hundred and nine."

A movable caisson or floating dock, made of wrought iron, of such capacity that the largest first-rate of that day, in fighting trim, might in a few hours be dry-docked, was an undertaking worthy of Trevithick.

The internal figure of the floating dock resembles that of a square-sided boat, made of wrought iron, half an inch thick, 220 feet long, 54 feet wide, and 30 feet deep, with a flange 6 feet wide, extending outwards from the upper edge, giving strength and a convenient platform for workmen to stand on. Large semicircular air-chambers, extending the whole length of the external sides, sufficed to float it when full of water, and gave strength to its sides. It would weigh about 400 tons, and when floated to the ship, valves allowed the dock portion to fill with water, which brought it down to about the top line of air-chambers; the bottom line of air-chambers was then filled until the dock was almost betwixt sinking and floating, when by the use of chains it was hauled under the bottom of the floating vessel, and again raised as high as convenient, the vessel being

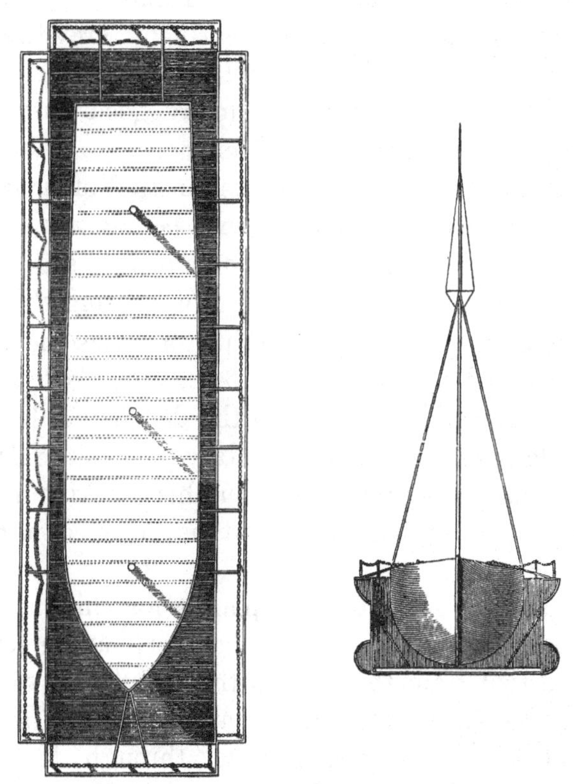

PLAN AND SECTION OF TREVITHICK'S IRON FLOATING DOCK, 1809.

inside; the water was then pumped out of the air-chambers, causing the top of the dock to rise above the water-line, after which the dock and vessel might be floated to shallower water. A steam-engine pumped the water out of the dock in three hours, raising its top about 8 feet above the floating water-line of the vessel.

One end of the dock might be removed, thus allowing the vessel to be more easily floated into it.

Iron had before been used in constructing small craft for canals and inland navigation; but not for ships of war, East Indiamen, or other large vessels fit to navigate the ocean. The reasons why iron should be used in preference to wood were seen clearly by Trevithick at first, as they are now known with sixty years of experience. His last reason of the many good ones was, perhaps, better understood by him then than it is now, and the day may still come when, instead of trying to keep out an enemy's shot, it will be allowed to go through with the least amount of damage, soft tough iron being used in the construction of fighting ships.

The invention of hollow wrought-iron masts and yards and the method of sliding one into another, like a pocket telescope, bringing all down to the deck level, has not been made practical during the sixty years since Trevithick recommended their use; but his working model and detail description show that he believed it practicable.

The bending of timbers or planks for ship-building was to be performed in an iron chamber, having screws so placed that from the outside they might be worked to force the timber into the required curve. Hot air and coal smoke or gas, passing through the iron chamber, heated the wood, causing it to bend freely, and at the same time the sap to evaporate.

The doing away with the ribs or bent framing of ships, substituting a double layer of planks, placed diagonally, was a method of ship-building since known as "diagonal planking," and much used in the present day.

The sixth invention, that of wrought-iron buoys,

has since come into general use, especially where, as Trevithick points out, buoyancy is required; and non-liability to injury from worms.

The seventh invention was an improvement on the nautical labourer of the year before, and might be called the "steam-arm."[1] The engine and boiler complete, weighing a ton, was moved about the deck on wheels. That part of the piston-rod extending beyond the cylinder was a rack, and worked a toothed wheel, on the axis of which an iron arm or lever was fixed, which could be lengthened or shortened at pleasure, probably by sliding on the axle with tightening screws, making a sort of steam-crane. Suppose coal was being discharged in the ordinary baskets, the basket in the hold would be hooked on to the rope attached to the end of the steam-arm, then about horizontal; turning on the steam caused the toothed wheel to make a portion of a revolution, causing the arm to rise to the required height with the attached load. Should the height to be lifted be greater than the length of the arm, the bight of the whip-rope worked on a pulley in the end of the arm, so that the space moved through by the basket was twice as much as that moved through by the end of the arm. The toothed wheel on the same axis as the arm could also be changed in size, to suit the speed and power required. If the pumps had to be worked, the steam-arm was shortened to suit the stroke of the pump. If the windlass had to be worked for lifting the anchor, warping the vessel, or other purposes, the steam-arm was attached to the ratchet-lever on the windlass; if the vessel had to be rowed, the steam-arm moved a paddle or rowing board of other construction.

---

[1] See patent, 1808, Propelling Vessels.

The eighth invention was devoted to the method of making this rowing arm effective by preventing the slip or escape of the water from the face of the oar or float-board, by causing it to work like a piston in a pipe, the sides of the pipe preventing the slip or dispersion of the water, except in the direction most suitable for propelling the vessel. The pipe was attached to the side of the vessel below the water-line, the steam-arm moving the piston backwards and forwards; the piston was a kind of thin hoop with a large valve, like a pump-bucket, except that the valve was larger. When the piston moved toward the bow of the vessel, the large valve opened, allowing the piston to move on its forward stroke without resistance; on the return or back stroke the valve in the piston closed; then the solid piston being forced against the water in the pipe, caused the vessel to advance.

The real object and value of these two last inventions was to make his high-pressure portable engine so simple and manageable, that it should do almost any of the ship's work more effectually and at a much less cost than manual labour. Sixty years have sufficed to bring into partial use the present donkey-engines, which are the growing representatives of the original nautical labourer and steam-arm, but they have not yet performed one-half of what Trevithick intended they should do.

The ninth invention in this remarkable patent proves the constant tendency of Trevithick's genius to elevate man, by making matter subservient to his wants.

The boiler of the steam-arm is to "conduce to the comfort and better subsistence of mariners," by inserting in its top an iron vessel having one or more compartments in which provisions are boiled in water, cooked in steam, or roasted in an oven, and the ship's company supplied with fresh water, hot, if required, from the steam given off by the salt water in the boiler.

A pamphlet was published, from which the following is extracted :—

"*Prospectus addressed to the consideration of His Majesty's Ministers, the Right Hon. the Lords of the Admiralty, and others. February* 10*th*, 1809.

"A BRIEF DESCRIPTION OF A WROUGHT-IRON MOVABLE CAISSON WITH A RUDDER, FOR DOCKING A SHIP WHILE RIDING AT HER MOORINGS.

"This plan may be practised in all countries, and must be particularly advantageous where there are no dry docks or flowing of tide. Ships on many foreign stations when requiring to be docked are now obliged to be sent home, at a great expense of money and waste of time, others being sent to replace them. This may be avoided in future. Docks made in England may be sent out in pieces of five or six tons, with the necessary rivets and bolts, ready to be put together where they may be wanted. Such a dock would last for twenty years without repair. When worn out it may be broken up, and will sell for one-third of its original cost.

"By constructing this caisson, adapted to local circumstances, ships of war and merchant ships, with all their stores and cargoes on board, can be carried to wharfs and storehouses up rivers, where the depth of water is not above one-half the ship's draught. For example, in the river Clyde the ships may be carried to Glasgow, instead of being obliged to unload at Port Glasgow, some miles lower down the river.

"For a long period the only means employed to effect the bending of ships' planks was by exposing them to the heat of open fires, and in most parts of Europe this is still the practice. As hitherto conducted, it has been found to be a tedious, slovenly process, attended with a great expense of fuel, and unequal in its effects, some parts being only partially heated, while others are burnt. Another system was therefore resorted to, that of employing steam; and it must be allowed that this mode of bending has been found to answer, so far as the interest of the ship-builder is concerned. But ship-owners have suffered from its effects. It is possible by means of steam to give the required degree of flexibility to planks; but steam of a degree of temperature high enough to destroy the vegetable sap cannot be confined in vessels of any reasonable strength. Wood so treated has been found liable to a sudden decay. The durability of charred wood is confirmed, by being found in burnt ruins free from decay after several hundred years.

"The process now recommended is for heating both planks and timbers without steam, and in such a manner that they may be enveloped and equally surrounded on all sides with hot air and smoke, the coal-tar contained in the latter entering at the same time the pores of the wood, and acting as a feeder.

"This process is so conducted as to prevent the wood from being burnt by it; all the air that reaches the timbers while under process being previously obliged to pass through the fire, and being by that means deprived of its oxygen, on that principle which maintains combustion.

"The means employed to effect this are horizontal curvilinear flues, made of cast iron, adapted to the forms intended to be given to the wood, and furnished with a powerful but simple apparatus for applying the force requisite to bend the timbers into the required form. The heat can be so managed as to give at the same time the degree of charring requisite for the preservation of the timber against wet and dry rot."[1]

Mrs. Trevithick spoke of a very pretty model at a scale of 1 inch to 8 feet, showing a first-rate ship, fully rigged, enclosed and floated in a dock of thin iron. It was sent to various places to make the plan known, and so was lost. A rudder was fixed to the dock for steering it, or it might be made in pieces, for a foreign country, and be there riveted together. During 1869 a floating dock, with a rudder, was towed to Bermuda; and a steamer of 150 tons burthen was

[1] Pamphlet, 10th February, 1809.

constructed in England, in separate pieces, or plates, to be sent to the White Nile, there to be riveted together.

The particular advantages of floating docks in countries not having the rising and falling of tides, arose from vessels in such situations having to be sent great distances in search of a dry dock.

Timber had been bent before the date of Trevithick's patent by exposing it to the heat of an open fire and to steam, the latter plan not destroying the sap, while the open fire was unmanageable in the uniform heating of the wood and in the difficulty of bending, both of which were given in the iron flue; while the coal-tar contained in the smoke entered the pores of the wood, preserving it, and may be called the origin of the modern most approved method of creosoting wood by heated coal-tar.

In addition to the pamphlet a long statement of the advantages to flow from the patent was sent to "The Honourable the Commissioners of His Majesty's Navy."

"We presume that our plan goes to obviate most of the evils complained of in Lord Melville's letter to Mr. Perceval.

"By our process we give a greater degree of curvature than is required for ships' ribs to a piece of oak fourteen inches square and fourteen feet long.

"Ships' ribs standing perpendicularly, as they now do, instead of holding by each other, require, *absolutely require*, to be held together by other and collateral means, namely, the planks and lining. Hence the longitudinal strength of a ship depends on these same planks and lining, and not at all on the ribs, whose disposition of themselves is to diverge and tumble to pieces, instead of leaning against and bracing each other, inch by inch, throughout the whole ship. Ships' ribs, as we purpose to place them, are to be split, and cross each other at all points, the one sett stretching diagonally towards the head, and the other towards the stern, each forming a diagonal stay, not at right angles, but forming the lozenge, diamond, or lattice, or close, if preferred, in which case the starting of a plank would be of no consequence. We have a model and drawings exemplifying these facts, which we desire to be permitted to exhibit."

The winner in the race of homeward-bound clipper tea-ships in 1869 was built on the diagonal principle. The experience therefore of sixty years has proved the correctness of Trevithick's plans.

In 1835 the writer was employed in putting engines on board the 'Diamond,' a new steamboat on the Thames, competing for the traffic below London Bridge, and specially built to excel the others. The outer planking was horizontal, backed by two thicknesses of diagonal planking; thus the whole thickness of the ship's side was but 3½ inches, while a few iron knees tied the deck timbers to the ship's sides. The vessel was 145 feet long. Engines of 40-horse power gave her a speed of 12⅓ miles an hour, the greatest that had then been attained on the Thames.

Diagonal planking is now generally adopted in building life-boats; its usefulness is still extending, and the writer believes one of Her Majesty's favourite yachts was so constructed.

A memorandum in Trevithick's writing, without date, but evidently of about the period of the patent of 1809, gives his views of the detail of an iron sailing ship. The necessary calculations for the parts are jumbled up with the writing, as he proceeds with the description. Unfortunately the drawing he refers to has disappeared. The vessel was for a trader, carrying 300 tons, drawing 8 feet 3 inches. The length 70 feet, width 35 feet, depth of hold 10 feet. The keel 1 inch thick, the vessel's bottom $\frac{3}{4}$ of an inch thick, the sides $\frac{1}{2}$ an inch, the deck $\frac{3}{8}$ths of an inch, and the gunwale $\frac{1}{4}$ of an inch thick, the whole riveted together and strengthened with iron ribs in the hull. Three iron keels of 2 feet in depth and 2 feet apart gave strength to the bottom and helped sailing qualities on the wind, while the great breadth enabled her to carry canvas without taking ballast. The weight of the vessel was 70 tons. The masts and yards were wrought-iron tubes, the shrouds and stays were iron chains; each of the two masts was supplied with a square sail 50 feet high by 44 feet wide, its larger half being aft of the mast, as a lug sail. This aft or long end of the fore-yard was connected by a chain with the fore or short end of the yard on the main-mast, thus making the two sails balance one another to save labour in trimming them.

The yards were centred on the masts, and could be lowered down, while the sails fixed to the yards reefed themselves by rolling on the yards as a roller-blind. Studding sails had two-thirds the spread of the square sails, the yards for which slid into the hollow main-yards when not wanted. Such an iron vessel would not disgrace a builder of the present day. The peculiarity of reefing by rolling on the yard has since been tried with some success.

The pamphlet of 1809 thus speaks of the tubular iron mast, and establishes the date of the drawing and memorandum referred to:—

"Sliding tubular masts, made of iron, and so constructed that the upper ones slide into the lower in a manner somewhat like a pocket telescope.

"A model of one for a first-rate ship, on a scale of one-eighth of an inch to a foot, is now ready for inspection.

"This tubular mast, being half an inch thick, and the same height and diameter as a wood mast, will be lighter, considerably stronger, much more durable, and less liable to be injured by shot, and will cost less money.

This mast is made to strike nearly as low as the deck, to ease the ship in a heavy sea. Yards and bowsprits may also be made of wrought iron, and chain shrouds and chains will not cost half the expense of rope."[1]

The description in Trevithick's handwriting gives such close detail, that a practical intention of constructing such a ship of iron, with tubular masts and yards, and self-reefing sails, was certainly made public, if not really acted on in 1809, as is evidenced by the workman-like grasp of the whole design, expressed in few words, interspersed with calculations establishing his practical deductions, resulting in the greater speed, carrying power, and economy of management of iron sailing ships as compared with wooden sailing ships.

The figures in his detail description are omitted, as the calculations were frequently broken off, when he mentally saw the approximate result.

"A plan, side and cross section of a wrought-iron trading ship. Scale one-sixth of an inch to a foot.

"Vessel to be 70 feet long, and 35 feet wide on the deck, and 10 feet deep in the hold, with three iron keels of 2 feet wide, with iron rods for strengthening the hull, the whole riveted together the same as a steam-engine boiler, the keel to be 1 inch thick, the bottom three-quarters thick, the sides half an inch thick, the deck three-eighths thick, and the gunwale one quarter of an inch thick. The mast and yards wrought-iron tubes, and the shrouds and stays iron chains.

"Weight of the ship 70 tons, width of the sails 44 feet by 50 feet high, two-fifths before the mast, and three-fifths aft the mast.

[1] Pamphlet, February 10th, 1809.

"Studding sails two-thirds of the width of the large sail. Deep water-line with 300 tons, 8 feet 3 inches, including the keels of 2 feet in depth.

"The yards fall down near the deck, and the sails reef themselves without going aloft, and are always drawn tight in a chain frame, that holds the sails close to the wind, and prevents their flopping.

"The yards being centred on the masts, and the after-end of the fore-yard chained to the fore-end of the after-yard, balances the sails, that they go about easily.

"The iron keels act as ballast; and, together with the extreme width of beam, enable the required canvas to be carried without taking in additional ballast. This vessel would be lighter, sail faster, stow more cargo, be more durable, and worked with a less number of hands than a wood vessel."

The news of the day describes a race between the English yacht 'Livonia' and the winning American yacht 'Columbia,' a large centre board schooner, built in 1871, 98 feet long, 26 feet beam, 8½ feet hold. Trevithick's iron trader of 1809 was 70 feet long, 35 feet beam, 10 feet hold, to which add the sliding keels, described in the following chapter; and the two schooners, allowing for the different requirements of racer and trader, exhibit similar general principles, though the vessel of sixty years ago had spars and hull of iron.

From the time of his work on the Dredger in 1803 to his leaving London in 1810 his occupations caused him to be almost daily within sight of the ships and mercantile operations on the Thames—iron buoys, iron tanks, iron docks, iron steamboats, iron sailing vessels, iron fighting ships, designed, patented, and more or less brought into commercial use, were results of those seven years of labour.

Our small and imperfect knowledge of Richard

Trevithick has caused him to be regarded merely as the inventor of the high-pressure steam-engine, while in truth that one machine, among his numerous useful discoveries, was but as a foot or hand to a perfect man; just as it took its place of relative usefulness in the improved sailing ship made of plates of iron, giving greater safety and comfort to the sailor, reducing his labour by the use of sliding keel, water ballast, and self-reefing sails; the steam-engine giving auxiliary propelling power, discharging cargo, and weighing the anchor, while the steam-boiler gave facilities for cooking food, and the iron tanks a larger supply of more wholesome water.

Many men scheme, take patents, and are no more heard of. Trevithick's working models, made known by his written description, show on what a different basis his schemes were built. Every detail is practically described, as though it was a history of a thing of the past rather than of a thing to come.

After a lapse of sixty years the greater portion of those schemes have become valuable realities; a paper of the day,[1] on the great increase of iron steamships, states that "about 11,000,000*l.* worth of shipping is now in course of building;" and this marked branch of national prosperity took its rise from the designs that we have traced in their original working form in 1809; the sliding tubular masts and yards remain to be made practicable.

---

[1] 'The Western Morning News,' October 26th, 1871.

CHAPTER XV.

PROPELLING VESSELS BY STEAM.

A LETTER, in 1803,[1] speaking of his high-pressure engine, with 45 lbs. of steam to the inch, says, "I was sent for to explain the engine at the Admiralty Office. They sent to inspect, and say they are about to erect several for their purposes, and that no other shall be used in the Government service." Again, in 1804,[2] "An engine is ordered for the West India Docks, to travel itself from ship to ship." "An engineer from Woolwich was ordered down, and one from the Admiralty Office, to inspect."

With this intimate knowledge of Trevithick's schemes, and the use of his engines by the Admiralty, it is the more strange that such events are ignored, or but slightly referred to in the history of early steamboats, for in the latter part of 1804 several gentlemen had consulted him on the feasibility of destroying the French fleet by the use of high-pressure steam fire-ships.

"MR. GIDDY, "COALBROOKDALE, *October 5th*, 1804.

"Sir,—Several gentlemen of late have called on me to know if these engines would not be good things to go into Boulogne to destroy the fleet, &c., in that harbour by fire-ships. They told me that a gentleman from Bath was then in London trying experiments under Government for that purpose, but whether by engine or by what plan I do not know.

[1] See Trevithick's letter, May 2nd, 1803, chap. ix.
[2] See letter, Feb. 22nd, 1804, chap. ix.

"A gentleman was sent to speak to me yesterday on the business, from a marquis; the name I am not at liberty to give you. I put him off without any encouragement, because I would much rather trust to your opinion in bringing this business forward than any other man.

"I have two 10-inch cylinders here completely ready; they are exceedingly well executed. I will not part with them until I hear further from you. If you think you could get Government to put it into execution, I would readily go with the engines and risk the enterprise. I think it is possible to make these engines drive ships into the middle of the fleet, and then for them to blow up.

"However, I shall leave all this business to your judgment, but if you give me encouragement respecting the possibility of carrying it into force, I am ready to send off these two engines on speculation.

"I believe if you do not bring this business forward, some other person will, and it would not please me to see another person take this scheme out of our hands.

"Be silent about it at home, for I should not like my family to know that I was engaged in such an undertaking.

"I would thank you for your sentiments on this business soon, as these two engines will be left unsold until I receive your answer.

"There are orders for several engines since I last wrote to you.

"I am, Sir,

"Your very humble servant,

"RICHARD TREVITHICK."

"TRENTHAM HALL, NEAR NEWCASTLE,
"*October 20th*, 1804.

"MR. GIDDY,

"Sir,—I received your last letter, and am really much obliged to you for your goodness in offering to give me your assistance in promoting my schemes. I am ordered by the Marquis of Stafford to leave this place for London on Monday morning to meet Lord Melville and Mr. Pitt. It is likely they may wish to be satisfied respecting the possibility of the plan I shall lay before them. I shall refer them to you for your opinion and calculations. I will write to you when I have seen these gents, stating every particular. There are many things I wish to communicate to you, for I cannot satisfy myself from my own figures. I hope you will be called on in the House in the course of next month. I expect to be in town at that time. If I am ordered to proceed with this plan for Government, there will be several things that I shall not be able to get through without your assistance. I must beg your pardon for so often troubling you for these things, for I am not master of myself, but hope you will have the pleasure of seeing these plans carried into execution.

"I am, Sir,

"Your very humble servant,

"RICHARD TREVITHICK."

These traces of proposals to propel vessels by his high-pressure steam-engines, in 1804, to destroy the French fleet in Boulogne Harbour, show him as the pioneer of marine engineers. A messenger from a marquis—"a profound secret"—caused a few minutes' cautious reflection, and a wish to consult his friend, followed by,—"If you think you could get Government to put it into execution, I would readily go with the engines and risk the enterprise." He thought it possible with those engines to drive ships into the middle of the fleet, and was prepared, without a moment's delay, to place two of them in a vessel, and by them propel her to France.

Within a fortnight he was at Trentham Hall, plotting with the Marquis of Stafford, who sent him to London to see Lord Melville and Mr. Pitt. He gave more than usual thought to this compound question of propelling a vessel by steam, steering it without a helmsman, and exploding it at the required moment;

for he admits, "I am not master of myself." Probably this was the only time in his life that he doubted his ability to make his high-pressure steam-engine do everything.

"SOHO FOUNDRY, MANCHESTER, *January* 10*th*, 1805.

"MR. GIDDY,

"Sir,—I have answered Mr. Gundry's letter. Their engine will cost about 170*l*., and will draw one hundred kibbals with about three bushels of coal.

"I fear that engine at Dolcoath will be a bad one. I never knew anything about its being built, until you wrote to me about Penlerthy Croft engine, when you mentioned it.

"I then requested Captain A. Vivian to inform me the particulars about it, and I find that it will not be a good job. I wish it never was begun. I was sent from the Marquis of Stafford to Lord Melville. I was at the Admiralty Office, and was ordered to wait a few days, before they could say to me what they wanted. I called five or six days following, but never received a satisfactory answer, only to still wait longer. But I left them without knowing what they wanted of me, for I was tired waiting, and was wanted much at Coalbrookdale at the time. When they send for me again, they shall say what they want before I will again obey the call. There was an engine, a 10-inch cylinder, put into a barge to be carried to Macclesfield, for a cotton factory, and I tried it to work on board. We had a fly-wheel on each side of the barge, and a crank-shaft across the deck. The wheels had flat boards, 2 feet 2 inches long and 14 inches deep, six on each wheel, like an undershot water-wheel. The extremity of the wheels went about 15 miles per hour. The barge was between 60 and 70 tons burthen. It went in still water about 7 miles per hour. This was done to try what effect it would have. As we had all the apparatus of old material at the Dale, it cost little or nothing to put it together.

"I think it would have been driven much faster with sweeps.

"The engine that was made for the London Docks, for discharging the West Indiamen, is put to work in a manufactory. They would not permit fire within the walls of the dock; there is an Act of Parliament to that effect.

"There is a small engine making in Staffordshire, for the London coal-ships to carry with them for unloading.

"The boiler is 2 feet 6 inches diameter and 5 feet long, with the cylinder horizontal on the outside of the boiler. The cylinder 4 inches in diameter, 18-inch stroke. I think it will be equal to six men.

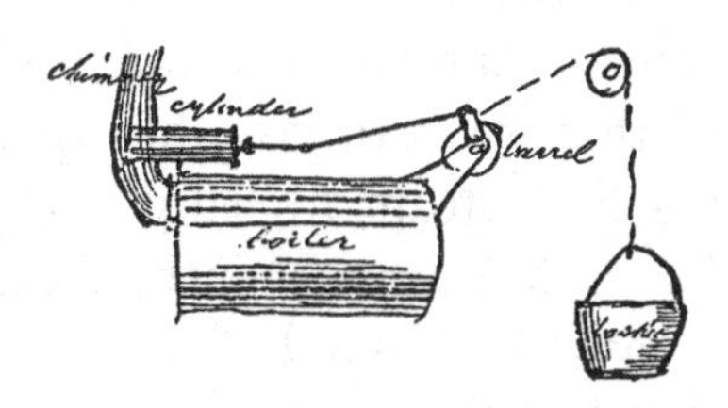

"The engine will always go one way. The man that stands at the hatchway will have a string to throw out a catch, which will let the barrel run back with the empty basket.

"I shall go to Newcastle-on-Tyne in about four weeks, and most likely shall return to London. By that time the little engine will be sent to London for the coal-ships. A great number of my engines are now working in different parts of the kingdom. There are three foundries here making them.

"They finish them in very good style; all the wrought iron is polished, and ornamented with brass facings. They are the handsomest engines I ever saw. I expect there are some of the travelling engines at work at Newcastle. As soon as I get there I will write to you.

"This day I received a letter from Jane [his wife]; sad lamentations on account of my absence. I am obliged to promise to return immediately, but shall not be able to fulfil it at this moment. I should be wrong to quit this business, as there are now seventeen or eighteen foundries going on with those engines, and unless I am among them, the business will fall to the ground, and after such pains as I have taken, I am sorry to quit it, until I get it established. I am sorry to hear that your sister is about to go to London to reside, for I fear that you will now spend but a small part of your time in Cornwall, and I shall be deprived of your good advice in future.

"I am, Sir,

"Your very humble servant,

"RICHARD TREVITHICK."

This account of a 60 or 70 ton barge, driven by a high-pressure steam-puffer engine, having a cylinder 10 inches in diameter, at a speed about seven miles an hour, the float-boards going at fifteen miles an hour, proves that had Lord Melville spared five minutes out of that week to keep his appointment with Trevithick, the first steam fire-ship would have dated from 1804; for within three months from that time he had driven a barge at a speed of seven miles an hour by an engine that had been constructed for another purpose, and was adapted to the steamboat on the spur of the moment by a crank-shaft across the deck, having at each end a paddle-wheel, with six paddle-boards 14 inches deep by 2 feet 2 inches long.

Trevithick's steamboat was an ordinary canal barge, into which one of his steam-engines had been placed as goods for conveyance to Manchester from Coalbrookdale, as mentioned in his note of the 5th of October, 1804, and 10th January, 1805. The fact came under the notice of the Prime Minister, and the makers of the engine were among the best-known manufacturers of the day, yet no one has recorded that Trevithick constructed this first practical steamboat.

Napoleon I. is said to have exclaimed, about 1804, when reviewing his army and ships at Boulogne, "The English do not know what awaits them; if we are masters of the Channel for a few hours, England has lived her time." Napoleon certainly did not know of Trevithick's proposed steam fire-ships, or of his courage in offering to guide them into the middle of his fleet.

Lord Dedunstanville had formed a troop of Cornish volunteers to drive Napoleon from the shores, and among them was Trevithick, who, his wife said, appeared well pleased with his red coat. One night a beat of drums in the Camborne streets startled the sleepers; Trevithick awoke his wife, and asked what all the noise could be about. "Oh! I suppose the French must be come; had you not better put on your red coat and go out?" "Well, but Jane," suggested the volunteer, "you go first and just look out at the window, to see what it is!"

"MR. GIDDY, "BLACKWALL, *July 23rd*, 1806.

"Sir,—This day I set the engine to work on board the 'Blazer' gun-brig.

"It does its work exceedingly well. We are yet in dock, and have lifted up mud only. I hope to be down at Barking shaft in a few days, at our proposed station, when I will write to you again. I think there is no doubt of success.

"A gentleman has ordered an engine for driving a ship. It is a 12½-inch cylinder. I am at a loss how to construct the apparatus for this purpose; therefore am under the necessity of troubling you for your advice on the subject.

"The plan I have is as under, unless you condemn it, or suggest a better plan. I propose to put a horizontal engine below the deck, and to put a wheel of 14 feet diameter in the hold. This wheel is to work in an iron case, air-tight; the axle to work in a stuffing box, and a pump to force air into this case to keep down the water from flooding the wheel, so that only the floats on the extremity of the wheel shall be in the water, and then only extend about 15 inches below the keel of the ship.

"The cutter is about 100 tons burthen.

"I wish to know the size of the floats on the wheel, and the velocity you think they should be driven.

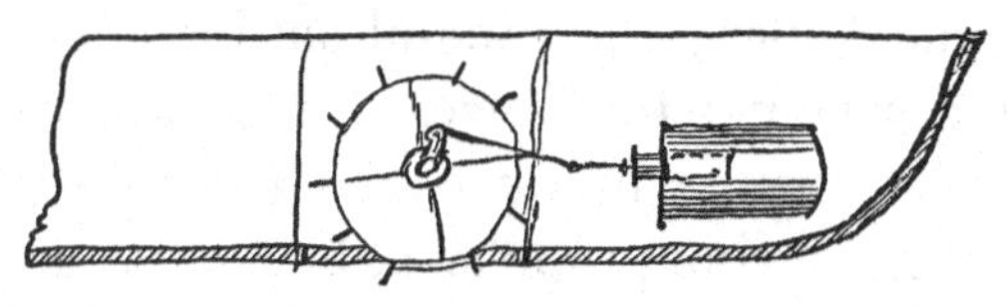

"I think the power of the engine is equal to 400,000, 1 foot high, in a minute, from which you will be able to judge what size of floats, and what velocity will be best. The air that is forced into the wrought-iron case will always keep the water down in the case to the level of the bottom of the ship. A space will be left on each side of the wheel, so that the air will never be displaced by the working of the wheel.

"Your answer to this will very much oblige

"Your very humble servant,

"RICHARD TREVITHICK.

"Direct to me at the Globe Tavern, Blackwall, near London."

In 1806 he had undertaken the still difficult task of placing a high-pressure puffer steam-engine of about 12-horse power in a gentleman's sailing yacht of 100 tons. Side wheels and paddle-boxes would destroy her appearance and sailing qualities. One centre paddle-wheel, enclosed in an iron case, was therefore to be fixed in the hold, the float-boards passing through an opening in the vessel's bottom. Water was prevented from rising in the case by a pressure of air forced into it by the engine, sufficient to depress it to near the level of the vessel's bottom, that the wheel might not be flooded. The paddle-shaft, where it passed through the two sides of the wheel-case, was made air-tight by stuffing-boxes. The bottom of the float-boards was 15 inches below the keel of the ship, attached to a wheel 14 feet in diameter. The engine to drive it was a 12½-inch cylinder high-pressure puffer direct-action engine, of about 12-horse power, with cylindrical tubular boiler; the general outline of engine and boiler being very like the Newcastle locomotive of 1804.

About that time, while producing changes and improvements in marine propulsion, he became acquainted with Mr. Robert Dickinson, who had been a West India merchant, and they became partners in the following patent:—

Propelling Vessels, etc.

Trevithick and Dickinson's Specification of 1808.

"In a ship or vessel properly constructed for the purpose, and to which we give the name of a nautical labourer, we place a rowing wheel shaped like an undershot water-wheel furnished with floats or pallets, but which we call our propelling boards, and of a size proportioned to the vessel; the said rowing wheel is placed vertically in a box or casing fitted to receive it in such a manner that its axis shall be at right angles to the length way of the ship or vessel, while the edge of the propelling board that is lowest shall be even with the keel of the ship or vessel, or rather with the keels, for we prefer having two, and placed at such a distance from each other as just to allow the propelling boards of the rowing wheel to pass freely between them. The foresaid box or casing which contains the rowing wheel is made air-tight, and is open only at the bottom, where the floats are intended to act in propelling the vessel, and the water is prevented from rising in the said box or casing by an air-pump forcing air into it, so as to keep the water always down at the level intended. The end or ends of the axle of the rowing wheel works or work in a collar or collars of leather or any other substance fitted to prevent air from passing out; and the said wheel, as also the fore-mentioned air-pump, is worked by means of a crank, a wheel, or any other suitable contrivance at one or both ends of the axle, and connected with a steam-engine, which is the power we employ to move the rowing wheel, or, in other words, to give the required movement to the vessel. In the wheel itself there is no novelty.

"The novelty of this part of our invention consists in working the wheel in an air-vessel, and by means of the contained air keeping the surface of the water at such a lower level than the general surrounding surface as may prevent the water from flooding the wheel. When we mean to employ such a vessel or ship and apparatus as the above for the purpose of towing other vessels, the propelling boards of the rowing wheel and also the steam-engine must be larger in proportion to the work intended to be performed; the vessels are then fastened to each other by a towing rope or towing ropes, the headmost of them being made fast to the rowing vessel, that being the name which we have adopted for the above-described ship or vessel and apparatus. When strong tides and currents oppose the progress of the rowing vessel encumbered with those she has in tow, we facilitate the business of towing in the following manner:—To the headmost or to all of the vessels in tow we attach a pair of long poles, or what bargemen call setts, for the purpose of being dropt into the water from time to time, as is now the practice in shoal stream water, to answer the purpose of an anchor. The vessels in tow being thus secured from going along with the stream, we then propel our rowing vessel forward alone, allowing the rope that is made fast to the headmost of them to run off from a windlass placed near the stern of the rowing vessel, or in any other convenient situation. When we have run off all the rope to the end that is made fast to the windlass, we then secure the rowing vessel in her place by dropping a pair of setts or an anchor; this being done, the vessels in tow take up their setts, the machinery of the towing vessel is then disengaged from the rowing wheel, and thrown into gear to work the foresaid windlass by the power of the steam-engine. By this means the rope is wound up on the windlass, and the vessels in tow are brought up to the rowing vessel, when the same method of proceeding is again repeated till the vessels have passed the heavy part of the stream or current, and are able to proceed without those interruptions. We need hardly add that anchors may be employed in place of setts, and that in deep water they must be used; but where the water is not too deep to admit of it, the employing of setts in place of anchors is attended with less trouble and delay. What we have stated respecting the use of our rowing vessel and apparatus in towing vessels against a stream or current applies also to the towing ships out of any harbour. If there are floating buoys outside the harbour we row our rowing vessel to one of them, letting off from the windlass in the manner before described the towing rope made fast to the ship that is to be towed out; having reached the buoy, or dropt another if there is no buoy, we then disengage the machinery from the rowing wheel, and put it into gear with the windlass, and work the ship out of harbour by the power of the steam-engine. In many harbours this will be of immense advantage to the public, and particularly to ship-owners, for it not unfrequently happens that the wind is fair for the voyage, but foul for getting out of harbour. Here it may be proper to mention once for all, that in every case where we drop anchor, when it is to be taken up again we weigh it by throwing its windlass into gear to be worked with the steam-engine. The benefit to be derived by the public from our contrivance for towing vessels or ships would be increased if chains of proper strength moored at each end were laid in difficult currents and mouths of harbours, to be used in the following manner, that is to say, by giving the said chain one turn round the end of the windlass connected with and worked by the steam-engine the vessel could be worked in either direction of the chain without making use of the rowing wheel, merely by the revolving of the windlass. The chain should have a buoy at each end for the purpose of being laid hold of. Nor are these all the uses to which we apply our contrivance. The apparatus or machinery which we employ to propel the vessel or vessels in the manner before described is so contrived that we can also employ it in loading or unloading not only the rowing vessel, but those she has towed, or any other that may want that assistance. Where a few articles only are to be taken on board or discharged from the rowing vessel, the common blocks and tackle may be worked by a rope attached to a windlass, and worked by the steam-engine. When more dispatch is wanted, as in discharging coals or grain from the rowing vessel or from others, and such like work where economy of time and labour is an important consideration, we facilitate the business by working a barrel in place of a windlass. This barrel, worked by the steam-engine, is kept in constant motion, while the business of discharging is constantly going on. On this barrel is placed as many endless belts as may be required for the business in hand. These belts pass respectively over a pulley of two grooves centred in a lever, the lower end of which lever is on a hinge. To the upper end of the lever, which is the longest from the centre of the pulley, a rope is attached, which goes to the vessel then discharging, where it passes through a common block in the rigging, whence the end of it descends to a man standing on the deck at the hatchway of the ship discharging. This rope is for the purpose of enabling the man, by pulling it, to lighten the endless belt by means of the longest end of the lever. From this it is plain that the pulley will then receive motion from the revolving barrel, and continue in motion so long as the man pulls the rope tight. By this motion another rope, which we call the discharging rope, made fast to the other groove of the pulley, that is, to the one not occupied by the before-mentioned endless belt, is wound up on the said groove. The discharging rope now mentioned is the one that brings up the bucket, sack, basket, or other article from the hold of the vessel that is discharging, no more being necessary for this purpose but to carry the rope first through a block in the rigging of the vessel containing the working apparatus, and then through another in the rigging of the vessel that is discharging, and making that end fast by a hook or otherwise to the bucket, sack, basket, or other article that is to be raised; while the man who holds the other rope keeps it tight, as before mentioned, the discharging rope keeps winding up on the fore-mentioned pulley; when the bucket, sack, basket, or other article has been raised to the height required, the man who holds the tightening rope slacks his hold, and the endless belt is thereby freed from action, so that it ceases to be carried round by the barrel before mentioned. The bucket, basket, or other article being emptied, is let go to descend again into the hold, and the endless belt being now slack, the weight of the descending bucket, basket, or other article makes the double-grooved pulley revolve the contrary way to what it did when winding on, so that the rope is unwound from the pulley, and remains so until the man who holds the tightening rope again applies a force sufficient to make the barrel give motion again to the double-grooved pulley by means of the endless belt, as before described. Instead of the barrel and endless belt before described being used for the above purpose, other common mechanical means may be employed to throw the pulley, to which is attached the discharging rope, in and out of gear, so as to wind or unwind the discharging rope.

"We have before mentioned several endless belts, though in our description we have confined ourselves to one; the other belts are for the purpose of working any required number of discharging ropes on board one or more vessels, for several vessels properly arranged may all be discharging at the same time; but all these discharging ropes being worked in the same manner by the same barrel, and by the same moving power, the description already given of one applies to any number, each having its own proper double-grooved pulley and lever.

"We must here add that we do not mean, by the description which we have given of the barrel and double-grooved pulley, and the method of working them, to claim such contrivances as new inventions, or be understood as meaning to confine ourselves to the use of these means exclusively, for any of the common mechanical means that are now employed to communicate or suspend motions at pleasure to apparatus connected with any maintaining power may be used to produce the effect. The novelty of our invention is simply this: employing such a vessel as we have described, furnished with a steam-engine as a moving power, and with proper apparatus to enable us to employ the said vessel and its contents as a labourer to assist in towing of vessels in the manner before described, and in loading and unloading them, in place of using the methods hitherto in use. Where towing of vessels and discharging of vessels may not both be wanted, the apparatus may be relieved of part of its load, and of part of the expense of construction, that is to say, where only towing may be wanted the machinery need not be loaded with those parts which apply to the discharging of vessels; and, on the other hand, where towing is not wanted, but only discharging and unloading, the parts that apply to towing may be dispensed with.

"In witness whereof, we, the said Richard Trevithick and Robert Dickinson, have hereunto set our hands and seals, this Twelfth day of July, in the year of our Lord One thousand eight hundred and eight.

"RICHARD TREVITHICK. (L.S.)
"ROBT. DICKINSON. (L.S.)"

The nautical labourer was to tow vessels up rivers or in smooth water. If the current was too strong the vessels towed were to be moored or anchored, while it steamed ahead a convenient distance, and dropping anchor, disconnected the paddle-wheel, connected the steam-windlass, and warped the vessels at a slower speed. A permanent chain might be laid to avoid the trouble of anchoring, passing it around the windlass of the towing boat, as is now done in steam-ferries crossing rivers. The labourer, when made fast alongside a vessel to be discharged, raised the cargo by the steam-windlass.

"When a few articles only are to be taken on board, or discharged from the rowing vessel, common blocks and tackle may be worked by a rope attached to a windlass, set in motion by the steam-engine. When more dispatch is wanted, as in the discharging of coal or grain from the rowing vessel, or *from others*, and such like work, where economy of time and labour is an important consideration, we facilitate the business by working a barrel in place of a windlass. On this revolving barrel are placed as many endless belts as may be required for the business in hand. These belts pass respectively over pulleys with two grooves, centred in a lever, the lower end of which lever is on a hinge."

The movement of the lever took the slack out of the endless belt, causing the pulley to revolve; a rope from the other groove of the pulley went to the vessel's hold and raised the load until the endless belt was again loosened.

The engine used was Trevithick's high-pressure portable puffer.[1] Several experiments were made in the Thames, with every prospect of an immediate and extensive use of the nautical labourer, when the Society of Coal Whippers protested against the encouragement of

---

[1] See Trevithick's letter, January 10, 1805, chap. xv.

such a rival, declined to work with it, and threatened to drown its inventor. Trevithick was guarded by policemen, two of them keeping watch at his house; but he succeeded in proving that his idea was economical and practical, though it was impossible to stem the strong current of prejudices of the daily labourer, and the nautical labourer retired from the contest.

A curious old drawing of Trevithick's without date or name, probably belonging to the time of the patent (1808), or the fire-ship of 1804, judging from the rude simplicity of the engine and the oval tube of the boiler, similar to the first high-pressure model, shows a vessel, even including the decks, of iron, to be propelled by sails or steam. Her form was like a deep saucer, with portions of her sides cut away to make room for a paddle-wheel on each side. At the deep water-line she was 43 feet wide by 58 feet long; stem and stern pieces, each 5 feet long, extended from deep water-line to the bottom of the keel, containing sliding keels of iron, each about 5 feet wide, and when down reaching 6 or 7 feet below the bottom of the vessel, acting as lee-boards and also as lever ballast when the ship was light. She was sea-going, but suitable for trading to shallow harbours, drawing 3 feet 3 inches without cargo, and 6 feet 6 inches when loaded with 150 tons. One strong mast indicates a large spread of canvas. The steam-boilers were curved to the shape of the hold, the funnel aft of the stern, the paddle-wheels 25 feet 6 inches in diameter, float-boards 3 feet 6 inches wide; two steam-cylinders 3 feet 6 inches in diameter, 8-feet stroke, with open tops and guide-wheel pistons, allowing the paddle-shaft to be kept very low, and doing away with parallel motion or guide-rods; the cranks were at right angles to each other. The whole of the bottom of the vessel for 1 foot in depth was double, or cellular, a means now much used to give strength, but by him turned to further account as a surface condenser, the detail of this portion is not fully shown, judging from three other sheets of drawing in pencil, all of them unfinished, modifying and improving the whole design. The two cylindrical vessels in the plan joined to the steam-cylinder by the nozzles, were a supplementary condenser and an air-pump.

The size and shape of the boilers, and of the cylinders with the long stroke, indicate the intention to work high steam very expansively, so that at the finish of the stroke on passing to the condensers it would be of a comparatively low pressure. Immediately after this period Trevithick, on his return to Cornwall, constructed an expansive steam-engine that worked with 100 or more pounds of steam to the inch in the boiler, expanding to low pressure and condensing.[1]

This stray waif (Plate XI.) of combined steam and sails, in an iron hull, unnamed and undated, known to be Trevithick's handywork, from his well-known handwriting in numerous explanations on the well-covered sheet of large paper, at a working scale of 1 inch to 4 feet, so full in detail drawing, and with a reference to a written list of descriptions of the various parts, warrants the belief that the construction of such a vessel was fully contemplated, if not really carried out. *a*, two open-topped cylinders; *b*, crank-axle and paddle-shaft; *c*, pistons, with guide-wheels, thus avoiding parallel motion; *d*, paddle-wheels; *e*, boilers, with cylindrical exterior of cast iron, and interior wrought-iron oval fire-tube; *g*, iron chimney at the stem of vessel; *h*, surface condensers, on which he wrote, "space between the

[1] See Wheal Prosper Engine and Herland Engine, chap. xix.

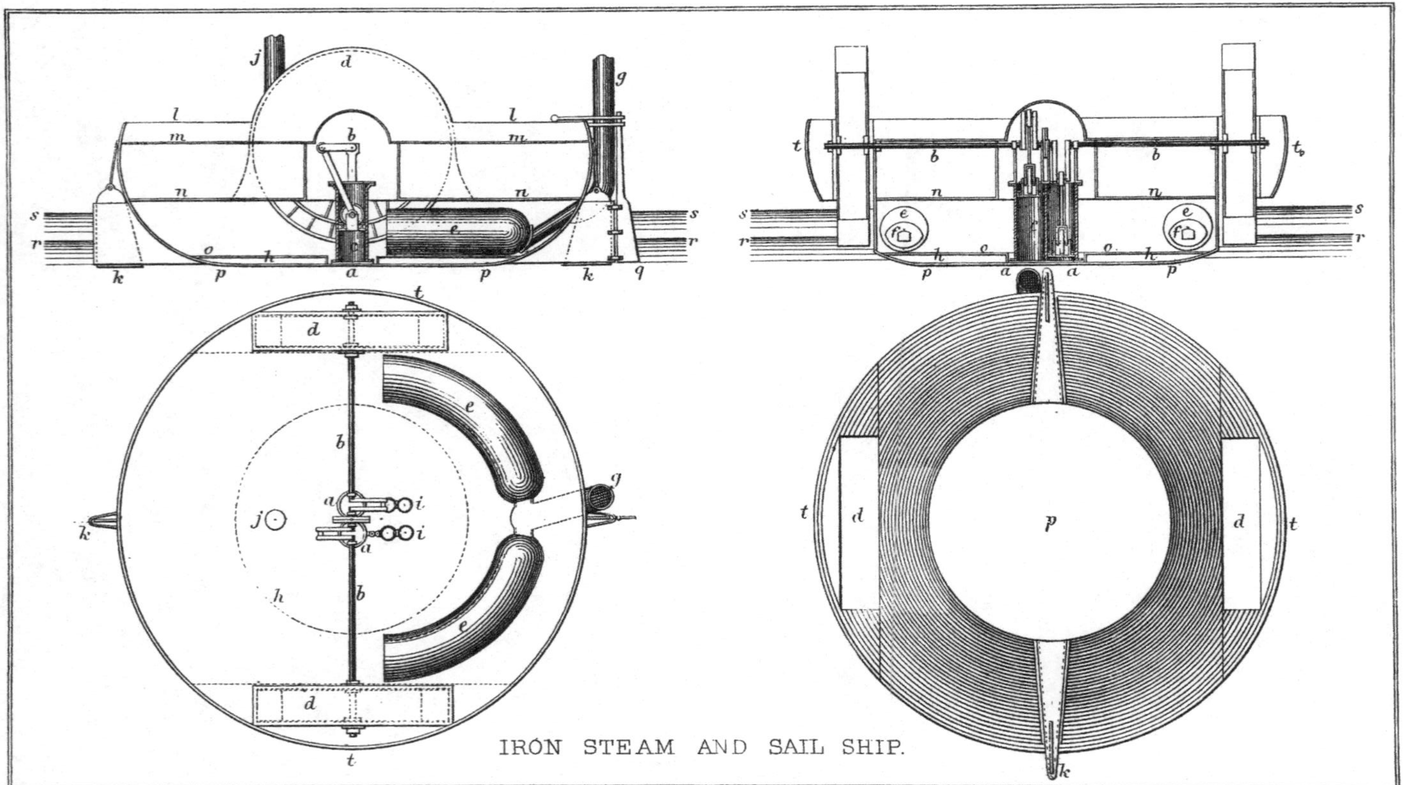

IRON STEAM AND SAIL SHIP.

London. E. & F. N. Spon, 48, Charing Cross.

Kell, Bros. Lith. London.

ship's bottom, and the lower deck to cool the water;" *i*, small air-pump and condenser; *j*, iron tubular mast; *k*, sliding iron keels; *l*, iron gunwale; *m*, iron deck; *n*, iron middle deck; *o*, iron bottom deck; *p*, iron bottom of vessel; *q*, rudder; *r*, light water-line; *s*, deep water-line; *t*, supports and shelter for paddle-wheels. The design combines the earliest Newcomen steam-engine with that of Watt, also of Trevithick's high-pressure steam direct-action engine of the present day, extending over one hundred and fifty years of steam-engine experience. The open-topped cylinders have the simplicity of Newcomen's first engine; but Newcomen dreamt not of the guide-wheel piston, and upward thrust of a rigid piston-rod moved by expansive steam, propelling a crank; neither was it the vacuum engine of Watt, for it used no injection-water, the vacuum being produced by surface condensation, as in Hall's condenser. The chief power of the engine was from high-pressure expansive steam during the up-stroke of the pistons: the ideas are traceable in his letters of 1812.[1] "The cold sides of the condenser are sufficient to work an engine a great many strokes without injection-water." "If this engine is worked with steam of 25 lbs. to the inch above the atmosphere, cutting it off at one-twentieth part of the ascending stroke of the piston, the power will be as three to two of Boulton and Watt's single engine." "The length of the piston, and the small variation that the beam will give it, is so trifling that it will not be felt." The open-topped cylinder, with a deep piston in lieu of the guide-wheel, the expansive steam, and surface condensation, are evidenced in letters of about that period.

---

[1] Trevithick's letter, December 7, 1812, chap. xvii.

The iron vessel in form and in detail of construction is as remarkable as the engine; its growth may be traced from the patents of 1808 and 1809 for iron ships with iron mast and yards and self-reefing sails.[1]

The shallow draught of the vessel and its sliding keels are principles still relied on by American yachtsmen; and making the cellular bottom of the vessel a condenser is the original idea in its most simple form of modern surface-condensers.

Certainly this drawing, explained by Trevithick's letters, patents, and working engines constructed sixty years ago, combine and fully illustrate, under the most simple forms, all the leading principles of the magnificent ships of the present day, with their iron sides, iron masts, small but powerful direct-action engines, with high-pressure expansive steam and surface condensation.

Everyone thought highly of his inventions, but no one enabled him to bring them into immediate profitable use: such being the inevitable result of going too much in advance of the time. Ten years of incessant labour since his first experiments with high-pressure portable steam-engines, to bring them within the range of everyday use, had left him a lone man, bereft of liberty and broken down in health and strength.

Steel's letter of the 20th April,[2] 1810, was addressed to Trevithick in Cornwall, where he had gone to raise money by mortgage or sale of mine shares or property, to pay the cost of showing his inventions to the public.

On his return to London, everything belonging to him was seized for debt, and he was obliged to retire

---

[1] Patent of 1809, and description of iron ship, chap. xiv.

[2] See chap. xii.

to a sponging-house in a street of refuge for debtors, a halfway house between freedom and imprisonment.

All this was too much for the strongest man. Typhus and gastric fever during many weeks reduced him to a state of physical helplessness, followed by the loss of intellect, and brain fever, and the patient, before so weak, required the care and strength of keepers.

In this emergency Dickinson brought his medical man to assist; Dr. Walford, known to the family, disagreed with the new comer. A third medical man was called in. Anonymous letters were sent to Mrs. Trevithick on the probable result of the injurious medical treatment.

In an emergency, when it seemed a question of life or death with her husband, Mrs. Trevithick, scarce knowing what she did, sought through the streets of London for a medical man. A gentleman observing her need of help, asked what was the matter, begged her to at once return to her house, and promised to seek and send a doctor. Misfortune had reached the turning point. A kind letter from her brother advised Trevithick's return to Cornwall, and in the early part of September, 1810, being still too weak to move hand or foot, he was carried on board a small trading vessel, called the 'Falmouth Packet,' his eldest boy, about ten years old, keeping him company.

It was war time, and the 'Falmouth Packet' with other vessels sailed from the Downs, under convoy of a gun-brig. After three days they anchored off Dover. Trevithick went on shore and enjoyed the first short walk since the commencement of his illness, four or five months before. On getting under way again they were chased by a French ship of war. The 'Falmouth Packet' knew how to sail and when to hug the shore, so she showed her heels to the enemy; and in six days after leaving London landed him at Falmouth, about sixteen miles from his Cornish residence at Penponds. Taking his boy by the hand, they walked to his home, from whence two months before, when too ill to be informed of his loss, his mother had been carried to her grave from the house of his childhood.

PENPONDS, TREVITHICK'S RESIDENCE IN CHILDHOOD. [W. J. Welch.]

"MY DEAR JANE, "HAYLE FOUNDRY, 1st *June*, 1810.

"Our hearing by every other post from Mr. Blewett, of Trevithick's rapid recovery, and also by Dr. Rosewarne last Saturday, that the fever had quite left him, gave us great satisfaction; but we are much concerned for your situation.

"In your letter of the 25th ult. you seemed to be much

alarmed from Trevithick's weakness, but I think you cannot expect otherwise than that he will be very weak for some time, after so dreadful an attack. Do not be alarmed, I hope he will do very well; you must not say anything to him about his business, that is likely to hurry his mind, until he gets better. If Mr. Dickinson receives money, he must be accountable for it. I beg that you will not hesitate asking Mr. Blewett for what you want. It gives us great happiness to hear that you enjoy health in this great trial. If you think it necessary and you wish it, I will come to you, but I sincerely hope that your next will bring a more favourable account; I know Mr. Blewett will be very happy to do anything for you in his power, and I wish you would ask his advice in any business that you think is not proper for Trevithick to be told of until he gets better.

"I hope you find Mr. Steel honest: in that case it is not in Mr. Dickinson's power to cheat you.

"Do let me know how Trevithick's affairs stand, and what his prospects are. If he is not likely to do well where you are, do you think he would consent to return to Cornwall—if not to settle, for a little while? His native air might be a means of getting him about. Both my sisters join with me in love to you and family.

"I am, dear Jane,

"Yours very sincerely,

"H. HARVEY."

An official document of receipt and expenditure for the five years prior to his bankruptcy shows a loss of more than 4000*l.* by his labours; and that he had sold or mortgaged his property and borrowed from his friends.

Mrs. Trevithick, when in London, in 1808 to 1810, saw an iron steamboat of her husband's at Mr. Blewett's yard. People said that an iron boat could not float. Trevithick made a trip down the river from London Bridge in the iron steamer, and towed after him one of his iron tanks to show that they also would float.

Dickinson's[1] letter in the early part of 1810 states that Penn had three steamboat engines then in hand; one of them with a caboose, or the nautical arm engine, and one of them a towing machine to take the place of towing machines that required repair.

On his return to Cornwall, he devoted his time to high-pressure expansive engines, and to preparing for South America; still turning over in his mind the propelling ships by steam. Mr. Newton, who rode on the first Camborne locomotive, thus writes:—

"My dear Sir, "Camborne, *28th February*, 1870.

"In reply to yours of the 25th, I recollect hearing Captain Trevithick, in conversation with the late Captain Andrew Vivian, above sixty years ago, say that he could construct a boat, to be propelled by the application of steam.

"His idea was a boat having two keels, between which two rods might be attached, with propellers to each, similar to duck's feet, one moving forwards, whilst the other moved backwards, and so *vice versâ*. Another idea of his was (being then war time) the application of thick, iron sides, to our men-of-war ships, to repel the shots from the enemy; these were ideas then floating in Trevithick's brain. It was at a mine meeting (North Bininer Downs). How plainly do these ideas appear as the embryo of what has taken place since; namely, the steam- and the iron-clad war ships.

"I am, dear Sir, yours respectfully,

"L. Newton."

Mr. Newton, who aspired to be a pupil of Trevithick's nearly seventy years ago on the working of the first locomotive, still lives to bear witness of the great

[1] See Dickinson's letter, chap. xiii.

engineer's designs for steam navigation, and the keeping out an enemy's cannon balls by iron-plated ships of war.

"CAMBORNE, *26th March*, 1812.

"SIR CHARLES HAWKINS, Bart.,

"Sir,—Enclosed I send you a letter for Sir John Sinclair, without a seal, for your inspection, and if you approve of it please to put a wafer and forward it; if not, please to leave it rest, and send me your remarks, that I may alter it. I have not gone so fully into the business as I might have done, in expectation of Sir John Sinclair's writing to me again, with particulars, the size engine he may want, and for what purpose. If I had money I would immediately construct an engine in a ship, at my own expense, much rather than be assisted by any person, but the misfortune I lately met with, left me without a shilling; therefore I am obliged to attend to other business for a maintenance for my family, instead of attending to purposes of twenty times the value.

"I think you have heard Mr. Giddy give his opinion on the value of my new plan for driving ships, and I am really sorry to part with it so cheap to strangers; but not having it in my power to carry it into execution, must submit. You see what a trifling sum will accomplish the experiment, and the value of it I believe is beyond calculation; besides the great utility to the public. I very much wish you would try the experiment, and take the advantages andhe merit to yourself, instead of its going to strangers.

"It might be done at Hayle Foundry, on one of their boats; the expense will not exceed 100*l*. If you wish, you can again consult Mr. Giddy on the plan, before I engage with any other. As you have land near the mine at Gwythian, I beg to inform you that I have begun the erection of a steam-engine in Wheal Prosper Mine. I was employed a few days at St. Ives, by the subscribers to the breakwater. I would be very much obliged to you to say if I must call on Mr. Hulse for payment.

"Your obedient servant,

"RICHARD TREVITHICK."

"CAMBORNE, CORNWALL, 26*th March*, 1812.

"TO SIR JOHN SINCLAIR,

"Sir,—I received from Sir Charles Hawkins a copy of Dr. Logan's letter to you, also a note from you to Sir Charles Hawkins, both respecting the driving boats by steam. This is a subject to which I have given a great deal of thought, as being a thing of immense magnitude and value, if it could be made to be of general utility.

"So far as ideas and the outlines of theory will warrant success, I have no doubt of making it fully answer the purpose on all sorts of vessels, and in the open sea; but this cannot be done by any plan that has yet been practised.

"I have been very fully informed respecting the plans and performance of the American steamboats, by practical men and others who have sailed on board them; therefore I think there can be no doubt of their performance in rivers; and am surprised that English engineers have not adopted it long since without waiting to be shown the way by Americans.

"The construction of steam-engines hitherto has made them very unfit for ships' purposes, on account of their immense weight, bulk, and complications.

"Also the method of applying the power by wheels over the ship's sides, would make them very unwieldy in gales of wind.

"The engine that I have lately invented is not one-tenth part of the weight in bulk of other engines of the same power; and so simple that they are managed by any common labourer.

"They are working with the cylinder perpendicular, others horizontal, and others diagonal, as circumstances require, and the fire, with all the apparatus, is enclosed within a wrought-iron boiler, so that no part could be injured from the rolling of the ship. Also the apparatus that I would use for propelling the ship should be fixed with the engine in her hold, so that no part of the machine should be exposed to injury from a heavy sea. Giving motion to a ship in the way that I propose is quite a new idea, and has never as yet been practised for any purpose whatever. When first I hit on this idea I communicated it to Mr. D. Giddy, whose theory on every occasion I have relied on, and found by practice to be correct. His opinion was with mine, that it was a thing of vast importance. I am so far convinced of its success, that if you wish to make an experiment I will lend you an engine for the purpose gratis, and also give my own time and invention; therefore the loss cannot be more than on the apparatus for propelling, which I will engage shall not exceed 100*l.*, which sum will prove it on a scale sufficient for the experiment both on my plan and that now in use in America.

"I have often observed horses employed to draw ships through the canal across the Isle of Dogs, from Limehouse to Blackwall, and found that three horses took two ships in tow, of three or four hundred tons burthen each, and drew them at two miles per hour.

"A 30-horse engine on my plan will not exceed 8 tons in weight, 16 feet long, 6 feet high, and 6 feet wide; very little more room than six hogsheads of sugar. I think less than half this power would be sufficient for driving a ship, pumping, loading, discharging, and cooking by steam.

"The steam from the salt water would be fresh for the ship's use, instead of burdening the ship on long voyages with, say six months' fresh water, weighing four times the weight of the engine and apparatus. One of the American steamboats is 160 feet long (a length equal to a 74-gun ship). The engine is called a 24-horse power; the boiler is from 8 to 10 feet long; from this I judge the engine cannot be above half the power stated.

"The wheels are on each side, similar to those of watermills, and under cover; her accommodation is fifty-two berths, besides sofas.

"There is one other steamboat of 85 feet long.

"Respecting the engine for thrashing, chaff cutting, sawing, &c., I am now making one of about two-thirds the size of Sir Charles Hawkins's, which will be portable on wheels, and may be removed from farm to farm as easily as a one-horse cart. By placing the engine in the farmyard, and passing the rope from the fly-wheel through the barn door or window, and around the drum on the machine axle, it may be driven.

"The steam may be raised, and the engine moved a distance of two miles, and the thrashing machine at work within one hour.

"The weight, including engine, carriage, and wheels, will not exceed 15 cwt.; about the weight of an empty one-horse cart.

"The size is 3 feet diameter, and 6 feet high. If you wish to have one of this size sent to the Board of Agriculture as a specimen, the price, delivered in London, will be sixty guineas. If you require further information, please write to me again, and I will furnish you with drawings, or models, or steam-engines of any size or for any purpose.

"N.B.—I have a portable engine nearly finished which will weigh about 30 cwt. The power is equal to 6 horses, or 24 horses every 24 hours. Price, delivered in London, 150*l.*

"I remain, Sir,

"Your most obedient servant,

"RICHARD TREVITHICK."

In March, 1812, the screw-propeller and plan for driving it were settled things in Trevithick's mind. He had seen that engines more powerful than the nautical steam-labourer of 1808, and steam-arm of 1809, were required for sea-going work, and his trials with paddle-wheels had shown that something more compact, and less in the way of wind and waves, should be devised before ocean-going steamers carrying canvas would be successful; for "applying the power by wheels over the ship's sides makes them very unwieldy in gales of wind. My apparatus for propelling the ship should be fixed with the engine in her hold, so that no part should be exposed to injury from a heavy sea. Steam-engines hitherto are unfit for ships' purposes, on account of their immense weight, bulk, and complications. My newly-invented engine is not one-tenth of their weight, a 30-horse engine complete being only

8 tons." The wisdom of those words is proved by their being still applicable after fifty years of rapid advancement.

The screw-propeller was offered by Trevithick to the Navy Board in 1812, with a proposal to test its superiority to paddle-wheels by trying both of them in the same boat, worked by the same engine, the apparatus to be fixed, so that no part be exposed to injury from a heavy sea: in other words, the screw was to be placed near the stern of the vessel, and to be under the surface of the water.

Sir John Sinclair having forwarded the proposal to the Navy Board, hoped that they would order a trial of the screw, and defray the cost; and on this account Trevithick postponed his application to Mr. Praed to test it on his canal boats. After waiting two or three months, he wrote to his friend Sir Charles Hawkins:[1]—

"Sir John Sinclair has taken a useless journey by calling on the Navy Board; for nothing experimental will ever be tried or carried into effect, except by individuals."

Acting on these views, he proposed, at his own expense, to place a 6-horse-power high-pressure steam-puffer engine, weighing 30 cwt., andcosting but 150*l*., in a barge, that the public might see the superiority of the screw-propeller and high-pressure engine over the paddle-wheel and low-pressure vacuum engine.

"DEAR SIR, "*April 7th*, 1812.

"I received your letter, enclosing one to Sir John Sinclair, which I forwarded. I wrote him that if he recommended your engines for propelling vessels, and made known your steam-engines to agriculturalists, I had no doubt but

[1] See letter to Sir Charles Hawkins, 13th June, 1812, chap. xvi.

they would be universally adopted. You may write Mr. Praed offering your engines on his canal for moving barges—you may say you take the liberty to write him through me—and offer him your engines for moving barges on canals by the means of steam. Give the cost and expense of the engines and barges worked by coals instead of horses, and refer yourself to me for the advantage of a steam-engine for thrashing corn. There can be no harm in your writing Mr. Praed, who is the chief manager of the Grand Junction Canal, and certainly has it in his power to adopt your plans. I think myself pledged to carry into effect, at least to do everything that can be expected of me, to effect the small breakwater at St. Ives. It appears to me that if the small breakwater was laid down from No. 3½ by the inner bamber loose to either No. 8, or about No. 9 . . . . that it would effectually secure a great number of shipping afloat. That even to No. 9 would not cost much above 20,000*l*. . . . . within the breakwater the tide would wash out the sand. . . . . If the breakwater was to be laid down as placed in the map . . . ."[1]

It does not appear that Mr. Praed encouraged the screw-propeller in canal barges, though Trevithick offered to defray the cost of the experiment;[2] and the idea remained in abeyance until 1815, when he constructed a rotary engine at Bridgenorth, hoping to make it and the screw-propeller a suitable combination for marine propulsion.

"I have drawn the cylinder 9 feet in diameter, in which the screw is to work. The screw is 8 feet 10 inches in diameter, and has two turns in 4 feet long. Of course it will gain 2 feet forward to each turn. The steam-arms revolve the same as the screw, being on the same first motion, at a rate of about 350 feet per second at the end of the arm, will give the screw about ten miles forward per hour. I have not worked the engine to

[1] Trevithick (in his copy of this letter) probably could not decipher the writing of Sir Charles Hawkins, marked by omissions, and toward the end of the letter gave it up as impossible.

[2] Trevithick's letter, June 13, 1812, chap. xviii.

any advantage since I wrote you last, because before I worked with very high steam I wished to force the boiler with cold water, which I did with 300 lbs. to the inch, which it stood exceedingly well."[1]

This description of an upright boiler, for the rotary engine for driving the screw-propeller, was referred to in a letter a week before,[2] and was also followed in four days by another letter, describing a tubular boiler, intended also for a steamboat:—"The boilers making for this towing engine are all tubes of 3 feet in diameter, five-eighths of an inch thick (with circular ends), of wrought iron. There are three horizontal tubes, and to each of the horizontal tubes three upright tubes are suspended."[3]

On the 6th June, 1815, a patent was taken for various improvements:—

"The fifth part consists of a mode of propelling, drawing, or causing ships, boats, and other vessels, to pass through the water, which purpose I effect by constructing a worm or screw, or a number of leaves, placed obliquely round an axis. The obliquity of the thread of the worm, screw, or leaves, admits of considerable variety, according to the degree of velocity given to it, and speed required, and according to the power with which it is driven. It may be made to revolve in the water at the head of the ship, or at the stern, or one or more worms may revolve on each side of the vessel, as may most conveniently suit the peculiar navigation on which the ship, boat, or vessel is to be employed.

"In order to make the boiler of a high-pressure steam-engine of very light materials, for portable purposes, and at the same time strong for resisting the pressure, as well as for exposing a large surface to the fire, I do construct the said boiler of a number of small perpendicular tubes, each tube closed at the

---

[1] See letter, 12th May, 1815, chap. xvi., pp. 368, 369.
[2] See letter, 7th May, 1815, chap. xvi., p. 364.
[3] See letter, 16th May, 1815, chap. xvi., p. 370.

bottom, but all opening at the top into a common reservoir, from whence they receive their water, and into which the steam of all the tubes is united."[1]

This tubular boiler, patented with the screw-propeller more than fifty years ago, is like a boiler now being tried in 1871 as a new thing. "The first of a series of official trials of one of the Cochrane boilers and one of the common boilers of the 'Audacious,' 14, double-screw iron ship, was commenced on Wednesday. The Cochrane boiler has the water contained in vertical tubes, which the fire surrounds, whereas in the common boiler the water surrounds the tubes through which the fire passes."

During twelve years he had at various times shown the practicability of steam propulsion; in 1804 with a seven-miles-an-hour steamer; in 1806 the single paddle steam-yacht. In 1808 the patent for universal steam on board ships. In 1812 the proposing to the Admiralty the use of the screw-propeller. In May, 1815, a screw 8 feet 10 inches in diameter had been made, and in June, 1815, a patent for the screw was applied for; and when on the verge of success, he, in 1816, sailed away to the mountains of an unknown land.

"Dear Trevithick, "East Bourn, *July* 18*th*, 1815.

"I really do not know what to say respecting the engines, beyond what we have frequently discussed in conversation about it. Your figures I believe are right. You have very much surprised me by stating that one of Mr. Woolf's is stopped. If the accounts of his engine at Caenver are correct, it does the most duty of either one now working in Cornwall. But what is very curious, his winding engine at Great Wheal Fortune is among the very worst of that sort. I hope that your experi-

[1] See patent, 6th June, 1815, chap. xvi.

ment with the recoil engine for moving boats is in progress. Nothing occurs to me respecting them, except that there is not any theoretical limit to the most advantageous velocity for the propelling screw. If you are disposed to try various sizes of these screws, moving with different velocities, you must recollect that the propelling zone, with the new screw, increases as the squares of the number of revolutions in a given time; and that supposing the velocities of rotation (that is, the number of revolutions in a given time) to be the same in different screws, the propelling zone will vary as the 4th power, or as the square squared, of the diameter; or thus, suppose D = the diameter and N = the number of revolutions in a minute of one screw, D = the diameter, and N = the number of revolutions in a minute of the other screws. Then will their propelling power be to each other in the proportion $D^4 \times N^2$ to $d^4 \times n^2$. The reason why no limit can theoretically be put to the velocity most advantageous for the propelling screw is this, that the faster it goes, the less will the loss be, occasioned by the motion of the vessel through the water. The centre of propulsion (as it may be termed) is distant from the centre of the screw (supposing the radius one) $\frac{1}{\sqrt{2}}$; that is, if a square piece, equal in size to the triangular vane  , were to move with the same inclination and the velocity which the vane has, or $\frac{1}{\sqrt{2}}$ (or about *c*), their effect would be equal. Suppose now, two vessels moving at the same rate through the water, each propelled by screws, one large, and moving so that the centre of propulsion has a velocity double to that of the vessel; the other smaller, but moving with a velocity ten times that of the vessel. In the first one, three parts out of four of the power will be lost; in the second, only nineteen out of a hundred, not one-fifth.

"I shall remain here for at least a fortnight.

"I am, dear Sir,

"Yours most respectfully,

"DAVIES GIDDY."

"GENTLEMEN, "PENZANCE, CORNWALL, *Nov.* 30*th*, 1815.

"Enclosed you have a drawing for the towing engine for London, which you will execute as soon as possible. The payment you will receive from Mr. John Mills, both he and myself being equally interested in this machine; therefore, you may call for payment on either, or both, which you please.

"On receipt of this drawing I wish you to inform me when you can execute it, and also whether you can understand the drawing.

"I expect that the screw will be too large for the barges to take to London. If so, leave it in two parts, and we will rivet them together in London.

"I wish you to send to me Herland invoice, and I will send you the balance. Let me know how you are getting on with Beeralstone engine. I hope you will forward it this next spring-tide down to Bristol, and ship it by the Plymouth trader. The adventurers are extremely uneasy on account of your delay. I was at Bristol the 22nd inst., and chartered a vessel of 90 tons burthen to take down Herland engine, which is only 11 tons, for which I am to pay 130*l.*, instead of 20*s.* per ton, which is the usual freight from Bristol to Cornwall.

"Therefore, I have made a sacrifice of 119*l.* entirely on account of your own neglect. Unless the Beeralstone engine is sent down this next spring (which is now nearly at hand), I shall get into further trouble and expense on that account also. Write to me by return of post to Penzance, and not to Camborne, as usual.

"I remain, your humble servant,

"RICHARD TREVITHICK.

"To HAZELDINE, RASTRICK, and Co.,
"*Bridgenorth, Shropshire.*"

Drawings had been sent for a steamboat on the new plan, and the screw or screws were actually on their way to London before the patent right had been sealed.

"MR. JOHN MILLS, "PENZANCE, *February* 28, 1816.

"Sir,—On the other side you have an account of the expenses of boilers, &c.

"I wish you would say what money I have received from you in the whole for Peru shares, as I do not exactly recollect, that I may enter it to your account all together.

"The castings I expect you have long since received from Bridgenorth. The performance of the great engine astonishes every person here; she will do the full work of a 40-horse engine, with one bushel per hour. The boiler I have sent you will do more than a 200-horse engine.

"You may get the boilers and case together in the barge immediately. Put the round end of them as near the stern of the barge as you conveniently can, because that will give more room for the machinery.

"As soon as you write me that all the engine is on board the barge, I will come up to town.

"The price I agreed for the freight was 10*l.* for each tube, and 25*s.* per ton for the other iron, which you will find about 9 tons.

"I have received news from Lima, dated 8th July last, and am informed by the Englishmen that went out, that they are getting on exceedingly well. They have been offered, and refused, at the rate of 8000 dollars for the same share as you hold. I would thank you to write and let me know whether you have received the boilers, and how you get on. Mr. Smith is returned to Greenwich; if you call that way, he can tell you the particulars of the engine, as he saw it at work several times.

"I remain,

"Your humble servant,

"RICHARD TREVITHICK."

The adaptation of the screw was a great difficulty, and the addition of a newly-designed tubular boiler for driving it was a complication of difficulties that Trevithick might have found his way out of; but it was too much for Mr. Mills, or for Mr. Smith, Trevithick's

London agent, and engine manufacturer, who had been to Cornwall to see what was being done by the great high-pressure steam engineer.

"DEAR SIR, "GREENWICH, *February 29th*, 1816.

"I have had some conversation with Mr. Marshall about the small engines for Manchester; he says that the power of two men will be quite sufficient. They expect to have them small enough to go in at a common door, to work them in the room where the machines are fixed. I will thank you to inform me what size boiler and cylinder you consider right for such an engine; you need not trouble yourself to make a drawing. I should like to know how the thrashing machine works, soon as you have got it finished. Penn called on me a day or two ago, and said he was going to assist Mr. Mills in fitting up the towing barge, and inquired very particularly about the Herland engine.

"He said Hawkins told him that your engines would do away with all the piston engines. Let me know how you are going on at Herland. I tried the engine which you sent me from Cornwall, yesterday, which works very well, except the large cock, which Field is grinding in this morning. I rather think it got damaged in the carriage. I have painted the fly-wheel, and all the parts, and it looks very well.

"I have been thinking of selling it for you (which I have no doubt may be soon done), and fit up one on the new principle for working the press. I hope to receive your drawing soon for the engine for the sugar-mill. I have answered Mr. Melvill's letter to you, and Mr. Quin's also, and from what Gordon and Merpey said, expect some orders from them.

"You have the enclosed in a parcel to Mr. Page. Have you any letters from Lima? Mr. Marshall says the engines are at work.

"Hoping to hear from you soon as convenient, I remain, with best wishes for your health, and that of Mrs. Trevithick and family,

"Yours sincerely,

"JAMES SMITH.

"P.S.—I have just got a letter from Carpenter and Smith's Wharf, saying the 'Jane,' of Penzance, is arrived."

Mr. John Penn, who assisted Mr. Mills in putting together Trevithick's screw-propeller and improved steam-engine, was a millwright, and probably the founder of the present eminent marine engine-building firm, Penn and Sons.

Field, who also helped, has probably since been known as a partner in the firm of Maudslay, Son, and Field, equally eminent with Penn and Sons. Harry Maudslay, also one of the firm, was then known as a clever smith employed by Trevithick.

James Nasmyth, then a boy, working for Maudslay, has since been the maker and patentee of Nasmyth's upright or chimney boiler, for which the writer, who had the pleasure of his personal acquaintance, paid him patent right, though it was almost an exact copy of Trevithick's boiler made and used in 1815,[1] and formed part of the engine that Field, and Penn, and others, probably Maudslay and Nasmyth, saw and worked on.

In 1833, the writer, seeking his fortune in London, and knowing that Maudslay, then a leading marine-engine builder, had benefited by Trevithick's patent tanks, and steamboat discoveries, and hoping a return for benefits received, sent in his name as the son of Trevithick, the engineer, and ask permission to go through the works. The answer was, "We are very sorry, but it is against our rule;" but he gained admittance by one of the leading mechanics, his relative, schoolfellow, and early shop companion, now the leading mechanic in Penn's factory, and the grandson or grandnephew of old Sam Hambly, who, as Trevithick's helper, had worked side by side with Penn and Maudslay in putting together the tanks, nautical labourer, and screw-propeller, during 1808 to 1815.

[1] See Trevithick's letter, 7th May, 1815, chap. xvi.

"Dear Sir, "Greenwich, *July* 26*th*, 1816.

"I have just seen Mr. Mills. He wishes me to inform you that there will be a great many engines wanted in Holland for the boats. He has seen the small engine which I have nearly finished, and wishes to try it on the water with your screw. I shall, therefore, be obliged to you to say by return of post what you think should be the diameter of the screw for this size engine, and in what way it would be best to work it: perhaps with a universal joint, if the engine will go fast enough this way; but let me have your opinion on this particular. I expect to have it finished in a week, when Mr. Mills will be ready with a proper boat for the purpose. He seems to have some doubts of the screw, but says that if this should not answer, the old way with the wheels will do very well. I have not seen any of the sugar people, but told Mr. Penn to make out his bill for them. Mr. Strattan, Gutter Lane, called on me the other day, and I promised him a sight of the engine at work very soon. Don't make any arrangement with him about engines till you hear from me again. Mr. Penn told me something about him which I will explain in my next. We had a letter from Mr. Page this day; shall write him in a day or two. I have not time for more now, as it is four o'clock. You shall hear from me again the beginning of the week, but expect to hear from you by next post.

"Yours, &c.,

"J. Smith.

"P.S.—I will give you the size of the oven in next."

Mills liked the look of Trevithick's new engine in Smith's shop, and wishing to have a screw-steamer, offered his boat that the engine might be placed in it, Smith engaging to discover from the great necromancer, then in Cornwall, how the screw should be made, how it should be fixed, and how it should be worked, to which an answer was sent by return of post. Mr. Smith had married a sister of Mr. Page, who, with his partner Day, were solicitors for the South American mining scheme.

"DEAR SIR, "GREENWICH, *August* 1*st*, 1816.

"I have just received your letter dated July 29, and shall proceed with the screw, according to your direction, but am much afraid that I shall not have the engine finished before you come. I wish you would stop another *week*, that I may get it done before you see it.

"Mr. Mills says that engines of 30 and 40 horse-power will be wanted, and a great many of them, and is very anxious to see the small one tried.

"I am very forward with it now; but, being a new thing to me, have had many tools to make, and Mr. Penn has been rather unfortunate in the casting the fly-wheel, which you saw on his premises was broke in the turning, and the first cylinder was good for nothing, and we had to cast another, which is pretty good. Horton charges 46*l*. for the boiler, which appears to be well made. I have got it on the wheels, and have finished the box at the bottom for the chimney, and have ordered the pipe for the chimney in town. The stuffing box for the valves is screwed on the top of the boiler, and the double cylinder is screwed to the side. The standards for the shaft of the fly-wheel are also fixed, and Penn is turning the shaft from a wood pattern, which I have given him. I am going to town this afternoon about the copper tubes or pipes for steam and discharging. The fire-door is finished, and the bars are in place. All this is done, and yet I think it will take another week to complete it, and I very much wish to finish it before you see it.

"The sugar-bakers have not got Mr. Penn's bill yet, but this I suppose does not much signify.

"I have asked the plumber to give his charge for what little he did in this business, but have not got it yet. We received the fish, hams, &c., yesterday evening, which are not in the least injured. We are much obliged to you for them. I shall not trouble you about the oven yet, as I cannot ascertain the size exactly, and am too much engaged at present to think about it. We are very glad that you are all well, and shall be happy to see you at Greenwich, and perhaps Mrs. Trevithick or Miss Trevithick will take a trip with you. I need not add that we shall be happy to see them.

"I am, dear Sir,

"Yours sincerely,

"JAMES SMITH."

Trevithick's leaving for Lima in two months from the date of those experiments prevented his perfecting the screw-propeller.

"A Select Committee of the House of Commons commenced its sittings May 8th, 1817, to inquire into the use and safety of Trevithick's high-pressure marine engines.

"Mr. Bryan Donkin examined:—Witness went down to Norwich as a volunteer, to inquire into the cause of the explosion of a steamboat; was accompanied by Mr. Timothy Bramah and Mr. Collinge. Was of opinion that the immediate cause of the explosion had been the use of steam of a very high expansive force. The approximate cause was a deficiency in the strength in the end of the boiler. It was cylindrical. One end was wrought iron, the other end cast iron. It appeared to have been previously of wrought iron, and had been cut out, and cast iron substituted in its place. Would not choose to use a high-pressure engine, from the danger which arose from their use. Scarcely ever saw the low-pressure engine beyond 6 lbs. to the inch. Thought it just to state to the Committee that there was an advantage to be derived from the use of high-pressure engines on board of boats. Had likewise been told, though without having seen one, that Trevithick had invented a method of making boilers by increasing the length, and decreasing their diameter, so as to render them capable of maintaining pressure to a much greater degree than heretofore. Had no doubt but Cornwall had derived incalculable advantages from the use of high-pressure engines.

"Report of the Select Committee of the House.—Your Committee find it to be the universal opinion of all persons conversant in such subjects, that steam-engines of some construction may be applied with perfect security, even to passenger vessels; and they generally agree, though with some exceptions, that those called high-pressure engines may be safely used, with the precaution of well-constructed boilers and properly-adapted safety-valves; and further, a great majority of opinions lean to boilers of wrought iron or metal in preference to cast iron."[1]

At the time of the inquiry by the House of Commons into the propriety of allowing the use of the high-pressure steam-engine in England, Trevithick was successfully working it on the top of the Cordilleras. The new boiler referred to by Mr. Donkin was the multitubular boiler, constructed for the screw-propeller engines.

On Trevithick's return from America ten or twelve years afterwards, he wrote:—"The boilers in use prior to your petitioner's invention could never, with any degree of safety or convenience, be used for steam navigation, because they required brick and mason work around them to confine the fire."

This statement, made fifteen or twenty years after his numerous applications of steam on board vessels, beginning in 1803 with the dredger-boat in the Thames, and then the steam-barge in 1804, prior to which no boiler, except his own, had been erected without a casing of brickwork flues, proves him to have been the first practical applier of the marine steam boiler and engine.

In 1788 Patrick Miller, Taylor, and William Symington, tried to propel a boat by a 1-horse-power steam-engine,[2] and after thirteen years of consideration, Symington, hoping he had made the idea practical, took a patent in 1801 for "the application of such

---

[1] Abstract of evidence before a Select Committee of the House of Commons on Steam.

[2] 'Distinguished Men of Science,' by H. Walker.

an engine, to propel a vessel by paddles;" and in 1803 the 'Charlotte Dundas' was propelled by steam in the Firth of Clyde, but after a short time was laid aside. "In 1807, Fulton, of New York, launched the 'Claremont,' said to be the first steamboat that successfully carried passengers for hire. In 1812 the 'Comet' steamboat, by H. Bell, plied between Glasgow and Helmsburgh, and in 1813 the 'Elizabeth' steamboat, on the Clyde, was probably the first remunerating steam-vessel in the world;"[1] so that twenty-four years elapsed between the idea of Miller and Symington, and its first practical use.

The first attempt, on any scale worthy of notice, at navigation by steam in Britain was made about this period on the river Clyde. A boat of about 40 feet keel and 10½ feet beam, having a steam-engine of 3-horse power, began to ply on the Clyde as a passage-boat between the city of Glasgow and Greenock in 1812; but owing to the novelty and apparent danger of the conveyance, the number of passengers was so very small that the projectors for some time hardly cleared their expenses.[2]

It is curious that while this experiment was going on in Scotland, Trevithick was corresponding with Sir John Sinclair on the subject, and on the 26th March, 1812, stated: "I have found that three horses towed two ships, of 300 or 400 tons burthen, at two miles an hour," thus intimating the probable power of their 3-horse 'Comet'; and added, "a 30-horse steamboat-engine on my plan will not exceed 8 tons in weight." But Trevithick was far in advance of such small work; he had then the screw-propeller in hand. "The construction of steam-engines

---

[1] Woodcroft 'On Marine Propulsion.'
[2] 'History of the Steam-Engine,' by Stuart.

hitherto has made them very unfit for ship purposes, on account of their immense weight, bulk, and complications; also the method of applying the power by wheels over the ship's sides would make them very unwieldy in gales of wind; the engine that I have lately invented is not one-tenth part of the weight or bulk of other engines of the same power; the fire, with all the apparatus, is enclosed within a wrought-iron boiler, so that no part could be injured from the rolling of the ship."[1] This boiler, patented by Trevithick in 1802, and used in his steamboat of 1804, was precisely the one which enabled this Clyde boat of 1812 to work, and is thus described: "The boiler in which the steam is produced is made of strong wrought-iron plates, and the fire-place is an iron tube contained withinside; the smoke, after passing through two or three turns of the tube in the boiler, passes off through the chimney, which is an iron tube. The boat was about 80 tons burthen; she went about six miles per hour in smooth water; the circumference of the paddle-wheels went at a speed of thirteen miles per hour."[2] Not only was the Clyde boat of 1812 a close copy of Trevithick's boat of 1804 in size of boat, speed of wheels and of boat, and kind of boiler, but the words used by Rees in its description read almost as extracts from Trevithick's letters of 10th January, 1805, and 26th March, 1812, thus making good his statement that "the boilers in use prior to your petitioner's invention could never, with any degree of safety or convenience, be used for steam navigation, because they required a protection of brick and mason work around them."[3]

---

[1] See Trevithick's letter, 26th March, 1812, p. 344.
[2] See Rees' 'Cyclopædia,' "Steam-Engine," published 1819.
[3] See Petition to Parliament, chap. xxv.

According to Rees, the American first steamboat experiments had the same origin, from Trevithick through Symington, though he says that the high-pressure engines ought not to be allowed in steamboats, and fails to mention the name of Trevithick.

The Scotchmen feared to go by the Clyde boat of 1812, because it was worked by Trevithick's high-pressure boiler; and if, as has been said, articles in Rees' 'Cyclopædia' were written by Dr. Robison, we can understand why, as the supporter of Watt, he did not trouble to trace the particulars of Trevithick's steam-barge of 1804, which worked with steam of from 50 to 100 lbs. on the inch, being just the pressure now much approved of in steamboats.

"The 'Egypt,' which is, we believe, except the 'Great Eastern,' the largest steamship afloat, is fitted with engines constructed on the compound principle, with boilers giving steam of 75 lbs. to the square inch. The ship has five steam-winches, and a steam capstan and windlass of Napier's patent. The steam-winches supersede a vast amount of manual labour; they work the pumps, hoist the sails, and discharge and load the cargo. The 'Egypt' has four masts; all the lower masts are of iron, while the lower yards and lower top-sail yards are made of steel."[1]

The principles illustrated in this modern iron steam and sail ship, built one hundred years after the birth of Trevithick, may be traced in his own description of his inventions sixty years ago. It is not meant to compare the 'Egypt' of to-day with the iron ship of 1809, or the Thames experiments and screw-propeller and multitubular boiler of 1816; but it may be said of each

[1] 'The Illustrated London News,' November 18, 1871.

of them that they were iron sailing ships, with iron masts and yards, propelled by a screw, driven by direct-action expansive engines, with surface condensers, and tubular boilers giving steam of about 75 lbs. on the square inch, and that in each of them the cargo, the sails, and the anchor were to be lifted by the steam-engine. The acts of Trevithick in the steam-dredger engine of 1803, the steamboat of 1804, the cargo and anchor lifting engine of 1808, the iron steamship with iron masts, direct-action expansive steam-engines, and surface condensers of 1809, and the screw-propeller and multitubular boiler of 1812 to 1816, added to the modern improved detail of construction, constitute the iron steam fleets of the present day.

CHAPTER XVI.

RECOIL ENGINE AND TUBULAR BOILER.

"MR. GIDDY, "BRIDGENORTH, *7th May*, 1815.

"Sir,—Yesterday I fixed the pumps to the new engine which goes exceedingly well.

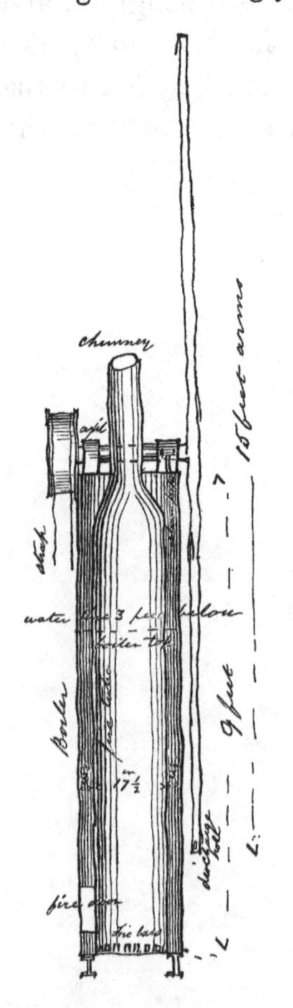

"The bucket is 24½ inches diameter, 18½ feet high, stroke 3 feet, works fifteen strokes per minute, with about 28 lbs. of coal per hour. The arms are 15 feet from point to point, but only one of them gives steam. They run twenty rounds for one stroke of the engine, which is worked by a strap, like a common lathe. The steam was about 100 lbs. to the inch. The opening in the arm was about half an inch long, by a quarter of an inch wide. I intend to work the steam much higher. The fire-place is one tube in another. The fire-tube is 17½ inches diameter, the outer tube 24 inches, 9 feet high. It stands perpendicular. I shall continue to make several experiments, and will give you the results, but as yet cannot say much about it, as it was late yesterday before I finished it.

"I am convinced that it has by far the greatest power for the consumption of coal that was ever made. When I consider the friction in this small machine and the small fire-place, where there is not above three gallons of coal, with the work it performs, I can scarcely believe my own eyes.

I have written by this post to Cornwall to desire the Foxes to send up their engineer to inspect it. The large engine is begun, according to the drawings.

"It will be about thirty-five times the fire sides and fire-place of this small engine. I think it will do fifty times the work, because the friction will be much less in a large than in a small machine. The expense will be about 800*l*., and the power will be equal to a 70-inch cylinder double power. I shall be able to carry every part of this powerful engine on my back (boiler excepted). I would be much obliged to you to inform me what speed you think would be the best for the ends of the arms to travel.

"I remain, Sir,

"Your very humble servant,

"RICHARD TREVITHICK."

This overcrowding of invention on invention was beneficial to the country, prompting others to complete and make useful the ideas so freely scattered, but it had the effect of retarding the perfection of any one in particular.

The screw-propeller, in itself complete, was tied to the incomplete recoil engine, which in its turn was hampered with the new multitubular boilers, of novel construction, and giving steam of greatly-increased pressure. The main object of the recoil engine was to avoid complexity of machinery, and consequent weight in marine steam propulsion.

The recoil engine consisted of two arms, in form like the scabbard of a cavalry sword, each arm 7 feet 6 inches long, projecting from an axis. Steam passed through those hollow arms, and escaping through a small hole near the end, its impact on the surrounding air caused the arms to be forced into circular movement. The propelling screw in the steam-vessel was to be attached to the same axle. A belt-wheel fixed on the shaft drove a pump for the purpose of practically testing the economical power of the engine. The boiler, of novel construction, though its simplicity almost leads to its being called not at all new, was his comparatively old tubular boiler shifted from the horizontal position to the vertical. Yet the two boilers in operation were totally distinct things, the new one being suitable and convenient for the hold, taking up little room, and not liable to have portions injuriously exposed to the fire from the roll of the vessel. That portion of the fire-tube above the water-line was an effective substitute for what has since been patented as the superheating steam apparatus, and at the same time served as a portion of the chimney. The working steam-pressure was 100 lbs. to the inch. Several experiments were tried by altering the size of the escape-hole in the end of the revolving arm from a quarter of an inch long, by one thirty-second of an inch wide, through various widths up to half an inch wide. The speed of the arm was from 10,000 to 15,000 feet a minute. Richard Preen, who saw it work, gave the following recollections:—

"I am now seventy years old, and was working in 1809 in John Hazeldine's engine foundry at Bridgenorth, and have been there most of my time. About that time Mr. Trevithick came very often to the foundry. The engines Hazeldine was building were called Trevithick's engines. The outer boiler was a cast-iron cylinder like a barrel, about 5 feet in diameter. The fire was inside in a wrought-iron tube. The cylinder was let into the boiler, the four-way cock had a handle that was knocked up and down. The piston-rod had a cross-head, and two side rods went down, one to the pin in the fly-wheel, the other to a pin in the cog-wheel; some of them had a crank on one end of the shaft instead of the cog-wheel.

"Mr. Trevithick made another kind of engine, called the Model, some people called it the Windmill, and said it was intended to throw balls against the French. There were two great arms, each of them 10 or 12 feet long, placed opposite one another on a hollow shaft or axle, which had a nozzle in it When steam was turned on it puffed out at the ends of the arms, and they went around like lightning, with a noise like shush! shush! so then it was called by that name.

"This engine was made just before Mr. Trevithick went to South America. He did not know what to do with it, and so gave it a present to Jones, the foreman in the works.

"Master and Jones had a pretty quarrel about who should have it afterwards.

"Mr. Rastrick was considered the engineer. He quarrelled with Hazeldine about putting up the Chepstow Bridge, and set up for himself at West Bromwich, to construct portable engines, the same as they made for Mr. Trevithick.

"He sent word to me to come to work for him. I was then working one of Mr. Trevithick's engines in Mr. Sing's tan-yard. They said they would put me in prison if I left them. That was about 1818.

"Mr. Hazeldine's brother (William Hazeldine) lived at Shrewsbury. He built Bangor Bridge.

"A Spaniard came once or twice with Mr. Trevithick. I was getting the core out of one of the cast-iron boilers before it was cold, for fear of straining the iron. The Spaniard sent for quarter of a cask of beer because we worked hard. Several engines and plunger-pumps were made, no piece to be larger than a mule could carry."[1]

"Hazeldine's Foundry, Bridgenorth,
"Davies Giddy, Esq., M.P., "*May* 12*th*, 1815.

"Sir,—I have received both your favours, which I cannot clearly understand. You say that the point of the arm ought to go five-sixths of the speed of the fluid, which under one atmosphere would be 1070 feet per second, and to increase as the pressure and speed of the fluid increases. Suppose one

[1] Richard Preen's recollections, taken at Bridgenorth, 1869.

atmosphere, or 15 lbs. to the inch, to be 1070 feet per second for the point of the arm to go, four atmospheres, or 60 lbs. to the inch, would be 2070 feet per second for the point of the arm to go; 16 atmospheres, or 240 lbs. to the inch, would be 4140 feet per second; this would be a speed more than double a cannon-ball, and such as can never be obtained in practice and while working in fluid instead of vacuum.

"The speed of the fluid does not increase after the pressure of one atmosphere.

"I cannot make out why the speed of the arm should ever increase beyond 1070 feet per second; but as you have given it so much thought, I doubt not your calculations are right, and I only wish to understand you, whether the speed of the arm should never exceed 1070 feet per second, under all pressures, or whether it is to increase as I have stated above. If so, it will go so far beyond practice that I must lose a part of the power for a less speed, so as to accommodate it to general use. What do you suppose the loss was in working 225 feet per second, which was about the rate in the data that I gave you in my first letter?

"Suppose I make the large engine for which I am now making the drawings to go with a speed of 350 feet per second, what part of the whole power should I get in that case, or what deficiency would there be to deduct from the $\frac{3}{8}$ths which you call the best data? As I have not yet begun that part of the machine, I can alter it to any speed you think proper, and I will do so; but that is a speed that is suitable to the screw on the first motion, and a speed that the centres will stand well. I am doubtful whether the speed is so great on steam as the theory of air gives, because I find that the arm that contains the steam holds about 30 feet long of 1 inch square of steam, and when the discharge-hole at the end of it has an opening of $\frac{1}{2}$ inch by $\frac{3}{16}$ths, which is about $\frac{1}{10}$th of a square inch, when the throttle-cock is suddenly shut, the steam continues to discharge for four seconds, first strong, and gradually falling down from 100 lbs. to the inch to atmosphere strong. I have not worked the engine to any advantage since I wrote to you last, because before working with very high steam I wished to force the boiler with cold water, which I did with 300 lbs. to the inch. It stood exceedingly well; only in a few critical places, in angles where the smiths could not come to close the rivets well, the cement has been forced from between the plates, but no alteration in the shape, nor has any part given way. 999 parts of the boiler out of the 1000 did not leak a single drop. I am now getting out the few leaking rivets, and shall then try it again. I find a great deal of difficulty in keeping the water in the boiler from getting away with the steam; the one tube being 18 inches diameter, and the other only 2 feet diameter, leaves but 3 inches space on each side, or but 2 superficial feet of surface of water to give out the steam, which comes with such violence that it brings the water with it, unless it is 5 feet below the boiler-top, obliging me to lose near one-half of my fire sides in the inner tube.

"About a foot and half down from the top I put a quantity of hemp, and loaded it with small pieces of iron like a packing, to serve as a sift for the water, but though it mends it, yet it is not a cure. I have taken a patent, and shall put in my specification as soon as I return to London, which I expect will be in the course of ten or twelve days.

"I remain, Sir,

"Your very humble servant,

"RICHARD TREVITHICK.

"P.S.—I have drawn the cylinder 9 feet diameter, in which the screw is to work, which will be 8 feet 10 inches in diameter; having two turns in 4 feet long, of course it will gain 2 feet forward to each turn and to each round of the steam-arms, the same as the screw, being on the first motion. At a rate of about 350 feet per second on the end of the arm, will give the screw about ten miles forward per hour."

The recoil engine was not quite satisfactory, for the steam issued from the aperture for four seconds after the supply had been shut off, showing a loss of power during its passage through the arm: one hoped-for means of obviating it was to get stronger steam. The gossip that the new engine was to throw balls against the French, and Trevithick's calculation coupling the speed of the arms with that of a cannon-ball, are as floating straws indicating that the current of conversation still ran on the possibility of attacking the French by steam, though the method was shifted from the steam fire-ship to the centrifugal steam-catapult.

"DAVIES GIDDY, ESQ., M.P., "BRIDGENORTH, *May* 16*th*, 1815.

"Sir,—The boilers making for this towing engine are all tubes, of 3 feet diameter, $\frac{5}{8}$ths of an inch thick, with circular ends, of wrought iron.

"There are three horizontal tubes, to each of which three other tubes are suspended, which will all be fixed or hung up in a large chimney, thus:—

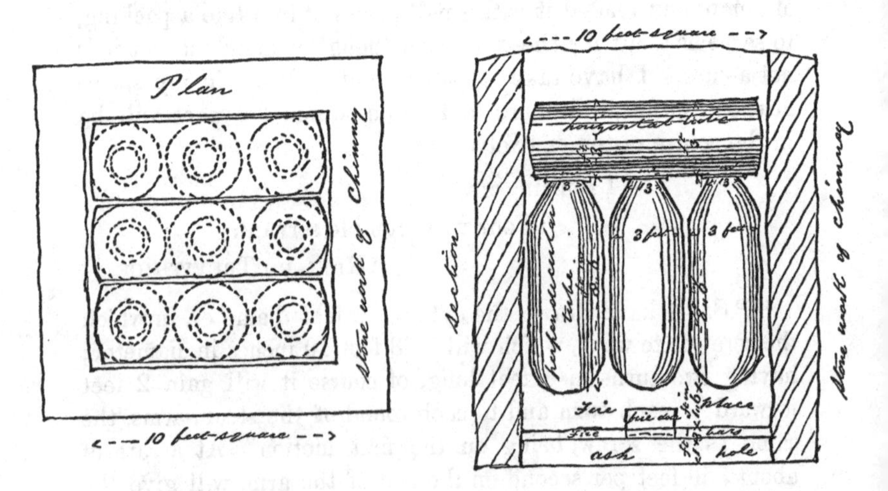

"Though I have called it stone-work, it will have for portable purposes an outside case of iron luted with fire-clay, but for fixtures, stone-work. I am satisfied that for rotary engines it will do well, but for engines lifting water the speed must be so reduced that it will be more expensive for that purpose than other machinery. I have been thinking of a method to use high steam in another way for that purpose, of which I sent a sketch for your inspection. Suppose I work with 200 lbs. to the inch pressure, which, in theory, the tubes on the other side will stand, according to the experience in proving wrought-iron cables, which stand 50,000 lbs. to the square inch of iron, above 2000 lbs. to the inch on the tubes; therefore the pressure I intend to work with will be only 10 per cent. of the real strength. Suppose a common plunger piece and pole to be placed over the shaft, or a little below the surface, to stand on a beam of wood across the pit, on each side the iron plunger-pole side rods connect to the shaft rods below. From the

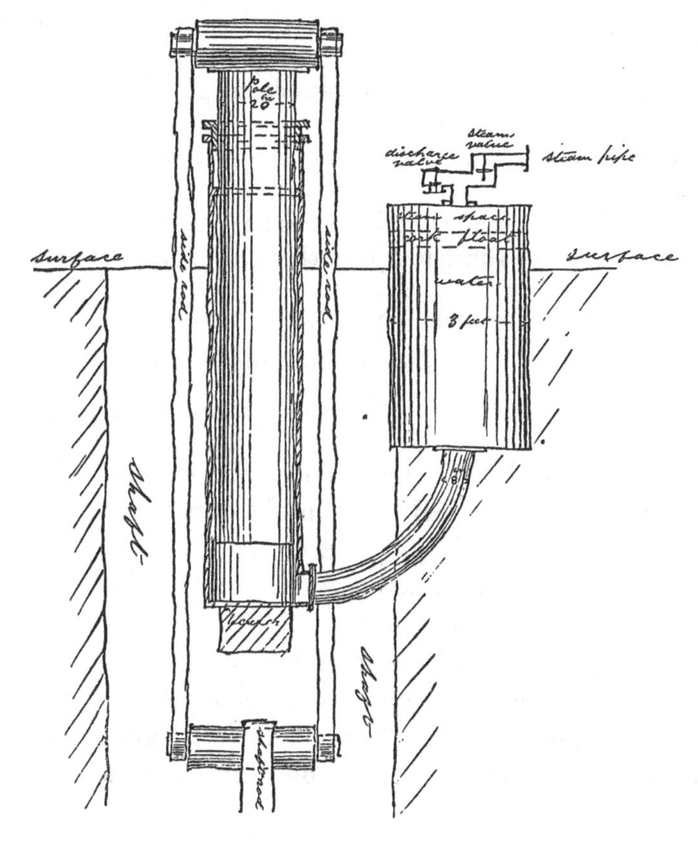

bottom of the plunger-piece a small pipe conveys water forward and backward to a steam-vessel, which water would pass and repass every stroke by the steam pressing on the surface of the water in the steam-vessel.

"This would not need an engine-house or beam. The whole could be fixed complete (boiler excepted), of the power of the Watt 63-inch double, for 250*l.* or 300*l.*, because a 20-inch plunger-pole would give that power.

"By this plan the expansive steam might be used. Would the steam on the surface of the water heat it above the boiling point, and then imbibe heat and again discharge it, as the pressure of the steam is thrown on and off, or would it not answer to put a float of cork on the water?

"By this plan the packing could be kept tight with any pressure.

"The heavy rods would take the advantage of the expansion of strong steam, and I cannot see but that the engine would be manageable with 3-inch valves, which would be easily worked.

"The whole expense of the machine would not exceed 1000*l.* for a power equal to a 63-inch double-power engine of Watt.

"I remain, Sir,

"Your very humble servant,

"RICHARD TREVITHICK."

Within a week from the trial of the simple vertical wrought-iron boiler with fire in the tube, he had designed and was putting into execution an entirely new boiler, composed of numerous tubes, and giving steam of 200 lbs. to the inch.

To avoid the loss of power by loss of heat which had been observed in the recoil arm, he thought of forcing water through it in lieu of steam; but to make it more intelligible to his friends, he applied the idea to one of his pole-engines.

The following letters give indications of experiments with various forms of tubular boilers:—

"DEAR TREVITHICK, "TREDREA, *December* 31*st*, 1815.

"I have really been uneasy on one point since I saw your engine at Herland, and that in respect to the raising of sufficient steam. There is without doubt sufficient surface, but then a great part of the long tubes will be remote from the fire, and consequently, I fear, of but little power for raising steam. I should very much have preferred using four or five shorter tubes arranged in this manner.

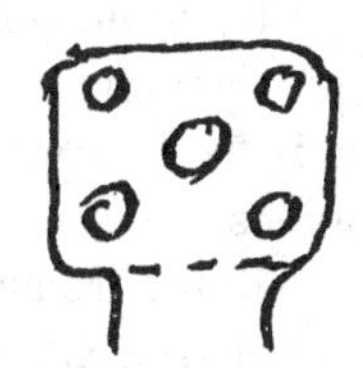

"It is difficult to give any reason for the greater power of fire applied in this manner, but there seems to be no reason for doubting the fact.

"No alteration can, however, now be made, and I shall wait with much anxiety for an account of the engine's performance.

"You will not forget the Geological Society. I am not even desirous of beautiful or costly specimens. What I request is that you will furnish us with samples of the different countries producing the silver ore, and specimens of the lodes in their bulk and in depth. I shall leave Tredrea on some day of the second week in January; but it may be as well to direct to me here, as I shall leave directions with the Post Office at Marazion about forwarding my letters.

"Yours, ever most faithfully,

"DAVIES GIDDY."

Mr. Davies Giddy's sketch is very like the fixing of tubes in modern locomotive fire-boxes; his words do not fully explain his views. My friend and railway companion in work, the late Mr. Henry Booth, has been called the inventor of the tubular boiler, used in Stephenson's trial locomotive, the 'Rocket,' in 1829; while Trevithick and the President of the Royal Society freely corresponded on the principles of such boilers and their practical construction fifteen years before, when they were not only used with higher pressure steam than Booth dreamt of, but were also patented.

"MR. GIDDY, "HERLAND, *3rd January*, 1816.

"Sir,—I have your favour of the 30th December. You are mistaken in calculating the tubes as not having so much fire exposed to them as in the usual way. The fire-place is 6 feet 9 inches wide, and 8 feet long, and the two tubes have nearly two-thirds of their surface, as the flue copes in a great way above the centre of the tubes. There will be about 96 feet exposed immediately over the fire-place. The surface of the fire is 54 feet, which is as much as is usually given in the largest engines.

"All the remainder of the heat, in the old way, is given through the sides of the boiler, when passing round perpendicular flues; but in this way the fire, after it has passed the fire-place, will descend to the lower end of the tubes, and give heat that will ascend instead of going horizontally, as is the usual way of perpendicular flues.

"If you leave home before the engine works, I will write to you every particular. If I do not chance to see you again before I leave Europe, I shall not forget to furnish you with information and specimens from America. I did not recollect to inform you when at Herland that the Foxes' engineer was here the day before you were, and said that he had worked a model about a week before with water, in the same way as I first proposed, and that the water did not get so hot as condensing water in four hours' working; neither did it escape, but worked exceedingly well, and far beyond any other engine.

"I remain, Sir,

"Your very humble servant,

"RICHARD TREVITHICK."

The results of these experiments for using water as a medium for conveying motion and reducing the amount of loss of steam and heat in the ordinary steam-engine, are not traceable.

His use of a screw for causing a draught in the chimney, in the fourth clause of the patent, where from the character of the engine he could not apply his former steam-blast, has been taken by some as a proof that Trevithick had used the steam-blast without understanding its value, when in truth it was only an additional proof of his inventive resources and power of working, not only with the blast but also without it.

Propelling Vessels, &c. 6th June, 1815.

"Instead of a piston working in the main cylinder of the steam-engine I do use a plunger-pole similar to those employed in pumps for lifting water, and I do make the said plunger-pole nearly of the same diameter as the working cylinder, having only space enough between the pole and the cylinder to prevent friction, or, in case the steam is admitted near the stuffing box, I leave sufficient room for the steam to pass to the bottom of the cylinder, and I do make at the upper end of the cylinder, for the plunger-pole to pass through, a stuffing box of much greater depth than usual, into which stuffing box I do introduce enough of the usual packing to fill it one-third high. Upon this packing I place a ring of metal, occupying about another third part of the depth of the stuffing box, this ring having a circular groove at the inside, and a hole or holes through it communicating with the outside, and with a hole through the side of the stuffing box; or, instead of one ring containing a groove, I sometimes place two thinner rings, kept asunder by a number of pillars to about the distance of one-third of the depth of the stuffing box, and I pack the remaining space above the ring or rings, and secure the whole down in the usual manner. The intention of this arrangement is to produce the effect of two stuffing boxes, allowing the space between the two stuffings for water to pass freely in from the boiler or forcing pump through a pipe and through the hole in the side of the stuffing box, so as to surround the plunger-pole and form the ring of water for the purpose of preventing the escape of steam by keeping up an equilibrium between the water above the lower stuffing and the steam in the cylinder. By this part of my said invention I obviate the necessity of that tight packing which is requisite when steam of a high pressure is used, and consequently I avoid a greater proportion of the usual friction, because a very moderate degree of tightness in the packing is quite sufficient to prevent the passage of any injurious quantity of so dense a fluid as water. And I do further declare that I use the plunger-pole, working in a cylinder and through a double stuffing, either with or without a condenser, according to the nature of the work which the steam-engine is to perform.

"The second part of my said invention consists in causing steam of a high temperature to spout out against the atmosphere, and by its recoiling force to produce motion in a direction contrary to the issuing steam, similar to the motion produced in a rocket, or to the recoil of a gun. The mode of carrying this into effect will be readily understood by supposing a gun-barrel to be bent at about a quarter of its length from the muzzle, so that the axes of the two limbs shall be at right angles to each other, and the axis of the touch-hole at right angles to the axis of the short limb, or the limbs containing the muzzle; then in the top of a boiler suitable to the raising steam of a high temperature make a hole, and insert the muzzle of the gun-barrel into that hole, so that the gun-barrel may revolve in the hole steam-tight, and let the short limb of the gun-barrel be supported in a vertical position by a collar, which will permit the breech of the gun-barrel to describe a horizontal circle, the touch-hole being at the side of the barrel. If steam of a high pressure be then raised in the boiler, it will evidently pass through the gun-barrel and spout out from the touch-hole against the atmosphere with a force, greater or less, according to the strength of the steam; and as the steam is also exerting a contrary force against that part of the breech which is opposite to the touch-hole, the barrel will recoil, and because the other end is confined to a centre, the breech end will go round in a circle with a speed proportionable to the pressure given, and may readily be made to communicate motion to machinery in general. This simple form of my invention is given merely to convey to the mind a clear idea of its nature; but in practice I do make the revolving limb, which for the sake of distinction I call the arm, much greater in its horizontal breadth than in its perpendicular thickness, in order that the arm may meet with but little resistance from the surrounding medium during its revolution, and I do introduce between the muzzle of the barrel and the boiler a steam-pipe with a cock or valve for the purpose of occasionally stopping the passage of the steam, and thereby preventing the motion of the arm, without wasting or lowering the pressure of the steam in the boiler, as well as to regulate the speed of the arm by allowing more or less steam to pass through it; and I do attach another arm of equal weight with that first ascribed in an opposite direction from the centre of motion to equalize the centrifugal force, and also by acting as a balance to avoid the great friction which would otherwise take place between the barrel and the supporting collar. I do also construct this improvement with two or more arms, having similar perforations and apertures for the steam with the arm first above described, and I do, when more convenient, place the axis of revolution horizontal, or even inclined, instead of vertical; and I do sometimes make the communication from the steam-pipe into the middle of the supporting collar, and from thence into the barrel, through a hole in that part which turns in the collar, making a groove round the inside of the collar opposite to the hole in the barrel, or a groove round the barrel opposite to the hole in the collar, in order to allow the steam free passage during the whole of the revolution; and I do make the aperture through which the steam is to spout capable of being increased or decreased in its dimensions by means of a sliding piece, marked with a screw, in order to apportion the size of the aperture to the strength of the steam, the size of the boiler, and the work to be performed. In cases where great speed is required, the revolving arms should be made short, and applied immediately on the axis of motion, without the usual intervention of multiplying wheels. It is unnecessary to mention any more variation of the mode of carrying this second part of my invention into effect, since I claim as my invention the various forms of causing steam of a high pressure to pass from a boiler through one or more perforated revolving arms, and to spout out against the atmosphere from an aperture at the side, near the extremity of the arm or arms.

"The third part of my invention consists in causing steam of a high temperature to act upon water, and that water to operate upon or under a piston, without permitting the steam to come in contact with the piston; by this contrivance, also, I obviate the necessity of that tight packing which is required when steam of a high temperature acts immediately on a piston. To effect this purpose I use two vessels, standing side by side, which, for the sake of bearing high pressure, I make of a cylindric form, closed at the bottom, but connected at the lower part by a tube. One of these cylinders, which I call the steam-vessel, has a close cover, into which are inserted a steam-pipe and a discharging pipe, and in each pipe there is a valve or cock. The other vessel, which I call the piston-cylinder, is furnished with a piston, the rod of which may be attached to a pump or other machinery. To put this apparatus into operation the piston is to be placed in and near the bottom of the piston-cylinder, and the steam-vessel filled with water; then, the discharging valve being shut, and the steam-valve opened, the steam will pass on the surface of the water, or upon a piece of cork or other substance floating on the surface, and by that pressure force the water down to the lower part of the vessel, through the tube, into the bottom of the piston-cylinder, and drive up the piston, without admitting any steam to pass out of the steam-vessel into the piston-cylinder. As soon as the steam has arrived near the orifice of the tube the steam-valve is to be shut and the discharging valve opened; the steam will then escape through the discharging pipe, and allow the piston to descend by its own weight, and drive the water back again into the steam-vessel, and by that means make a complete stroke, which may be repeated as often and with as much force as the size and construction of the boiler will supply the requisite quantity of steam. By the alternate passage of the same water from one cylinder to the other the water may be kept nearly to boiling heat, and consequently there will be no considerable condensation of the steam on its coming in contact with the water; and, if the little condensation which will take place should add to the quantity of water in the vessel, that extra quantity may be ejected at every stroke through a pipe and regulating cock communicating with the piston-cylinder. A double engine is to be constructed by using two steam-vessels with their valves, to force water alternately on each side of the piston, in which case the piston-cylinder must be closed at the top, and the piston-rod worked through a stuffing box, in the usual way; or the steam-vessels may be applied to force water alternately under two single pistons, working in cylinders open at the top, to produce the effect of a double engine. And I do further declare that this part of my said invention is not capable of being used with a condenser, because the water in the steam-vessel would, if exposed to a vacuum, evaporate into the condenser, and destroy the vacuum. By lengthening the pipe of communication between the steam-vessel and piston-cylinder the power of this engine may be conveyed to a great distance from the boiler, without intermediate beams or connecting shafts. And I do further declare that, although I have described this part of my said invention as acting with a piston in the piston-cylinder, I sometimes use the plunger-pole, working through a double stuffing box instead of a piston.

"The fourth part of my said invention consists in interposing the steam-vessel of the third part between the boiler and revolving arms described in the second part, so that, instead of the steam passing into the revolving arms, it shall pass on the surface of the water in the steam-vessel, and force the water through the arms and out at the apertures near their extremities against the atmosphere, and produce the revolution of the arms with more or less velocity, according to the strength of the steam; in this case, also, when the steam has forced nearly all the water out of the vessel the steam-valve must be shut and the discharging valve opened for the steam to escape, and at the same time another valve or cock opened for the purpose of letting a fresh portion of water from a reservoir into the steam-vessel, when the steam is again to be admitted from the boiler, in order to force the water through the revolving arm or arms. The arms should be enclosed in a case, to prevent the steam or water being thrown at a distance, to the annoyance of the by-standers, the bottom of which case may become the reservoir for supplying water to the boiler and steam-vessel. The top of the steam-vessel should be placed at a lower level than the bottom of the reservoir, in order that the water may flow into the steam-vessel when the communication from the boiler is cut off and the steam discharged. Various modes of opening and shutting the cocks or valves are so commonly known to all persons conversant with steam-engines, that a description of them is quite unnecessary. This fourth part of my said invention is in some respects similar to the contrivance described in many books or machines by the name of Barker's Mill; but in Barker's Mill no greater power can be obtained than what arises from the perpendicular altitude of the reservoir or head of water above the place where it spouts out from the revolving arms, whereas in my invention the water acted upon and driven out by steam of a high pressure, which is well known to pass through tubes with very little friction compared with water, a much greater power will be obtained from the same quantity of water; and as the water, after it has passed through the revolving arms, may be made to flow back again into the steam-vessel, to be driven through the revolving arms by a succeeding portion of steam, a small quantity of water may be made to work the engine a long time. By putting flat plates or leaves upon the revolving arms within the case I produce a current of air in the manner of a winnowing machine for blowing the fire, and I do sometimes place in the flue a screw or set of vanes, somewhat similar to the vanes of a smoke-jack, which screw or vane I do cause to revolve, by connection with the steam-engine, for the purpose of creating an artificial draught in the chimney, always proportioned to the size of the fire-place and situation of the chimney. By either or both of these means I obviate the necessity of a tall chimney when the engine is used for portable purposes.

"The fifth part of my said invention, and that which comes under the second part of my title, consists of a mode of propelling, drawing, or causing, ships, boats, and other vessels to pass through the water, which purpose I effect by constructing a worm or screw, or a number of leaves placed obliquely round an axis, similar to the vanes of a smoke-jack, which worm, screw, or vane shall be made to revolve with great speed, having the axis in a line with the required motion of the ship, boat, or other vessel, or parallel to the same line of motion; the obliquity of the thread of the worm, screw, or leaves, admits of considerable variety, according to the degree of velocity given to it and speed required, and according to the power with which it is driven, but as a general medium I by preference contrive that the thread of the screw at its outer edge shall make with its axis an angle of about thirty degrees. This worm is in some cases to revolve in a fixed cylinder, in others to revolve together with the cylinder, similar to the screw of Archimedes, but generally to revolve in the water without any cylinder surrounding it. This worm or screw may be made to revolve in the water at the head of the ship, boat, or other vessel, or at the stern, or one or more worms may revolve on each side of the vessel, as may most conveniently suit the peculiar navigation on which the ship, boat, or vessel is to be employed. In some cases when the screw is to work at the head of a ship, it is to be made buoyant, and move on a universal joint, at the end of an axle, turning in the bow of the ship, in order that the screw may accommodate itself to the unevenness of the waves. And I do further declare that, in order to make the boiler of a high-pressure steam-engine of very light materials for portable purposes, and at the same time strong for resisting the pressure, as well as for exposing a large surface to the fire, I do construct the said boiler of a number of small perpendicular tubes, each tube closed at the bottom, but all opening at the top into a common reservoir, from whence they receive their water, and into which the steam of all the tubes is united.

"In witness whereof, I have hereunto set my hand and seal, this Twentieth day of November, in the year of our Lord One thousand eight hundred and fifteen.

"RICHARD (L.S.) TREVITHICK."

The variety and extreme range of inventions made known in this patent specification cannot be other than superficially spoken of in this chapter, including, as they do, multitubular boilers, expansive high-pressure steam, the screw-propeller, and the recoil engine.

The absence of a drawing illustrating his claims is to some extent made good by reference to other drawings. Claim 1 is for an application of the pole,—either in a puffer or vacuum engine, shown in Wheal Prosper drawing, chapter xix.,—to marine purposes, modified if necessary by the water medium of communicating power in Claim 3, shown in Trevithick's sketch, page 371. Claim 2 is for the recoil or rotary engine, sketched by Trevithick, page 364, worked by the rush of strong steam against the atmosphere, or by the water medium described in Claims 3 and 4. Claim 5 is for the famous screw propeller, described in Trevithick's letter, page 348, and includes the no less famous multitubular boiler, as sketched by Trevithick, page 370: as the customary puffer-engine blast-pipe could not be applied, draught was produced by a screw in the chimney.

INDEX TO VOLUME I.

D.

F.

F

G.

M.

N.

O.

P.

Q.

R.

V.

W.

Y.

END OF VOLUME I.

LONDON: PRINTED BY W. CLOWES AND SONS, STAMFORD STREET AND CHARING CROSS.